Imperial College London
2406588901
AF804974

OPTIMAL PORTFOLIOS

OPTIMAL PORTFOLIOS

Stochastic Models for Optimal Investment and Risk Management in Continuous Time

Ralf Korn

Johannes Gutenberg-Universität Mainz

World Scientific

Singapore • New Jersey • London • Hong Kong

Published by

World Scientific Publishing Co. Pte. Ltd.

P O Box 128, Farrer Road, Singapore 912805

USA office: Suite 1B, 1060 Main Street, River Edge, NJ 07661

UK office: 57 Shelton Street, Covent Garden, London WC2H 9HE

Library of Congress Cataloging-in-Publication Data

Korn, Ralf.

Optimal portfolios : stochastic models for optimal investment and risk management in continuous time / Ralf Korn.

p. cm.

Includes bibliographical references and index.

ISBN 9810232152

1. Portfolio management -- Mathematical models. 2. Options (Finance) -- Mathematical models. 3. Risk management -- Mathematical models. 4. Stochastic processes. I. Title.

HG4529.5.K674 1997

332.6'01'5118--dc21 97-36352

CIP

British Library Cataloguing-in-Publication Data

A catalogue record for this book is available from the British Library.

First published 1997

Reprinted 1998, 1999

Copyright © 1997 by World Scientific Publishing Co. Pte. Ltd.

All rights reserved. This book, or parts thereof, may not be reproduced in any form or by any means, electronic or mechanical, including photocopying, recording or any information storage and retrieval system now known or to be invented, without written permission from the Publisher.

For photocopying of material in this volume, please pay a copying fee through the Copyright Clearance Center, Inc., 222 Rosewood Drive, Danvers, MA 01923, USA. In this case permission to photocopy is not required from the publisher.

Printed in Singapore by Regal Press (S) Pte. Ltd.

PREFACE

Nearly 25 years ago, Fischer Black and Myron Scholes came up with their famous option pricing formula which is regarded today as the cornerstone of mathematical finance. However, why should a formula for option pricing be the starting point for writing a book on continuous-time portfolio optimisation? The reason for this lies in the totally different success in the application and science of the Black-Scholes formula and the continuous-time portfolio methods. While both rely on the same type of underlying continuous-time model, the Black-Scholes formula is today common knowledge among practitioners, financial economists and financial mathematicians; everyone at the stock exchange is familiar with it, but only very few people are actually applying continuous-time portfolio methods. For most practitioners and also in large parts of academic circles, the simple one-period mean-variance approach of Markowitz is still state of the art! But this approach was developed 45 years ago! To show that something has really happened in portfolio optimisation theory during the last 45 years is one of the main intentions of this book. Especially, we will concentrate on a continuous-time framework which is the same as the one underlying the Black-Scholes formula.

The whole idea underlying portfolio optimisation is totally natural. One has got a certain amount of money and tries to use it in such a way that one can draw the maximum possible utility from the results of the corresponding activities. This principle covers nearly every situation of daily life. After a hard day's work, you are hungry, thirsty and bored. You quickly walk over to your news agent, choose something to drink, some snacks and a magazine, but a quick look into your wallet tells you that either one drink, one chocolate bar or the magazine is too much. You are facing a problem of portfolio optimisation! Imagine that you are thinking of buying a house and are offered two different ones you can afford. One close to your office with public transport connections, but without a garden and close to a crowded motor way, the other one with a beautiful landscape but requiring you to commute a long distance to work everyday. The decision about which is more convenient for you is in principle (you have definitely guessed it!) a portfolio problem. There are a lot of further examples which you will face without even thinking of applications in finance.

This book should give you an introduction to modern, continuous-time portfolio optimisation starting with the basic (discrete-time) Markowitz mean-variance approach, covering Merton's pioneering work of the late sixties and early seventies, presenting the martingale method as alternative to Merton's method of stochastic control, looking at various constrained portfolio problems and portfolio problems in the presence of transaction costs and finally, it closes with some alternative approaches to the expected utility maximisation. A more detailed description of the topics and a guideline on how to read through the book is given in "Suggestions to the reader".

As with every book, there are a lot of people without whose help it would not have been possible for me to write this book. It is therefore a great pleasure for me to take the opportunity to thank them all. In particular, I would like to thank my wife Elke and my son Uwe for being so patient with me during the time of preparing the manuscript. Also, Elke prepared figures for the cover design and helped me with implementing some algorithms to produce various numerical examples. I really appreciated that Ian Buckley read parts of this manuscript to check my written English (I hope that you will see some progress in my style, Ian!). Thanks go to Thomas Wiesemann who taught me nearly everything I know on discrete-time models of value preserving portfolio selection. The stochastics group of the mathematics department at the Johannes Gutenberg-Universität Mainz under the leadership of Wolfgang J. Bühler provided a nice and stimulating atmosphere which helped much in writing this book. I have also benefited from a lot of discussions with various friends and colleagues at different meetings and conferences over the last years. To mention them all would bear the danger of forgetting someone. So a global thanks to you all! The staff at World Scientific never tired of answering all my questions in a friendly and accurate way. In particular, I would like to thank Elena Nash for initiating the book and Caroline Ching for guiding me patiently through the final stages of the project. Finally, I should not forget to thank my students who definitely helped to improve this book with all their questions and comments during my lectures.

It's now time for you to work and start with reading and I hope that you will also find some pleasure in doing so!

Hallgarten, July 1997 Ralf Korn

CONTENTS

SOME GUIDELINES AND GENERAL NOTATIONS

There are different ways to read (parts of) this book. The way which is most suitable for you depends on your interests and your knowledge in finance, stochastics, differential equations and optimisation. As continuous-time portfolio models require knowledge on stochastic processes, Itô-calculus and optimisation, the necessary results are provided in the Appendix. As on one hand, it is not possible to give a detailed introduction in these different fields in one single book and on the other hand not necessary to know all the proofs of the relevant results, the Appendix can also serve as an introduction to the above mentioned areas. It is therefore regarded as an essential part of the book which will also open non-experts in stochastic calculus and stochastic control the gate to continuous-time portfolio optimisation. This role of the Appendix is the main reason why it is written in a rather narrative form. However, to allow for a quick way to the central parts of portfolio optimisation, I have decided to leave it in the back of the book and you can judge it on your own if you need to recall some of the basic facts before reading the main subject.

It is definitely recommended to read Chapter 2 (description of the continuous-time model) and at least one of the descriptions of the main approaches to the continuous-time portfolio problem, the stochastic control approach or the martingale method, in Chapter 3. One can then jump to various parts of the book. Only Section 4.4/5 will sometimes require knowledge of the martingale method. Chapter 4 on constrained portfolios is not necessary to understand Chapter 5 on portfolio problems under transaction costs and Chapter 6 on alternative models for portfolio selection.

At some points, it was necessary to use some results which are not given in the index. There, the additional introduction of technicalities and further mathematical theory would have taken much more pages than it could be justified by their actual application. As a compromise, the arguments behind the relevant proofs are sketched and the interested reader is referred to the original sources.

We have used the following hierarchical system for numbering and referring to equations, theorems, remarks and definitions: In each chapter, important equations are numbered consecutively starting with (1). By referring to such an equation given in the same chapter, we only use this bracketed number. If it is referred to an equation occurring in a different chapter then we add the chapter number inside the bracket, i.e. equation (3.12) is equation (12) of Chapter 3. Theorems, remarks, lemmas, definitions, ... are all put together in one group with respect to numbering.

As with the equations, they are numbered consecutively in each chapter and are referred to by their number without chapter number if they are in the same chapter or referred with chapter number and number if they are from a different chapter. For example, Remark 4.5 has the "group number" 5 in Chapter. Chapters are decomposed into sections (numbered by Arabian digits) which are — sometimes — further split up into sub-sections (numbered by small Latin numbers). The Appendix is made up of five different parts, A – E, each of them having sub-parts which are numbered consecutively. Also, equations, theorems, remarks,... are numbered as in the main chapters. Results from the Appendix are always referred to in the book by putting the letter of the relevant part of the Appendix in front of their number.

Most of the notation used in the book is explained where it is used first. However, we will explain some frequently used notation and abbreviations below.

Abbreviations

iid	independent, identically distributed
ode	ordinary differential equation
pde	partial differential equation
qvi	quasi-variational inequality
sde	stochastic differential equation
w.l.o.g.	without loss of generality
s.t.	such that

Some sets and spaces

$\mathbf{N}$	$:= \{1, 2, 3, \ldots\}$
$\mathbf{R}^d$	d-dimensional space of real numbers
$C^{i,j}(A, B)$	space of functions f: $A \times B \to \mathbf{R}^d$ which are i times continuously differentiable with respect to the variables corresponding to A and j times continuously differentiable with respect to the variables corresponding to B
$\sigma(\mathcal{A})$	the smallest σ-algebra containing the collection of sets $\mathcal{A}$
$F \otimes G$	the product σ-algebra of the two σ-algebras F and G

Notations

x'	denotes the transposed of the vector x
$\|x\|$	$:= \sqrt{\sum_{i=1}^{n} x_i^2}$, i.e. the Euclidean norm of $x \in \mathbf{R}^n$
$a \wedge b$	$:= \min\{a, b\}$, i.e. the minimum of the real numbers a and b
x^+	$:= \max\{0, x\}$, i.e. the positive part of $x \in \mathbf{R}$
x^-	$:= \max\{0, -x\}$, i.e. the negative part of $x \in \mathbf{R}$
$\mathcal{L}$	Lebesgue measure

$L \otimes P$	product measure of the two measures L and P
$\underline{1}$	$:= (1,\ldots, 1)'$ (of appropriate dimension)
e_i	$:= (0,\ldots 0, 1, 0\ldots,0)'$ the i th unit vector (i.e. the one having zero entries for all components but number i where it has "1" as entry)
I	the identity matrix of appropriate dimension
$1_A(x)$	$:= \begin{cases} 1, & x \in A \\ 0, & x \notin A \end{cases}$, i.e. the indicator function of the set A
$\max\limits_x f(x)$	maximise the function f(x) with respect to x
$\max\limits_x f(x)$ s.t. $x \in A$	maximise the function f(x) over all $x \in A$

CHAPTER 1

INTRODUCTION AND DISCRETE-TIME MODELS

1. General and Historic Remarks : A Short Survey

The portfolio problem of an investor trading in different (financial) assets is to choose an optimal investment and consumption strategy. To be more precise, an investor endowed with a given initial capital x has to decide **how many** shares of **which** asset he should hold at **what** time instant to **maximise** his (expected) utility of consumption during the time interval [0, T] and/or of his total wealth at the time horizon T. The main object of portfolio management theory (for brevity: portfolio theory) is to deliver solution methods for the portfolio problem.

In principle this approach can also be applied to general decision problems under uncertainty, but the models we will consider in this book rely crucially on some features which are typical of financial markets (such as infinitely divisible assets, trading possibilities in continuous time and absence of storage problems,...).

As a scientific topic, portfolio theory is part of the (mathematical) theory of financial markets, the central aspect of which is to explain the mechanism of the capital markets. We will not follow this topic here, but instead make some short remarks on the history and development of portfolio theory up to and including some of the most recent innovations.

The earliest approach to solving the portfolio problem is the so called mean-variance approach. It was pioneered by H. Markowitz (see (Markowitz 1952), (Markowitz 1959)) and J. Tobin ((see (Tobin 1958), (Tobin 1965)) and is only suited for one-period decision problems. It consists of a one-off decision at the beginning of the period (t = 0) and no further actions until the end of the period (t = T). It still has great importance in real-life applications and is widely applied in the risk management departments of banks. The main reasons for this is being the ease with which the algorithm can be implemented and that the method demands no special knowledge beyond very basic stochastic models (in fact, it is enough to know what expected values and covariances of random variables are). As a tribute to the importance of his contribution, in 1990 Markowitz gained the Nobel prize in economic sciences for his work on the mean-variance approach. However, the whole approach extremely simplifies reality.

To overcome the limitations and problems raised by modelling the portfolio problem in a discrete time setting (and especially a single-period one) a continuous-time approach for modelling the stock prices and the actions of the investors was proposed in the late fifties and early sixties by (among others) Fama, Mandelbrot and Samuelson (see also (Samuelson 1973)). Proposed models for stock prices ranged from lognormally distributed ones to models having stable Pareto and other stable distributions. However, it is remarkable that the first known continuous-time model was introduced far earlier when at the turn of the century Louis Ferdinand Bachelier wrote his dissertation "Théorie de la speculation" (Bachelier 1900). Amongst his achievements, he was the first the first person to conceive of the stochastic process nowadays known as Brownian motion. By modelling stock prices as Brownian motions with drift his aim was to compute theoretical prices for various options and to compare them with the empirical prices observed at the Paris stock exchange at that time. The basic flaw of Bachelier's approach was his use of Brownian motion as his model for prices, a process which always has a positive probability of attaining negative values, a characteristic which is not exhibited by real world stock prices. That Bachelier's continuous-time approach of was so far ahead of his time did not count in his favour: his thesis was rejected !

The work of Robert Merton (see (Merton 1969), (Merton 1971),...) must be regarded as the real starting point of continuous-time portfolio theory. By applying standard methods and results from stochastic control theory to the portfolio problem he was able to obtain explicit solutions for some special examples. However, the crucial point in his approach is that the whole problem reduces to solving the Hamilton-Jacobi-Bellman Equation of dynamic programming (for brevity: HJB-Equation). This typically leads to the problem of solving a highly non-linear partial differential equation for which even a numerical solution may prove elusive. Despite these limitations, the Merton approach is still very popular in finance, and much recent work deals with refinements of this approach to include transaction costs or market imperfections in the model (we will come back to this shortly).

With the growing application of stochastic calculus to finance in the early eighties (initiated in (Harrison and Kreps 1979), (Harrison and Pliska 1981), (Harrison and Pliska 1983)) a more elegant method — the martingale approach to portfolio optimisation — was developed by Karatzas e.a. (see (Karatzas e.a. 1987) or (Karatzas 1989)) and Cox and Huang (see (Cox and Huang 1989)). Its main idea is to establish a separation of the portfolio problem into a static optimisation problem ("Find the optimal terminal wealth and the optimal consumption process") and a representation problem ("Compute the corresponding trading strategy"). The underlying market model differs only slightly from the one used by Merton, but instead of using stochastic control theory, the martingale approach is based on deep results of stochastic calculus (such as the martingale representation theorem (see Appendix B.5)) and on convex optimisation.

In recent work on continuous-time models both these approaches have been refined to make the models more realistic. The most significant improvements have been the introduction of additional constraints and of transaction costs to the portfolio problem. The work on constraints can be roughly divided into work concerning constraints both on the trading strategies and on the terminal wealth of an investor. Typical constraints on the strategies include short-selling and leverage constraints, bounds for the wealth held in one asset or incomplete market constraints (see for example (Cox and Huang 1991), (Cvitanic and Karatzas 1992), (He and Pearson 1991) and (Xu and Shreve 1992 a, b)). Constraints on the terminal wealth are the subject of (Korn and Trautmann 1995) or (Korn 1997b). As a special case, one is now able to formulate and solve a continuous-time version of the mean-variance problem of Markowitz.

As rebalancing of the holdings is the essential action of an investor solving the portfolio problem, transaction costs and their impact on the form of the optimal strategy cannot be ignored; especially because the optimal trading rules in continuous-time models (without transaction costs) typically involve trading at every time instant. (Magill and Constantinides 1976) was among the first papers dealing with the subject of transaction costs. Later, in (Davis and Norman 1990) and (Soner and Shreve 1994), a pure consumption problem with an infinite time horizon (i.e. $T = \infty$) was solved explicitly in the case of proportional transaction costs (i.e. the transaction costs are always proportional to the value of the transacted shares). The difference between the two papers is the method used. Davis and Norman used the more traditional but straight forward approach to construct an explicit smooth solution to the variational inequality arising from the corresponding HJB-Equation. Conversely, Soner and Shreve used a viscosity solution approach to singular stochastic control. A very general transaction cost structure (including a fixed cost component) can be treated with the help of an impulse control model for the portfolio problem (see (Eastham and Hastings 1988) and (Korn 1994)). The main drawback of the approaches presented so far lies in their numerical complexity. It is difficult — if not impossible — to apply them in an efficient way to markets consisting of more than two assets. To overcome this problem, and to be able to work with a realistic number of assets (around 20–30), Atkinson and Wilmott used perturbation methods to solve a portfolio problem with a special transaction cost structure, introduced in (Morton and Pliska 1995). Their main idea is to solve the problem approximately via an asymptotic analysis of the optimality equation for small but non-zero transaction costs (see (Atkinson and Wilmott 1995)).

As alternatives to the expected utility maximisation approaches described so far, we should also mention non-utility based approaches to portfolio optimisation. We will explicitly present the approach of asymptotically optimal portfolios developed by T. Cover in a discrete time setting (see (Cover 1991)) which F. Jamshidian transferred to the continuous-time framework (see (Jamshidian 1992)). By rebalancing the holdings according to the past performance of the stocks, the authors showed that the portfolio so obtained outperforms (nearly) all constant portfolios when the time horizon approaches infinity. However, the most attractive feature of this approach seems to be that it does not require knowledge of the market coefficients. Being similar to empirical Bayes methods, it continuously updates its view towards the different stocks on the basis of their price movements.

Trading in index portfolios is motivated by the philosophy that it is very hard to beat the market. However, it should be possible to achieve the same performance as the market — measured by a market index (such as the Dow-Jones, FTSE, or Dax) — by holding a portfolio of shares in exactly the same fractions as they enter this index. The problem encountered by a fund manager who tries to track such an index, but is also faced with random in- and outflows to/from his fund, is treated in (Buckley and Korn 1997).

Another non-utility based approach dealt with in this book relies on the existence of value preserving portfolio strategies. The principle of value preservation was conceived by K. Hellwig (see (Hellwig 1987)). The main idea hinges on the definition of the economic value of a future cash stream. A value preserving investor acts in such a way as to maintain this economic value at a constant level. He is only allowed to consume those parts of his wealth that are not necessary for keeping the economic value constant. This very general idea was applied to portfolio problems in discrete time by (Wiesemann 1995) while (Korn 1997a) considered the existence of value preserving portfolio strategies in continuous time.

1.2 Mean-Variance Analysis in a One-Period Model: the Markowitz-Approach

The origin of modern portfolio optimisation is widely regarded to be the work of H. Markowitz, especially (Markowitz 1952). The crucial observation of Markowitz was that a pure maximisation of expected return would lead to putting all of the money in the stock with the highest expected return (assuming that there exists a unique such stock). As such a strategy represents a highly risky position, Markowitz claimed that such an investment rule must be rejected. His suggestion was to quantify the risk by the variance of the position and to follow only so called **mean-variance efficient strategies**. The corresponding mean-variance approach to portfolio theory is still highly valued which is reflected by the fact that Markowitz gained the Nobel prize in economics, the wide-spread use of this approach in practise, and its appearance and recommendation in textbooks on investment theory (see e.g. (Haugan 1993) : "The Tool is cool ..."). Therefore, we will briefly sketch the method and its underlying market model.

Assume that an investor can trade in d different securities at time t = 0. He will allocate his wealth to the different securities at time t = 0 and will hold this position until the time horizon t = T. Such a model is called a **one-period** or **static model** (independent of the actual length of the time period). We assume that there are no transaction costs and that the securities are perfectly divisible. The future prices of the securities are modelled via their returns R_i, i = 1,..., d, at time t = T. More precisely, if $P_i(0)$, $P_i(T)$ are the security prices at times t = 0,1 then we look at

$$R_i = P_i(T) / P_i(0)$$

(which indeed is equivalent to considering the relative returns $(P_i(T)-P_i(0))/P_i(0) = R_i-1$). As the returns are not foreseeable they are modelled as random variables with expectation and covariances given by

$$\mu_i = E(R_i),\ \ \sigma_{ij} = \mathrm{Cov}(R_i, R_j),\ \ i,j = 1,\dots, d.$$

Let π_i be the fraction of initial wealth x of the investor invested in security i at time t = 0, i.e.

$$\pi_i = \frac{\psi_i P_i(0)}{x}, \tag{1}$$

where ψ_i is the number of shares of security i held by the investor at time t = 0. We will call the vector $\pi = (\pi_1,\dots, \pi_d)'$ the **portfolio vector** of the investor. The choice of the portfolio vector π at the beginning of the trading period results in a total return $R = X^\pi(T) / x$ (or portfolio return) of

$$R = \sum_{i=1}^{d} \pi_i R_i \tag{2}$$

at time t = T, where $X^\pi(T) = \sum \psi_i P_i(T)$ is the wealth of the investor at time t = T and ψ_i is the number of shares of security i held by the investor related to π_i by equation (1). To see that equation (2) is valid note that we have

$$\sum_{i=1}^{d} \pi_i R_i = \sum_{i=1}^{d}\left(\frac{\psi_i P_i(0)}{x}\frac{P_i(T)}{P_i(0)}\right) = \sum_{i=1}^{d}\frac{\psi_i P_i(T)}{x} = \frac{X^\pi(T)}{x} = R\,.$$

We further deduce that the expected value and the variance of the portfolio return for the portfolio π be

$$E(R) = \sum_{i=1}^{d} \pi_i \mu_i =: \mu(\pi),$$

$$Var(R) = \sum_{i,j=1}^{d} \pi_i \sigma_{ij} \pi_j =: \sigma^2(\pi).$$

To avoid a negative wealth of the holdings in the time horizon $t = T$ (i.e. to avoid bankruptcy in the strict sense), we will restrict ourselves to portfolios with non-negative components (i.e. we forbid short-selling). Such portfolio vectors will be called **admissible**.

The essential parts of the philosophy of Markowitz are now summarised in the following two formulations of the **mean-variance principle** :

- For a given upper bound σ^2 for the variance of the portfolio return, choose an admissible portfolio π such that $\mu(\pi)$ is maximal under all admissible portfolios π, with $\sigma^2(\pi) \leq \sigma^2$.

- For a given lower bound μ for the mean of the portfolio return choose an admissible portfolio π such that $\sigma^2(\pi)$ is minimal under all admissible portfolios π, with $\mu(\pi) \geq \mu$.

A visualisation of these two rules is given by Figure 1, where we have plotted a typical region for the attainable pairs $(\mu(\pi), \sigma^2(\pi))$ for all admissible portfolios π. Only those pairs lying on the bold part of the boundary satisfy the requirements of either formulation of the mean-variance principle. These pairs (and also the portfolios leading to such pairs) are called **mean-variance efficient**. The bold line is called the **mean-variance efficient set**.

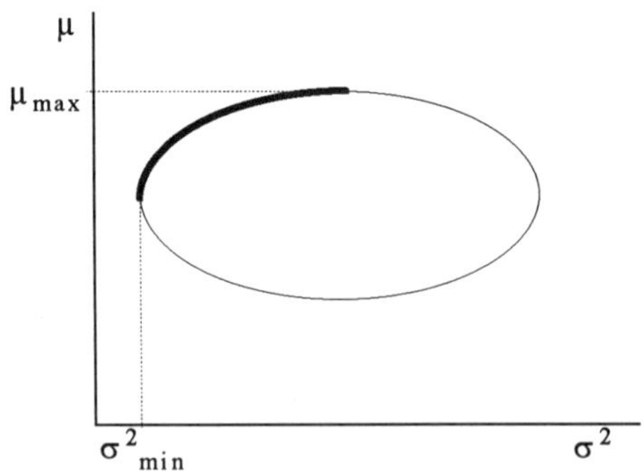

Figure 1: Mean-variance efficient set

Of course, if we have a market containing a riskless security, then the set of possible pairs $(\mu(\pi), \sigma^2(\pi))$ must have an intersection with the y axis. Furthermore, one could imagine other forms for the shape of the set of possible pairs $(\mu(\pi), \sigma^2(\pi))$ (see e.g. (Haugan 1993)). So, Figure 1 only serves as a visual aid. However, the main task is to compute the efficient set and the corresponding portfolio vectors. This can be achieved by solving a multi-objective programming problem or by solving a whole family of quadratic programming problems. We will concentrate on the latter method. Therefore, consider the problem of an investor who has fixed a lower bound $\mu_{low} \in [\mu_{min}, \mu_{max}]$ (where μ_{min}, μ_{max} are the smallest and highest value of the means of the returns for the different securities, respectively). According to the first version of the mean-variance rule he then tries to minimise the portfolio return variance $\sigma^2(\pi)$. His corresponding optimisation problem reads :

$$\min_{\pi \in \mathbf{R}^d} \sigma^2(\pi) \tag{3}$$

$$\text{s.t. } \mu(\pi) \geq \mu_{low}, \ \pi_i \geq 0, \ \sum_{i=1}^{d} \pi_i = 1 .$$

According to the definition of $\mu(\pi)$ and $\sigma^2(\pi)$, this problem is a quadratic programming problem (i.e. it has a quadratic objective function and linear constraints). The choice of μ_{low} ensures that the feasible set is non-empty. If we assume that the covariance matrix $\Sigma = (\sigma_{ij})$ is positive definite, then this problem possesses a unique solution π^* (see e.g. (Fletcher 1981)). It can be calculated with standard algorithms such as those by Gill and Murray (see (Gill and Murray 1978)) or by Goldfarb and Idnani (see (Goldfarb and Idnani 1983)) for quadratic programming problems.

The assumption of a positive definite covariance matrix Σ is violated either if (the return of) one security can be obtained as a linear combination of (the returns of) some of the remaining securities, or if there is a riskless security (which would correspond to a zero row and column in Σ). While the non-degeneracy assumption on the risky assets is a natural one, it is also a natural case to consider the presence of a riskless security. In such a situation, we number the securities in such a way that the riskless appears first, which yields

$$\Sigma = \begin{pmatrix} 0 & 0 \\ 0 & \Sigma^* \end{pmatrix}$$

where Σ^* is a positive definite $(d-1,d-1)$-matrix. If we now set $\nu = (\nu_1,\ldots, \nu_{d-1})' = (\pi_2,\ldots, \pi_d)'$ and note

$$\pi_1 = 1 - \sum_{i=1}^{d-1} \nu_i ,$$

then we could rewrite the optimisation problem (3) in the form

$$\min_{\nu \in \mathbf{R}^{d-1}} \nu' \Sigma^* \nu \tag{4}$$

$$\text{s.t.} \quad \mu_1 + \nu_1(\mu_2 - \mu_1) + \ldots + \nu_{d-1}(\mu_d - \mu_1) \geq \mu_{low}, \quad \nu_i \geq 0, \ \sum_{i=1}^{d-1} \nu_i \leq 1 .$$

As this problem is again a quadratic programming problem with a positive definite objective function matrix (or equivalently with a strictly convex objective function), it admits a unique solution ν^*, which then induces the optimal solution to our mean-variance problem (3) via

$$\pi^* = \left(1 - \sum_{i=1}^{d} \nu_i{}^*, (\nu^*)' \right).$$

In either of the two foregoing cases, we can calculate the mean-variance efficient set by solving the family of problems (3) for all possible values $\mu_{low} \in [\mu_{min}, \mu_{max}]$.

On the other hand, if an investor has specified an upper bound for the portfolio variance, $\sigma^2_{high} \in [\sigma^2_{min}, \sigma^2_{max}]$ (where σ^2_{min}, σ^2_{max} are the smallest and highest possible value of the variance of the returns for the feasible portfolios π, respectively) and tries to maximise his expected portfolio return, then he has to solve the optimisation problem

$$\max_{\pi \in \mathbf{R}^d} \mu(\pi) \tag{5}$$

$$\text{s.t.} \ \sigma^2(\pi) \leq \sigma^2_{high}, \ \pi_i \geq 0, \ \sum_{i=1}^{d} \pi_i = 1 .$$

This is a linear programming problem with one additional quadratic constraint. It cannot be solved in such a direct manner as problems (3) and (4), but we can deduce the following relation between the two problems (3) and (5) (where for simplicity we assume that Σ is positive definite).

Proposition 1

Let Σ be positive definite. Assume further that in the following parts a) and b) we always have $\mu_{low} \in [\mu_{min}, \mu_{max}]$ and $\sigma^2_{high} \in [\sigma^2_{min}, \sigma^2_{max}]$, respectively.

a) If π^* is a solution to problem (3) satisfying $\mu_{low} = (\pi^*)'\mu$ (i.e. the expectation constraint is active in the optimum), then π^* also solves problem (5) provided that there we choose $\sigma^2_{high} := (\pi^*)' \Sigma\pi^*$.
b) If $\hat{\pi}$ is a solution to problem (5) with $\sigma^2_{high} = (\hat{\pi})'\Sigma\hat{\pi}$ (i.e. the variance constraint is active in the optimum) then $\hat{\pi}$ also solves problem (3) provided that in this case we choose $\mu_{low} := (\hat{\pi})'\mu$. .

Proof:
a) π^* is feasible for problem (5) provided that we chose $\sigma^2_{high} = (\pi^*)' \Sigma\pi^*$. If there exists a solution to problem (5) with a higher expected value than μ_{low}, then it must have a variance of $\sigma^2(\pi) = \sigma^2_{high}$ (otherwise this would contradict the optimality of π^* for problem (3)). But π would then be feasible **and** optimal for problem (3). Uniqueness of the solution to problem (3) then implies $\pi^* = \pi$.
b) $\hat{\pi}$ is feasible for problem (3) provided that we chose $\mu_{low} = (\hat{\pi})'\mu$. If there exists a solution to problem (3) with a lower portfolio variance than σ^2_{high}, then it must have a mean $\mu(\pi) = \mu_{low}$ (otherwise this would contradict the optimality of $\hat{\pi}$ for problem (5)). But π would then be feasible **and** optimal for problem (5). Uniqueness of the solution of problem (3) again implies $\hat{\pi} = \pi$.

□

Remark 2
a) From one perspective, problem (5) ("bounding the risk and then maximising the gain") seems to be the more natural one for an investor. It is therefore noteworthy that with the help of Proposition 1 we can give an algorithm for solving this problem simply by solving a sequence of quadratic programming problems. The basic idea is to select a sequence $\mu^{(n)} \in [\mu_{min}, \mu_{max}]$ and to solve problems (3) with $\mu_{low} = \mu^{(n)}$. If the solution of the actual problem (3) has a variance that is bigger than σ^2_{high}, then one should choose a value for $\mu^{(n+1)}$ to be smaller than $\mu^{(n)}$. If the solution of the actual problem (3) has a variance that is smaller than σ^2_{high}, one is allowed to increase the lower bound on the mean of the return and thus choose a value for $\mu^{(n+1)}$ to be bigger than $\mu^{(n)}$. Under a suitable choice of the sequence $\mu^{(n)}$ this algorithm will converge. Of course, one should stop if the resulting variance of the optimal solution of problem (3) is close enough to σ^2_{high}.
b) Another widespread formulation of the mean-variance approach is the problem

$$\max_{\pi \in \mathbf{R}^d} \left(\mu(\pi) - \alpha\sigma^2(\pi)\right) \tag{6}$$

$$\text{s.t. } \pi_i \geq 0, \ \sum_{i=1}^{d} \pi_i = 1$$

i.e. to choose a weighted sum of portfolio return variance and portfolio return mean as objective function where α is a positive constant. Here, the variance constraint is only implicitly included into the problem. Although this formulation can easily be transformed into the other two formulations it has the disadvantage that their is a priori no control over either the magnitude of the mean or the variance. For its solution we have the same results and remarks as for problem (3).

At the end of this section we will give a small example that serves to illustrate the shape of the feasible region and the location of the optimal portfolio for the mean-variance problem.

Example 3
Assume

$$n = 2, \quad \mu = \begin{pmatrix} 1 \\ 3/2 \end{pmatrix}, \quad \Sigma = \begin{pmatrix} 2 & -1 \\ -1 & 2 \end{pmatrix}, \quad \sigma^2_{\text{high}} = 3/2 .$$

We show how the corresponding problem of type (5) can be solved graphically :

$$\max_{\pi_1, \pi_2} \{ \pi_1 + 3/2\,\pi_2 \}$$

$$\text{s.t. } \pi_1 \geq 0, \pi_2 \geq 0, \ \pi_1 + \pi_2 = 1, \ \sigma^2(\pi) = 2(\pi_1{}^2 + \pi_2{}^2 - \pi_1\pi_2) \leq 3/2 .$$

For the following arguments, we always refer to Figure 2, which shows the feasible region for the portfolios (π_1, π_2) and the contour lines of the objective function.

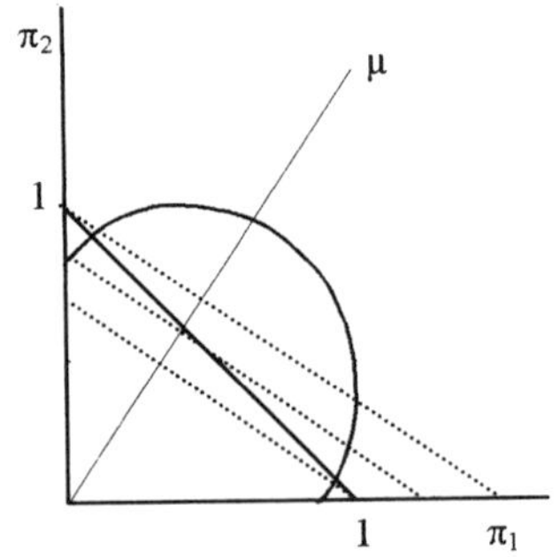

Figure 2: Feasible region and level sets for a mean-variance problem

Note that the feasible region for the portfolios (π_1, π_2) consists of that part of the line given by the equation "$\pi_1 + \pi_2 = 1$" that lies inside the region "$\sigma^2(\pi) \leq 3/2$ ".

As the objective function is linear, its contour lines are straight and are orthogonal to μ (the dotted lines in Figure 2). The corresponding levels increase in the direction of μ. Hence, the optimal portfolio lies at the upper end of the feasible set. One can compute it by determining the two intersection points of the line "$\pi_1 + \pi_2 = 1$" and the curve "$\sigma^2(\pi) = 3/2$". The optimal portfolio corresponds to the one that lies above the line "$\pi_1 = \pi_2$". Hence, we only have to solve these two equations for and choose the correct of the two possible solutions which is given by

$$(\pi_1, \pi_2) = \left(\tfrac{1}{2} - \sqrt{\tfrac{1}{6}}, \tfrac{1}{2} + \sqrt{\tfrac{1}{6}}\right).$$

1.3 More on One-Period and Discrete-Time Approaches to Portfolio Selection

Our supply of material concerning one-period and discrete-time models and corresponding methods for portfolio optimisation is far from exhausted. Instead of the mean-variance approach one could look at optimisation problems of the form

$$\max_{\pi \in \mathbf{R}^d} E(U(R)) \tag{7}$$
$$\text{s.t.} \sum_{i=1}^{d} \pi_i = 1, \quad \pi_i \geq 0, \quad i = 1,\ldots, d$$

with a general utility function U(.). For suitable choices of U(.) we could even drop the non-negativity constraints and still get a finite optimal solution. For a discussion of such problems, and especially the characteristics of reasonable classes of utility functions we refer the interested reader to (Merton 1990).

As a possible generalisation of one-period models, one immediately thinks of multi-period models in which an investor is allowed to change his holdings at some given trading dates 0, 1,..., T−1, before the time horizon $T \in \mathbf{N}$. Instead of a single portfolio vector the investor is now obliged to choose a **portfolio process** $\pi(t)$, t = 0, 1,..., T−1 where $\pi(t)$ has to be chosen at time t, without knowledge of the security prices at later trading dates. If we further assume that an investor behaves in a self-financing way (i.e. the wealth of his holdings at time t *before* choosing $\pi(t)$ equals the wealth of the holdings *after* the choice of $\pi(t)$) then we can set up a multi-period version of problem (7). Under the assumption that at each trading date t the set of possible security prices is finite, it can be solved using dynamic programming methods. We will present an example of such a multi-period model and its numerical solution in Section 3.2.

1.4 Criticisms and Limitations of Discrete-Time Models

Although the mean-variance approach has been widely accepted and appreciated by practitioners and academics for a number of years, one can raise some serious arguments against it. One criticism is that risk is only measured in terms of the variance of the portfolio return. Bounding or minimising the variance does not only lead to low deviations from the mean return on the down side but also on the up side. The symmetric form of the variance has the undesirable side effect of not only bounding possible losses, but also possible gains. To overcome this drawback, one should instead look at one-period problems of the form (7) that allow for a choice of a suitable (non-symmetric) utility function.

However, the main drawback of the mean-variance approach is the static nature of the problem (which is of course a common limitation of all one-period models). After the decision concerning the allocation of initial wealth to the different securities has been made at the beginning of the period, there will be no further action until the time horizon. Even if there are dramatic changes in price expectations, or in the whole market situation, the mean-variance approach does not allow for a reaction by the investor. Once the initial portfolio is chosen, the investor's job is complete and his only feasible action is to watch the prices move. In fact, the modelling of the security prices merely consists of looking at the means and covariances of the returns of the prices. This is an extreme oversimplification of reality and totally ignores the highly volatile behaviour and dynamic nature of the prices that is invariably observed in practise.

A discrete-time generalisation of the mean-variance approach in particular and of one-period models in general is no real solution to this problem. First of all, the number of trading events is still finite, and — more seriously — each trading event can occur at fixed time instants. To allow for a relatively fast reaction from the investor to changes in the stock market, the time between the trading dates must be small, and this requires a large number of trading times to be modelled. However, in order to solve such a discrete-time problem by dynamic programming methods, either the number of possible security prices at each trading date or the number of trading dates must be small. If this is not the case the computational effort required to perform the usual backwards induction to solve the problem will explode (see Section 3.2 for a short explanation and discussion of this argument). Thus we are forced to discretise not only time but also the state space (i.e. the set of possible security prices) in a rough way .

In contrast, continuous-time models offer more satisfying solutions to these problems. By allowing for the possibility of trading at every time instant (during the opening hours of the stock exchange) the investor can react immediately if the situation dictates this. Furthermore, the highly developed mathematical tools of stochastic control theory and stochastic calculus permit the development of models in which the stock prices can attain any positive number, i.e. there will be no limitation either in time or space. Although the mathematical models and methods are more complicated, the structure and form of the solution to the portfolio problem is typically much clearer and more explicit than in the discrete-time case. There is even an enormous gain in tractability and speed in computing the optimal solutions.

Of course, there are also drawbacks with continuous-time models. The biggest such problem still seems to be that of including transaction costs. To be tautological, since trading is the only action of a trader, these costs cannot be neglected and their impact on the form of optimal trading strategies in continuous–time models is dramatic (see Chapter 5). As shown in the examples in Chapter 5 of this book, there are some promising approaches to this problem but there is still a need for more theoretical results and algorithms that establish a (usually numerical) solution for portfolio problems under transaction costs in (segments of) markets of realistic size (i.e. where the number of traded securities is at least around 20–30, not to mention the enormous total number of securities traded at the major stock exchanges).

CHAPTER 2

THE CONTINUOUS-TIME MARKET MODEL

2.1 The Security Price Processes

The main criticism against the one-period mean-variance approach is that of the static modelling both of the stock prices and of the actions of the investors. To overcome the resulting problems and limitations we present a continuous-time framework for both prices and actions, which has become the standard model in mathematical finance (see (Karatzas 1989) or (Merton 1990)). In this section we will first state the (stochastic) differential equations governing the evolution of the security prices. This will be followed by an explanation of and motivation for their form and the statement of some properties of the so obtained price processes. The modelling of the behaviour of the investors and the presentation of the main characteristics of the resulting security market will be considered in subsequent sections.

i) The General Setting

We will consider a financial market consisting of one riskless asset with a rate of return $r(t)$ (the "interest rate") and n risky assets with mean rates of return $b(t) = (b_1(t),..., b_n(t))'$ and a volatility matrix $\sigma(t) = (\sigma_{ij}(t))$, $i = 1,..., n$, $j = 1,..., m$, at time t. The riskless asset will be called a "bond", although this is a bit misleading (Actually, it behaves like a bank account but we adopt the usual naming practise). Likewise, the risky assets are called "stocks". Let $P_0(t)$ be the price of the bond and $P_i(t)$ be the price of stock number i, $i = 1,..., n$, at time t. In our model, the price dynamics are given by the equations

$$dP_0(t) = P_0(t)\, r(t)\, dt, \qquad P_0(0) = 1, \tag{1}$$

$$dP_i(t) = P_i(t)\left(b_i(t)dt + \sum_{j=1}^{m} \sigma_{ij}(t)dW_j(t)\right), \quad P_i(0) = p_i \;, i = 1,..., n \tag{2}$$

where $W(t) = (W_1(t),..., W_m(t))'$ is an m-dimensional Brownian motion defined on a complete probability space (Ω, F, P). The information structure is given by the P-augmentation $\{F_t\}_{t\in[0,T]}$, $T < \infty$, of the natural filtration generated by W(t), $t\in[0,T]$ (see Appendix A). We will also assume $F = F_T$ and that the (components of the) market coefficients r(t), b(t), and $\sigma(t)$ are all F_t-adapted and uniformly bounded in $(t, \omega)\in[0,T]\times\Omega$. Furthermore, $\sigma(t)\sigma(t)'$ is required to be uniformly positive definite, i.e. there exists a positive constant δ with

$$v'\sigma(t, \omega)\sigma(t, \omega)' v \geq \delta \| v \|^2 \quad \forall (t, \omega)\in[0,T]\times\Omega,\ v\in\mathbf{R}^n. \tag{3}$$

This last assumption automatically implies $m \geq n$, i.e. the dimension of the Brownian motion W(t) should be at least as big as the number of stocks. This assumption ensures that there are no redundant stocks.

Of course, there are unique (in the sense of $L\otimes$P-a.e.), explicit solutions to the equations (1) and (2) which are given by (see Theorem B15 in the Appendix)

$$P_0(t) = \exp(\int_0^t r(s)ds), \tag{4}$$

$$P_i(t) = p_i \exp\left(\int_0^t (b_i(s) - \tfrac{1}{2}\sum_{j=1}^m \sigma_{ij}(s)^2)ds + \sum_{j=1}^m \int_0^t \sigma_{ij}(s)dW_j(s) \right), \tag{5}$$

but the differential representations via the equations (1),(2) will prove to be more convenient for our subsequent analysis.

Before we continue with our description of the continuous-time market model we will try to justify our way of modelling the security prices as the solutions of equations (1) and (2). For simplicity, we assume $n = m = 1$ and that the market coefficients r, b, σ are constant. For convenience, we will then write p, b, σ instead of p_1, b_1, σ_{11}.

ii) Motivation for the Form $P_0(t)$ of the Bond Price

Note first that if a riskless security (with initial unit price) earns interest at rate r, which is proportional to time but only paid at the time horizon T, it will have a value of $P_{0,1}(T) = 1+ rT$ at time $t = T$. If it earns interest at the same rate but if now interest is paid at times T/2 and T then it will have the prices $P_{0,2}(T/2) = (1+ rT/2)$ and $P_{0,2}(T) = (1+ rT/2)^2$ at these times. To understand the expression for $P_{0,2}(T)$, note that in the 2-period case on [T/2,T] compound interest is also paid on the interest payment rT/2 which has already accrued at time T/2. Generally, if interest payments occur at every time instant iT/n, the prices $P_{0,n}(iT/n)$ are given by $P_{0,n}(iT/n) = (1 + rT/n)^i$. Passing to the limit for n means a change to continuous compounding of the interest rate and results in the limiting price $P_0(t) = \exp(rt)$. This simple fact can be visualised with the help of Figure 3, in which we have taken $T = 1$, $r = 0.5$ (which is of course an unrealistically high value, but it helps visualising the differences between the various prices). Further, the prices $P_{0,n}(iT/n)$ for $n = 1, 2, 4$ have been joined by straight lines and the graph of $P_0(t)$ has been plotted. From the above construction the sequence $P_{0,i}(t)$ is monotonic in i with $P_0(t)$ as an upper bound.

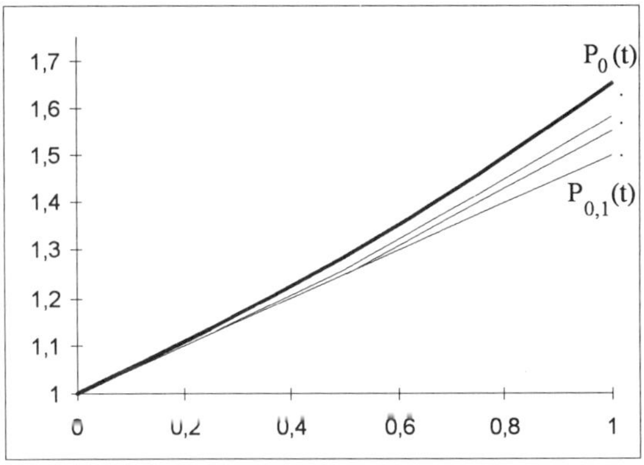

Figure 3: Bond prices with discrete and continuously compounded interest

iii) Motivation for the Form $P_1(t)$ of the Stock Price

The stock price model is based on the assumption that in average a stock behaves like a bond but possibly has a higher "interest rate" $\tilde{b}$, with the difference between the stock and bond interest rates regarded as a premium for the intrinsic risk in a stock investment. We assume that the whole trend of the price development lies in this part of the price. Thus, the risky part of the investment should be modelled by a stochastic process **without** tendency, i.e. by a process with zero mean at every time instant. If we further assume that the log-return $\ln(P_1(T))$ at time T be normally distributed around $\tilde{b}T$, it is also reasonable to assume that the deviation of the log-return from $\tilde{b}T/2$, at time T/2, should be normally distributed. But, because one thinks of the resulting deviations from the line $\tilde{b}t$ as a sum of small deviations over time, it seems natural to assume that the variance of the deviation at time 0.5*T is just half as big as that at time T. The same argument applied to every time instant in [0,T] and the additional requirement of a continuous price process then directly lead to the model

$$\ln(P_1(t)) = \tilde{b}t + \sigma W(t) = \left(b - \tfrac{1}{2}\sigma^2\right)t + \sigma W(t), \tag{6}$$

where $W(t)$ is a one-dimensional Brownian motion (and where we have assumed $P_1(0) = 1$). In Figure 4 we have plotted a simulated path of $\ln(P_1(t))$ of this form together with the line $\tilde{b}t$, the "bond part" of $\ln(P_1(t))$. As parameters we have chosen $\tilde{b} = 0.1$ and $\sigma = 0.2$.

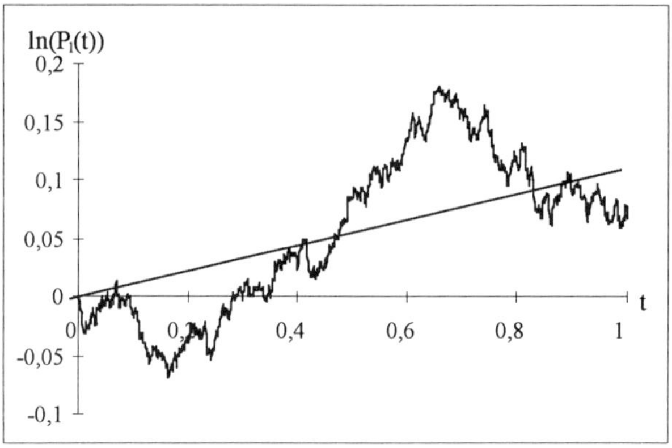

Figure 4: Simulated path of the logarithm of the stock price and its "bond"-part

So far, we have arrived at the form of a geometric Brownian motion,

$$P_1(t) = p\, e^{\tilde{b}t + \sigma W(t)} \tag{7}$$

for modelling the stock price. By computing the expected value and the variance of this stock price we get a self explanatory interpretation of the market coefficients.

Lemma 1

a) Let $P_1(t)$ be of the form (7). With the notation $b = \tilde{b} + \tfrac{1}{2}\sigma^2$ we have :

$$E(P_1(t)) = p\exp(bt),$$
$$Var(P(t)) = p^2 \exp(2bt)\left[\exp(\sigma^2 t) - 1\right].$$

b) The process $\{Z(t), F_t\}_{t\in[0,T]}$ with $Z(t) := \exp(cW(t) - \tfrac{1}{2}c^2 t)$ is a martingale for every $c \in \mathbf{R}$.

Proof :

a) Using the constant b introduced above, we obtain

$$E(P_1(t)) = p_1 \int_{-\infty}^{\infty} e^{(b-\frac{1}{2}\sigma^2)t+\sigma x} \frac{1}{\sqrt{2\pi t}} e^{-\frac{x^2}{2t}} dx$$

$$= p_1 e^{bt} \int_{-\infty}^{\infty} \frac{1}{\sqrt{2\pi t}} e^{-\frac{(x-\sigma t)^2}{2t}} dx = p_1 e^{bt} .$$

The proof for the representation of the variance is similar and therefore left as an exercise for the reader.

b) We have

$$E(Z(t)|\, F_s) = E\left(e^{cW(t)-\frac{1}{2}c^2 t} |F_s \right)$$

$$= e^{cW(s)-\frac{1}{2}c^2 s} E\left(e^{c(W(t)-W(s))-\frac{1}{2}c^2 (t-s)} |F_s \right)$$

$$= e^{cW(s)-\frac{1}{2}c^2 s} = Z(s) \quad \text{a.s..}$$

For the third equality note that the increment W(t)–W(s) is normally distributed with mean zero and variance t–s, and is independent of F_s. Hence, the conditional expectation is equal to the unconditional one, and we get the required equality with the help of part a) of the lemma.

□

Having this lemma in mind, we now see that the stock price

$$P_1(t) = p_1 e^{bt} e^{\sigma W(t)-\frac{1}{2}\sigma^2 t}$$

is simply given as the product of its mean and a positive martingale with mean equal to one. In other words, it is the product of a component reminiscent of a bond price and a process that causes the price to fluctuate randomly (but without tendency) around this "bond price". This representation (and Lemma 1) also justifies the name "mean rate of stock return" for b.

Figure 5 contains some simulated paths of such a price process where we have chosen $b = 0.14$ and $\sigma = 0.2$. Additionally, we have plotted the graph of its expectation, exp(bt), $t \in [0,1]$.

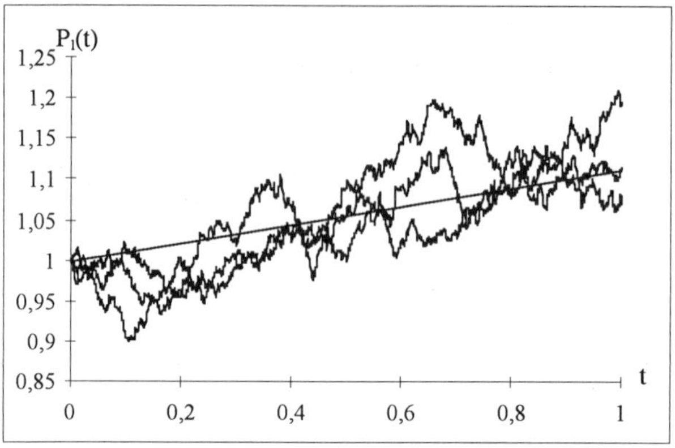

Figure 5: Some simulated paths of the stock price together with its expectation

Remark 2

a) Lemma 1 and the interpretation of the stock price(s) remain valid in the case of an arbitrary number of securities with constant coefficients. Then p and b have to be replaced by p_i and b_i, respectively. For σ^2 we have to put in the sum of squares

$$\sum_{j=1}^{m} \sigma_{ij}^{2}.$$

The proof of Lemma 1 goes through if we note that the sum of independent normal random variables with zero mean is again normally distributed with mean zero. For the same reason, part b) of the lemma stays valid if we make the analogous generalisations.

b) In the case of randomly fluctuating market coefficients b, r, σ that satisfy the assumptions made in subsection i) the interpretation of the stock price given above stays correct. The Novikov condition (see Section B.6 of the Appendix) yields that

$$Z_i(t) := \exp\left(-\tfrac{1}{2}\int_0^t \sum_{j=1}^{m} \sigma_{ij}(s)^2\, ds + \sum_{j=1}^{m} \int_0^t \sigma_{ij}(s)\, dW_j(s) \right)$$

is a martingale, and Theorem B13 gives us the existence of the relevant moments of $P_i(t)$. However, we then have

$$E(P_i(t)) = p_i\, E\left(\exp\left(\int_0^t b_i(s)\, ds \right) \right).$$

In Chapter 4 we will also encounter situations in which the market coefficients are not necessarily bounded and in which the Novikov condition is not satisfied. There, we cannot ensure that the process $Z_i(t)$ be a martingale (we can then only say that it is a positive local martingale).

2.2 The Wealth Process and the Actions of a Small Investor

After having modelled the security prices we now turn to the investors and their admissible actions. We assume that an investor is endowed with a given initial capital at time $t = 0$. His possible actions are rebalancing of his holdings (i.e. buying or selling of securities) and consumption of (parts of) his wealth. In this context, by **wealth**, we mean the value of current holdings. The investor is allowed to hold a negative amount of bonds, which corresponds to the purchase of shares of stocks that are at least partly financed by credit. Also we permit the possibility of a negative stock position, which corresponds to the practise of short-selling stocks (i.e. selling shares without owing them but having the obligation to deliver them later). We will consider short-selling constraints for both stocks and bond in Chapter 4. As a further idealisation, we assume that there are no transaction costs i.e. no brokerage fees for selling or buying that will reduce the investor's wealth (the impact of transaction costs will be the subject of Chapter 5). Our investor is assumed to be a **small** investor, meaning that his actions alone do not influence the price. Further, he is only allowed to act in a self-financing way. That is to say, all of his actions must be financed from his current wealth, which therefore has to be equal to the sum of his initial wealth, plus his gains or losses from investment, minus his consumption, up to the current time. Of course, all of these actions must be chosen without any knowledge of future security prices.

These requirements will be formalised in

Definition 3

Let $T > 0$ be fixed (the "time horizon").

i) A **trading strategy** is an $\mathbf{R}^n$-valued, F_t-adapted process $\varphi(t)$, $t\in[0,T]$, with

$$\int_0^T |\varphi_0(t)|dt < \infty \quad \text{a.s.},$$

$$\sum_{i=1}^{n}\sum_{j=1}^{m} \int_0^T (\varphi_i(t)P_i(t)\sigma_{ij}(t))^2 dt < \infty \quad \text{a.s.}\,.$$

ii) Let φ be a trading strategy. The process

$$X(t) := \sum_{i=0}^{n} (\varphi_i(t)P_i(t))$$

is called the **wealth process** ("value of the current holdings") corresponding to φ. X(0) is called the **initial wealth**.

iii) A non-negative, adapted process c(t), $t \in [0,T]$, with

$$\int_0^T c(t)dt < \infty \quad \text{a.s.}$$

will be called a **consumption rate process** (for brevity: consumption process).

iv) A pair (φ, c) consisting of a trading strategy φ and a consumption process c will be called **self-financing** if the wealth process X(t) corresponding to φ satisfies

$$X(t) = X(0) + \sum_{i=0}^{n} \int_0^t \varphi_i(s)dP_i(s) - \int_0^t c(s)ds \quad \forall t \in [0,T]. \tag{8}$$

Remark 4: "The self-financing condition"

The natural way to impose that a trading strategy be self-financing is to demand that the wealth of an investor before any action at time t should equal his wealth after this action, minus his consumption at this time as possible part of the action. However, there is no sensible, obvious analogue of this requirement in continuous time. To show that equation (8), given in part iv), of the definition really describes the type of behaviour we have in mind, let us look at the following discrete-time example with just one bond and stock with prices $P_0(i)$, $P_1(i)$, i = 0,..., n. Further let c(i) be the amount of consumption at time i and $\varphi_0(i)$, $\varphi_1(i)$ be the trading strategy. The natural self-financing condition is

$$X(i) = \varphi_0(i)P_0(i) + \varphi_1(i)P_1(i) = \varphi_0(i-1)P_0(i) + \varphi_1(i-1)P_1(i) - c(i). \tag{9}$$

However, the right hand side of equation (9) is equal to

$$\begin{aligned}\varphi_0(i-1)\,(P_0(i) - P_0(i-1)) + \varphi_1(i-1)\,(P_1(i) - P_1(i-1)) - c(i) \\ + \varphi_0(i-1)P_0(i-1) + \varphi_1(i-1)P_1(i-1)\,.\end{aligned} \tag{10}$$

If we assume that the investor was also acting in a self-financing way at the preceding time i−1, we can recast the last two terms of representation (10) in a similar way. Repeating the whole procedure i-times, using the notation $\Delta P_j(i) = P_j(i) - P_j(i-1)$,

$j = 0,1$, and noting $x = X(0) = \varphi_0(0)P_0(0) + \varphi_1(0)P_1(0)$, equation (9) acquires the form

$$X(i) = x + \sum_{j=1}^{i}\left(\varphi_0(j)\Delta P_0(j) + \varphi_1(j)\Delta P_1(j)\right) - \sum_{j=1}^{i} c(j)$$

which is exactly the discrete-time analogue of requirement (8), thereby demonstrating that the above definition is reasonable.

Apart from the above description of the investment strategy in terms of the **number** of shares of the different securities that the investor holds at time t, it is also possible to describe his strategy by the **fractions** of his wealth that he invests in the different securities, i.e. by the quotients

$$\pi_i(t) := \frac{\varphi_i(t)P_i(t)}{X(t)}, \quad i = 0,\ldots, n.$$

Given that we are provided with (the observable) wealth and price processes then we can deduce these quotients from the trading strategy and vice versa. The quotients and the trading strategies are in this sense equivalent representations of the investor's actions. The vector of these quotients will be called a **portfolio process**. For computational purposes, it will be convenient to use the name portfolio process only for the last n components of this vector. This will not constitute a loss of information as the quotient describing the fraction of wealth invested in the bond is given by $1 - \pi(t)'\underline{1}$ (where $\underline{1} := (1,\ldots, 1)'$ is the vector (of appropriate dimension) with unit components). We have thus defined a portfolio process with the help of a trading strategy φ. Further, with the help of the portfolio process, we can derive a stochastic differential equation as a representation for the wealth process of an investor. This will then allow to define such a portfolio process without referring to the corresponding trading strategy. To do so, we look at the differential representation of the wealth process of an investor who uses a self-financing pair (φ, c), as in part iv) of Definition 3. Using this part iv), the security price equations, and the relation between the trading strategy and the corresponding portfolio process we obtain

$$dX(t) = \varphi_0(t)dP_0(t) + \sum_{i=1}^{n}\varphi_i(t)dP_i(t) - c(t)\,dt$$

$$= \varphi_0(t)P_0(t)r(t)dt + \sum_{i=1}^{n}\varphi_i(t)P_i(t)\left(b_i(t)dt + \sum_{j=1}^{m}\sigma_{ij}(t)dW_j(t)\right) - c(t)\,dt$$

$$= X(t)[\{(1-\pi(t)'\underline{1})r(t) + \pi(t)'b(t)\}dt + \pi(t)'\sigma(t)dW(t)] - c(t)\,dt\,, \qquad (11)$$

i.e. we have derived a linear stochastic differential equation for X(t). Together with the initial condition X(0) = x, it has a unique solution whose explicit form is given by (see the "variation of constants" formula B15 in Section B.3 of the Appendix)

$$X(t) = Z(t)\left(x + \int_0^t \frac{1}{Z(u)} c(u)du \right)$$

with

$$Z(t) = \exp\left(\int_0^t \left(r(s) + \pi(s)'(b(s) - r(s)\underline{1}) - \tfrac{1}{2}\|\pi(s)'\sigma(s)\|^2 \right) ds + \int_0^t \pi(s)'\sigma(s)dW(s) \right).$$

If an investor uses a consumption rate of the special form c(t) = kX(t) (with k > 0), the corresponding wealth process has the simpler form

$$X(t) = x\, e^{\int_0^t \left(r(s) + \pi(s)'(b(s) - r(s)'\underline{1}) - \frac{1}{2}\|\pi(s)'\sigma(s)\|^2 - k \right) ds + \int_0^t \pi(s)'\sigma(s)dW(s)}$$

(which is also a direct consequence of the variation of constants formula). In particular, we see that for all those strategies of the investor that incorporate a consumption process of the above form, the wealth process will stay positive.

As the solution X(t) of the stochastic differential equations (11) is unique for any given portfolio process π(t) (satisfying some necessary integration requirements), we can define a portfolio process via the above equation for X(t). Note also that the form of this equation ensures that the trading strategy corresponding to the portfolio process will be self-financing. Thus, a portfolio process (as given in the following definition) is self-financing by construction.

Definition 5

A pair (π, c) consisting of a consumption process c and an $\mathbf{R}^n$-valued, F_t-adapted process π(t), t∈[0,T], such that the stochastic differential equation (11) with the initial condition X(0) = x has a unique solution X(t) satisfying

$$\int_0^T \|\pi(t)X(t)\|^2 dt < \infty \text{ a.s.},$$

will be called a **self-financing strategy**. π will be called a (self-financing) **portfolio process**.

For most of the time in this book, we will consider investors who trade in such a way that they achieve a non-negative wealth over the whole time interval $[0,T]$, i.e. those that successfully avoid bankruptcy (in the strict sense).

Definition 6
i) A wealth process with initial wealth $x > 0$ corresponding to the self-financing strategy (π, c) is called **admissible** if we have $X(t) \geq 0$ for all $t \in [0,T]$ a.s. .
ii) $A(x) := \{(\pi, c) | (\pi, c)$ self-financing strategy with admissible $X(t)$ with $X(0) = x\}$ is called the **set of admissible strategies** (with initial wealth $x > 0$).

2.3 Completeness of the Market Model and the Growth-Optimum Portfolio

If we are in the special case that the number of stocks is equal to the dimension of the underlying Brownian motion, then our market model has a remarkable feature: it yields a **complete** market. The meaning of the term completeness will be obvious after the statement of the following theorem (see for example (Cvitanic and Karatzas 1992) or Section 8.E of (Duffie 1992)). To state it, we introduce the notation

$$\theta(t) = \sigma^{-1}(t)(b(t) - r(t)\,\underline{1}),$$

$$H(t) = \exp\left(-\int_0^t (r(s) + \tfrac{1}{2}\|\theta(s)\|^2)ds - \int_0^t \theta(s)'dW(s)\right).$$

Theorem 7
a) For every $(\pi, c) \in A(x)$ with corresponding wealth process $X(t)$ we have

$$E\left(\int_0^t H(s)c(s)ds + H(t)X(t)\right) \leq x.$$

b) For every consumption process $c(t)$, $t \in [0,T]$, and every non-negative, F_T - measurable random variable B with

$$x := E\left(\int_0^T H(s)c(s)ds + H(T)B\right) < \infty,$$

there exists a portfolio process $\pi(t)$, $t \in [0,T]$, with $(\pi, c) \in A(x)$, such that the corresponding wealth process $X(t)$ satisfies

$$X(T) = B \quad \text{a.s.}\,.$$

Proof :
Note first that the process H(t) is the unique solution of the stochastic differential equation

$$dH(t) = -H(t)\{r(t)dt + \theta(t)\, dW(t)\}, \qquad H(0) = 1\,.$$

a) Let $(\pi, c) \in A(x)$ for $x > 0$ with corresponding wealth process X(t). Using the above differential representation of H(t) and the stochastic differential equation for the wealth process X(t), together with Itô's formula, we discover that

$$H(t)\, X(t) + \int_0^t H(s)c(s)ds$$

$$= x + \int_0^t H(s)dX(s) + \int_0^t X(s)dH(s) + \langle H, X\rangle_t + \int_0^t H(s)c(s)ds$$

$$= x + \int_0^t H(s)X(s)[r(s) + \pi(s)'(b(s) - r(s)\underline{1}) - r(s) - \pi(s)'\sigma(s)\theta(s)]ds$$

$$+ \int_0^t H(s)X(s)[\pi(s)'\sigma(s) - \theta(s)]' dW(s)$$

$$= x + \int_0^t H(s)X(s)[\pi(s)'\sigma(s) - \theta(s)]' dW(s). \qquad (12)$$

Due to the non-negativity of H(t), X(t) and c(t), the left side of the above equation is non-negative. On the other hand, the term after the last equality sign is a local martingale. But due to Proposition A17, non-negative local martingales are super-martingales. Thus, we have

$$E\left(\int_0^t H(s)c(s)ds + H(t)X(t)\right)$$

$$= x + E\left(\int_0^t H(s)X(s)[\pi(s)'\sigma(s) - \theta(s)]' dW(s)\right) \le x,$$

which proves part a).
b) Define the process

$$X(t) = \frac{1}{H(t)}\, E\left(\int_t^T H(s)c(s)ds + H(T)B \,\middle|\, F_t\right), \quad t \in [0,T]\,.$$

The process X(t) is non-negative with continuous paths (note that the path continuity of $E(H(T)B|F_t)$ is a consequence of the martingale representation theorem B22) and

$$X(0) = x, \quad X(T) = B \quad \text{a.s.}\,.$$

Further, define

$$M(t) = E\left(\int_0^T H(s)c(s)ds + H(T)B\middle|F_t\right).$$

The definitions of M(t) and X(t) yield

$$M(t) = X(t)\,H(t) + \int_0^t H(s)c(s)ds \quad \text{a.s.}\,, \quad t\in[0,T]\,. \tag{13}$$

Obviously, M(t) is a Brownian martingale. Using the martingale representation theorem, we establish the existence of an $\mathbf{R}^n$-valued, adapted process ψ satisfying

$$P\left(\int_0^T \|\psi(t)\|^2 dt < \infty\right) = 1,$$

$$M(t) = x + \int_0^t \psi(s)'dW(s) \qquad \forall\, t\in[0,T] \text{ a.s.}\,.$$

Equating this representation for M as a stochastic integral with the right hand side of equation (13), then, with the help of equation (12), we can conclude that X(t) is the wealth process corresponding to the strategy (π, c), where c is the given consumption process and $\pi(t)$ is given by

$$\pi(t) = (\sigma^{-1}(t))'\left(\frac{\psi(t)}{X(t)\,H(t)} + \theta(t)\right), \quad t\in[0,T].$$

In particular, the construction of $\pi(t)$ and of X(t) especially imply $(\pi, c)\in A(x)$.

□

Remark 8

a) Before interpreting this important result, let us take a closer look at the process H(t). It can easily be verified that 1/H(t) is the wealth process that corresponds to the strategy $(\pi, 0)$ with

$$\pi(t) = (\sigma^{-1}(t))'\theta(t) = (\sigma^{-1}(t))'\sigma^{-1}(t)(b(t)-r(t)\underline{1}) \ .$$

If we interpret $\pi(t)$ as the vector of the "relative local risk premia" gained for investing in the stocks (an interpretation, which is even more obvious in the case of n=1 and constant coefficients, in which we have $\pi(t) = (b-r)/\sigma^2$), then part a) of the proposition tells us that, compared to this investment, no other admissible strategy performs better (in the mean). H(t) seems to be exactly the right discount process to judge the advantage of a strategy. Therefore, it is also called a market numeraire. Moreover, we will see in Section 3.4 that $\pi(t)$ is the optimal portfolio process if an investor tries to maximise E(ln(X(T))). Therefore, $\pi(t)$ is called the **growth-optimum portfolio**.

b) In terms of H(t), part a) of Theorem 7 puts bounds on the ambitions of an investor regarding consumption and /or terminal wealth: For any given, desired consumption process c, the existence of a portfolio process π with $(\pi, c)\in A(x)$ is only possible if we have

$$E\left(\int_0^T H(s)c(s)ds\right) \leq x \ .$$

Any desired payoff B at the terminal time T (i.e. a non-negative, F_T-measurable random variable) can only be attained by following a suitable strategy $(\pi, c)\in A(x)$ if B satisfies

$$E(H(T)B) \leq x \ .$$

Put another way, given future aims, such as a payoff of B or consuming according to a consumption process c, part a) of the theorem yields lower bounds for the initial wealth that is needed to achieve these goals. However, part b) of the theorem ensures that these lower bounds are exactly the correct amount; more precisely, there exist admissible strategies (π, c) such that the goals can be reached with exactly the above mentioned initial endowments. As we are now able to reach every such goal exactly, provided we have a sufficient initial endowment, our market model is called a **complete market**. This feature of the market will play a central role in the martingale method for solving the portfolio problem in Chapter 3.

c) The portfolio process constructed in the proof of part b) of Theorem 7 is the unique admissible one with respect to $\mathcal{L}\otimes P$, such that the corresponding wealth process X(t) satisfies X(0) = x and X(T) = B a.s.. To see this, assume that $\nu(t)$ would be another such portfolio process with $(\nu, c)\in A(x)$ and corresponding wealth process Y(t). Then, the process Z(t) = X(t) – Y(t) would satisfy

$$Z(t) = E(Z(T)|F_t) = 0 \quad \forall \ t\in[0,T] \ .$$

Hence, X and Y coincide a.e. with respect to $L \otimes P$. From the uniqueness of the representation of an Itô-process (see Section B.2) we then get the uniqueness assertion for the portfolio process.

2.4 Option Pricing: A Short Introduction

As its title indicates, option pricing will play no central role in this book. However, this introduction serves as a reference, giving some ideas of option pricing that we will pick up again in Chapter 6 when we look at value preserving portfolio strategies.

There is now a huge variety of books on derivatives, their use and their valuation (see e.g. Baxter and Rennie (1996), Wilmott e.a. (1993), Hull (1993)) which also highlight the different mathematical approaches to the option pricing problem. As we will see shortly, the work we have put into showing the theorem on completeness of the market, Theorem 7, in the previous section, will now prove to be fruitful. Using this result in the market setting of Section 3 will make the solution of the option pricing problem extremely simple.

Definition 9
A non-negative, F_T-measurable random variable B will be called a **contingent claim** if it satisfies

$$E\left[(B)^{\mu}\right] < \infty \tag{14}$$

for some $\mu > 1$.

Example 10
a) "European Call Option"
The most popular example of a contingent claim is the so called **European call option**. It is a contract that gives its buyer the right (but not the obligation) to buy one share of stock at the agreed future time $t = T$ (the "maturity") for a fixed price K (the "exercise price") from the seller of the European call. If at maturity, the price per share $P_1(T)$ is above K, then the owner of the option would exercise his right and buy the share for a price of K. By selling it immediately at the market he will realise a profit of $P_1(T) - K$. However, if $P_1(T)$ is below K, then the owner of the option would not exercise his right and thus gain no profit but also suffer no loss. We can therefore identify the European call with its terminal payment

$$B = (P_1(T) - K)^+,$$

where x^+ denotes the positive part of the real number x.
b) "European Put Option"
In contrast to a European call, a **European put** gives its owner the right to sell one share of stock for the fixed price K at the terminal time t = T to the seller of the put. We can thus identify the European put option with the terminal payment of

$$B = (K - P_1(T))^+ .$$

There are many more examples of options (and also more complicated ones) such as Asian options, look backs, digitals, ... We refer the interested reader to the above cited literature. Also, there exist options for which the exercise time is not fixed in advance, rather, it is chosen by the owner of the option (so-called **American options**). For all of these claims, there is the question of how to determine the "fair price" of such a contract. The main principles underlying the calculation of this price are those of replication and no arbitrage. If one is able to replicate the payoff of a contingent claim by investing according to an admissible portfolio strategy, then the price of the contingent claim should not be above the initial amount of money, x, needed to set up this strategy. The reason for this is obvious: if we had an opportunity to sell an option for a price of y, exceeding x, then we could sell the option, take the amount x to set up the replicating strategy, and pocket the difference y – x at time t = 0. At time t = T the payment obtained from the strategy and the one we make to the option's owner cancel out. In that case, we have a deal containing no risk but yielding a positive payment at time t = 0, a so called **arbitrage possibility**. (a similar argument shows that, provided the existence of a replication strategy with initial wealth x, the option price cannot be smaller than x). The popular wisdom is that if such arbitrage possibilities were to exist, then the market would immediately react in such a way that they would disappear. Therefore, in theoretical models it is justified to assume absence of arbitrage opportunities (the "**principle of no arbitrage**"). Now the fair price of a contingent claim B will be defined as the infimum of the amounts of money needed to set up a replicating portfolio strategy for B. The following definition will put these considerations into a rigorous framework.

Definition 11

a) $(\pi, c) \in A(x)$ is called a **replication strategy** for the contingent claim B if the wealth process X(t) corresponding to (π, c) satisfies

$$X(T) = B \qquad \text{a.s.} .$$

Let $D(x) \subseteq A(x)$ be the set of replication strategies for B.

b) The real number

$$p := \inf\{\, x > 0 \mid D(x) \neq \varnothing \,\}$$

is called the **fair price** of the contingent claim B.

Using the boundedness of the market coefficients and Hölder's inequality, we can show that the requirement (14) implies

$$x^* := E(H(T)B) < \infty.$$

But then part b) of the theorem on complete markets (i.e. Theorem 7) already implies the existence of a replicating strategy with an initial wealth of x^* and a consumption process that vanishes identically. Hence, we must have

$$p \leq x^*.$$

Moreover, we have equality between p and x^*, which is the result of the following theorem.

Theorem 12
The fair price p of a contingent claim B is given by

$$p = x^* = E(H(T)B),$$

and there exists a unique replication strategy $(\pi^*,c^*)\in D(p)$ with $c^*(t)=0\ \forall\, t\in[0,\, T]$ a.s.. Its corresponding wealth process, $X^*(t)$, (the "valuation process for B") admits the representation

$$X^*(t) = \frac{1}{H(t)}\, E\,(H(T)B \mid F_t)\,.$$

Proof :
Let $(\pi,c)\in D(x)$ for some $x > 0$ with corresponding wealth process $X(t)$. Then by part a) of Theorem 7 and the replication property for B of (π,c) we have

$$x \geq E(H(T)X(T)) = E(H(T)B^*) = x^*\,.$$

Hence, every replication strategy needs at least an initial wealth of x^*. On the other hand, (π^*,c^*) as above is the unique replication strategy in $D(x^*)$ (see Remark 8 c)) which yields all the assertions of the theorem.

□

Remark 13
The compact and elegant form of the above proof demonstrates the usefulness and power of Theorem 7. Indeed, it yields that option pricing in a complete market is (in principle) simple, although the explicit computation of the expected values in Theorem 12 can be quite cumbersome.

Before discussing the theorem, let us apply it to our two examples. We will assume that the market simply consists of a bond and a single stock with constant market coefficients r, b, σ. This is exactly the setting underlying the famous Black-Scholes Formula (see (Black and Scholes 1973)).

Example 10 (continued): "The Black-Scholes Formula"
a) "European Call Option"
Computing E(H(T)B) for this particular setting, we get the fair price p_{call} of a European call as

$$p_{call} = \int_U^{\infty} e^{-\left(r+\frac{1}{2}\theta^2\right)T-\theta x}\left(p_1 e^{\left(b-\frac{1}{2}\sigma^2\right)T+\sigma x} - K\right)\frac{1}{\sqrt{2\pi T}} e^{-\frac{x^2}{2T}} dx$$

$$= p_1\Phi(d_1(0)) - Ke^{-rT}\Phi(d_2(0))$$

where $\Phi(.)$ is the distribution function of the standard normal distribution and where we have set

$$U = \frac{1}{\sigma}\left(\ln\left(K/p_1\right) - (b - \tfrac{1}{2}\sigma^2)T\right),$$

$$d_1(t) = \frac{\ln\left(p_1/K\right) + \left(r + \frac{1}{2}\sigma^2\right)(T-t)}{\sigma\sqrt{T-t}}, \quad d_2(t) = d_1(t) - \sigma\sqrt{T-t}\ .$$

The valuation process for the European call is given by

$$X_{call}(t) = P_1(t)\Phi(d_1(t)) - Ke^{-r(T-t)}\Phi(d_2(t))$$

which constitutes the famous Black-Scholes Formula. It is also easy to guess from its form that the trading strategy (φ_0, φ_1) that generate the valuation process is

$$\varphi_0(t) = -Ke^{-rT}\Phi(d_2(t)),\ \varphi_1 = \Phi(d_1(t)),$$

(where it is still necessary to check that this is a self-financing trading strategy !).

b) "European Put Option"
The valuation process of a European put has the form

$$X_{put}(t) = Ke^{-r(T-t)}\Phi(-d_2(t)) - P_1(t)\Phi(-d_1(t)) .$$

Discussion of Theorem 12 and of the Black-Scholes Formula
In a naïve approach to option pricing one might have suggested the discounted expected payoff

$$E(e^{-rT}(P_1(T)-K)^+), \text{ or } E\left(\exp\left(-\int_0^T r(s)ds\right)B\right),$$

in the general case, as the natural candidate for the option price. But in general, this is not the price given by Theorem 12. Moreover, a striking feature (and probably the main reason for the success) of the Black-Scholes Formula is that the mean rate of stock return, b, does not enter it. This stunning fact can be worked out with the help of Girsanov's theorem (see Section B.4 in the Appendix). By defining a new probability measure $\tilde{P}$ via

$$\tilde{P}(A) := E(Z(T)1_A) \quad \forall A \in F_T$$

with

$$Z(t) := \exp\left(-\int_0^t \theta(s)^T dW(s) - \tfrac{1}{2}\int_0^t \|\theta(s)\|^2 ds\right), \quad t \in [0,T],$$

we first note that we have

$$\tilde{E}\left(\exp\left(-\int_0^T r(s)ds\right)B\right) = E(H(T)B)$$

(where $\tilde{E}$ denotes expectation with respect to $\tilde{P}$). This implies that the naïve approach to option pricing is not totally wrong. It only needs a careful choice of measure. To interpret this change of measure from P to $\tilde{P}$, we note that by Girsanov's theorem B23, the process

$$\tilde{W}(t) = W(t) + \int_0^t \theta(s)ds$$

is a Brownian motion (of appropriate dimension) with respect to the new probability measure $\tilde{P}$. Thus, in the Black-Scholes model, the stock price $P_1(t)$ has the representation

$$P_1(t) = p_1 e^{(r-\frac{1}{2}\sigma^2)t+\sigma\tilde{W}(t)},$$

i.e. the discounted stock price $P_1(t)/P_0(t)$ is a martingale with respect to $\tilde{P}$. Because by construction, $\tilde{P}$ is also equivalent to P on F_T (i.e. the two probability measures have the same null sets when restricted to F_T), $\tilde{P}$ is called an **equivalent martingale measure** for $P_1(t)$. Changing from P to $\tilde{P}$ thus means changing from a subjective to a risk-neutral market, a market in which all securities have the same mean rates of return. This argument also applies in a general, complete market setting. Consequently, option pricing in a complete market as presented above, is simply option pricing via the naïve approach in a risk-neutral market. In such a market, b, the original mean rate of stock return, plays no role and therefore does not enter the option price !

However, in incomplete markets the situation is more subtle. While in complete markets there exists only one equivalent martingale measure (see (Harrison and Pliska 1981/3)), in incomplete markets we have many. In that setting, each equivalent martingale measure represents a consistent price system for the contingent claims, and we need additional arguments besides arbitrage reasons to justify the use of a particular one. We will return to this aspect in Chapter 6 when presenting the value preserving approach to portfolio selection.

2.5 Convergence of Discrete Markets to Continuous Markets

One of our main justifications for considering continuous-time market models (with a continuous state space) arose out of the limitations of discrete-time models (see Section 1.4). Although there is a gain in tractability and modelling possibilities by using a continuous-time model as in the previous sections, one could still argue against it in saying that due to our finite measuring scales in time, money and units of stock real world markets are discrete ones. Of course, one has to admit that the state space must be large and the time scale very fine such that a discrete-time model could be a realistic one. If such a model which is usually non-tractable from a computational point of view can be well approximated (in a reasonable sense) by a highly tractable continuous-time model then the use of continuous-time models is perfectly justified.

One could always argue that if time between trading dates in discrete-time models becomes smaller and smaller then a continuous-time model is a natural idealisation. However, such an argument would be a weak one if it would be the only justification. A much more important question is if structural properties of discrete-time models and their continuous-time limits coincide. Is it possible that properties we have in continuous-time models are also present in (at least certain) discrete-time ones. As an example, we could ask if for every complete continuous-time model there exists a sequence of complete discrete-time models approximating it. Other questions are of the type if convergence of discrete-time prices against continuous-time ones also implies the convergence of optimal trading strategies to corresponding portfolio problems in the different models. We cannot go into detail in this book and do not develop a convergence theory for stock market models.

The subject of convergence of finite market models against continuous-time ones is fully covered in (Willinger and Taqqu 1991). The topics treated there range from the basic approach of approximating continuous-time price models by discrete-time ones via approximating a Brownian motion by a random walk in the sense of weak convergence up to the treatment of the above mentioned structural questions.

In (Föllmer and Schweizer 1993) diffusion type models for stock prices (which include the model of this chapter) are derived as a limit of a sequence of prices obtained in (large scale) discrete markets by the law of supply and demand. As there is also a modelling of different types of traders, this paper offers a nice generalisation to the continuous-time model introduced in this section.

CHAPTER 3

THE CONTINUOUS-TIME PORTFOLIO PROBLEM

3.1 Introduction and Formulation of the Problem

The continuous-time portfolio problem consists of maximising total expected utility of consumption over the trading interval [0, T] and/or of terminal wealth X(T). We will formulate this problem in our continuous-time market setting introduced in the previous chapter. However, in this chapter we will often restrict ourselves to the complete market model . I.e. we will look at a security market with n+1 assets, one of which is a default-free bond, whose instantaneous rate of return r(t) may (possibly randomly) fluctuate, and the other n of which are stocks, whose prices are driven by n independent Wiener processes (i.e. by an n-dimensional Brownian motion) and have (randomly fluctuating) mean rates of return $b_i(t)$ and volatility coefficients $\sigma_{ij}(t)$.

There are two main approaches to solving the continuous-time portfolio problem:

- the **stochastic control approach** developed by Merton (see (Merton1969, 1971))
- the **martingale approach** (as e.g. presented in (Pliska1986), (Karatzas, Lehoczky and Shreve 1987), (Cox and Huang 1989), (Karatzas 1989),...)

The first approach is based on standard results of stochastic control theory. The optimal solution is computed by solving the so-called Hamilton-Jacobi-Bellman Equation in two steps. The first step consists of searching for the optimal portfolio strategy as a function of the (unknown) optimal expected utility. Inserting this portfolio and consumption strategy into the Hamilton-Jacobi-Bellman Equation results in a non-linear partial differential equation, whose solution forms the second step. In the special case of the Black-Scholes model and HARA utility functions, Merton was able to find closed form solutions for this problem. In general, however, it is very hard to get explicit solutions to the Hamilton-Jacobi-Bellman Equation. Even the numerical tractability of such problems is very limited.

In a complete markets setting, the second approach is based on martingale theory and stochastic integration. The optimal solution is again found in two steps. Here the first step yields both the consumption process and terminal wealth that maximise expected utility. This step corresponds to the solution of a static (in time) optimisation problem. In a second step, the corresponding portfolio strategy is derived by means of the martingale representation theorem. While this approach depends crucially on the completeness of the market model, it has several advantages. For instance, we are able to study problems with non-constant market coefficients and general utility functions. Moreover, if there are no closed-form solutions of the continuous-time portfolio problem, Monte Carlo methods offer a suitable way for numerical computations.

Both these approaches will be presented in Sections 3.3 and 3.4, respectively. However, we will give a first impression of the different philosophies behind them via a simple, discrete example in Section 3.2. Of course, prior to attempting this, we have to give a rigorous formulation of the portfolio problem. For this, we assume that we are considering an investor who makes consumption and investment decisions continuously in time. He is endowed with an initial wealth of x and tries to maximise his utility of consumption over a fixed time interval [0, T] and/or of terminal wealth at the time horizon T. The first ingredients needed for this problem are the utility functions:

Definition 1
A function $U: (0, \infty) \to \mathbb{R}$ such that $U \in C^1$ is strictly concave and satisfies

$$U'(0) := \lim_{x \downarrow 0} U'(x) = +\infty, \quad U'(\infty) := \lim_{x \downarrow \infty} U'(x) = 0 \tag{1}$$

will be called a **utility function**.

Remark 2
a) Typical examples of such utility functions are $U(x) = \frac{1}{\alpha}x^{\alpha}$ for $\alpha \in (0, 1)$ or $U(x) = \ln(x)$ for positive x.
b) The requirements in Definition 1 are in many ways natural ones. The concavity of U, in connection with requirement (1), implies that U is strictly increasing with a strictly decreasing derivative U'(x). In economic terms, an investor having such a utility function will always prefer a higher level x (of consumption or wealth) to a lower one. But his additional utility gained from an additional unit (of consumption or wealth) decreases with the actual level he is at. In particular, there will be a saturation effect if his level approaches infinity.
c) A set of functions U(t, .), $t \in [0, T]$, will also be called a **utility function** if for every fixed $t \in [0, T]$, U(t, .) is a utility function in the second variable (in the sense of Definition 1) and U(., x) is a continuous function in the first variable for every fixed positive x. An example of a utility function U(t, x) depending on $t \in [0, T]$ is

given by $U(t, x) = \exp(-\rho t)U(x)$ where $U(x)$ is a utility function in the sense of Definition 1.
d) We will look at slightly more general classes of utility functions in Sections 3.3 (in particular at the so-called HARA-functions (see Definiton 3 below)) and 3.4, depending on the method chosen to solve the portfolio problem.

As an investor is assumed to aggregate all his different attitudes towards risk of investments into the choice of his utility function, it is worth classifying the possible such functions via so-called risk-measures. The most commonly used of these is the **Arrow-Pratt measure of absolute risk-aversion** ARA(x), which is defined as

$$ARA(x) = -\frac{U''(x)}{U'(x)} . \tag{2}$$

Another widely used risk-measure is that of relative risk-aversion which differs from the absolute risk-aversion by an additional factor of x in the denominator.

Because by Definition 1 we restrict ourselves to strictly concave and increasing functions U, we only deal with utility functions with positive ARA(x) (if we also assume that we have $U \in C^2$). An investor having such a positive absolute risk-aversion measure is called risk-averse. To give this a more descriptive meaning, let X be an F_T-measurable, positive random variable with $E(U(X)) < \infty$, where U is a utility function as defined in Definition 1 (one should think of X as a possible terminal wealth at $t = T$). Define the positive real number z by

$$U(z) = E(U(X)) .$$

The investor with utility function U is today indifferent between having a constant terminal payment of z or a random one of X. But by Jensen's inequality we have $E(U(X)) < U(E(X))$ for a non-constant payment X, and as U(.) is strictly increasing, we obtain

$$z < E(X).$$

Hence, the investor would prefer a constant payment of $z + \varepsilon$ for $\varepsilon \in [0, E(X)-z)$ to the random payment of X, although the constant payment yields a smaller "expected wealth". The investor's distaste of risk foresakes the additional expected return.

An important class of utility functions are the so-called HARA-functions (see (Merton 1971).

Defintion 3

A function U(x) is said to be a **HARA-function** if it admits the representation

$$U(x) = \frac{1-\gamma}{\gamma}\left(\frac{\beta}{1-\gamma}x + \eta\right)^{\gamma} \tag{3}$$

for $\gamma < 1$, $\gamma \neq 0$, $\beta > 0$, $\frac{\beta}{1-\gamma}x + \eta > 0$.

Remark 4

a) The name of this class of functions (which is an abbreviation for **h**yberbolic **a**bsolute **r**isk-**a**version) is justified by looking at the Arrow-Pratt measure which then has the form

$$ARA(x) = \left(\frac{x}{1-\gamma} + \frac{\eta}{\beta}\right)^{-1}, \tag{4}$$

i.e. it is a hyberbolic function.

b) Usually, one includes U(x) = ln(x) into the HARA-family (which we will do, too) by formally setting $\gamma = \eta = 0$, $\beta = 1$. The inclusion of ln(x) into the HARA-family is a natural one, as we have

$$\frac{1-\gamma}{\gamma}\left(\left(\frac{1}{1-\gamma}x\right)^{\gamma} - 1\right) \xrightarrow{\gamma \to 0} \ln(x).$$

c) In Section 4, we will weaken the requirements of Definition 1 in such a way as to ensure that also $U(x) = 1 - e^{-cx}$ can be considered to be a utility function for positive c. This gives us an example with constant absolute risk-aversion. More precisely, we have ARA(x) = c for this choice.

Let $X^{x,\pi,c}(t)$, $t \in [0, T]$, be the wealth process of an investor who is endowed with an initial wealth of $x > 0$ and who follows an admissible strategy (π, c). The investor's (expectation of the) utility of using this strategy is defined as

$$J(x; \pi, c) := E\left(\int_0^T U_1(t, c(t))dt + U_2(X^{x,\pi,c}(T))\right) \tag{5}$$

where U_1, U_2 are utility functions in the sense of Definition 1 and of Remark 2 c), respectively (for the moment we will assume that the expectation in equation (5) exists). Of course, the goal of an investor will be to maximise this utility over all possible strategies (π, c). If we were now to restrict the set of admissible strategies to that subset of A(x) such that the expectation in equation (5) be finite then we would possibly exclude an optimal strategy. It goes without saying that the best possible strategy would be one from which the investor would gain an infinite utility. We will therefore consider the following sub-class of A(x):

$$A'(x) := \left\{ (\pi, c) \in A(x) \,\middle|\, E\left(\int_0^T U_1(t, c(t))^- \, dt + U_2(X^{x,\pi,c}(T))^- \right) < \infty \right\}, \quad (6)$$

(where z^- denotes the negative part of z) i.e. it is still allowed to draw an infinite utility from the strategy $(\pi, c) \in A'(x)$, but only if the expectation over the negative parts of the utility functions is finite. Clearly, if both utility functions are non-negative then A(x) and A'(x) coincide.

We are now ready to formulate the portfolio problem as

Definition 5

The (unconstrained) **portfolio problem** (for an investor with initial wealth $x > 0$) consists of the optimisation problem

$$\max_{(\pi, c) \in A'(x)} J(x; \pi, c) \,. \quad (P)$$

The rest of this chapter is devoted to the solution of this problem (P). As we will also consider portfolio problems with various additional conditions imposed on the terminal wealth and/or the portfolio processes, we will sometimes call (P) the **unconstrained portfolio problem**. In the literature the unconstrained portfolio problem is sometimes referred to as **Merton's problem** (see (Duffie 1992)).

3.2 Comparison between Stochastic Control and Martingale Method — A Preview via a Simple Discrete Example

Although the actual techniques used in the continuous-time setting will differ substantially from the computations given in the following simple example, the latter will serve its purpose of highlighting the main difference in philosophy between the two main approaches to solving the portfolio problem, as stated in the previous section. We have chosen the simpler discrete setting to present the main ideas behind the stochastic control and the martingale approaches, because the use of partial differential equation and stochastic calculus methods in the continuous-time setting often obscures the principles underlying both approaches.

Therefore, consider the following situation: our market consists of a bond and only a single stock. Trading is only possible at the times t = 0, 1, 2. The bond and stock prices $P_0(t)$, $P_1(t)$ at the different trading times are displayed in the following tree diagram:

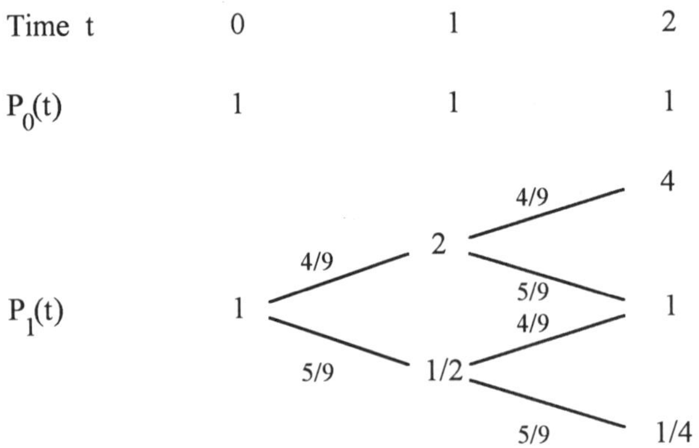

We observe that the bond price is assumed to be constant over time, whilst the stock price can only move to two different values at the next trading time. It can double or halve its value with probabilities of 4/9 and of 5/9, respectively.

Our investor is not interested in consumption; indeed, his only goal is to maximise his (expected) utility from terminal wealth at time t = 2. Therefore, his chosen strategy is completely described by the pair $(\pi(0), \pi(1))$ of the fractions of wealth invested in the stock at times t = 0, 1, respectively (i.e. by his portfolio process). Of course, he chooses $\pi(i)$ **after** having observed the stock price at time t = i. Further, $\pi(i)$ can depend on the actual price $P_1(1)$ of the stock at time t = 1 (although this will not be the case for the optimal solution in our example). Endowed with an initial capital of x, the investor's goal will now be to solve the optimisation problem

$$\max_{\pi=(\pi(0),\pi(1))} E\left(\sqrt{X^{x,\pi}(2)}\right). \qquad \text{s.t.} \quad X^{x,\pi}(2) \geq 0 \tag{P}$$

Before seeing the two solution approaches in action, let us point out that a "forward" optimisation is not possible without further information. More precisely, if we start at time t = 0 and try to determine the corresponding optimal strategy $\pi(0)$, it is not clear how to decide on its optimality. By fixing $\pi(0)$, we are only able to calculate the expectation of (functions of) the corresponding wealth process at time t = 1. However, maximising (a function of) this expectation by chosing a suitable $\pi(0)$ will in general not lead to a $\pi(0)$ that is part of an optimal strategy $(\pi(0), \pi(1))$. It is only possible to calculate an optimal $\pi(0)$ at time t = 0 if the optimal continuation $\pi(1)$ in all possible states of the world at time t = 1 (i.e. in the two different states given by the two possible values of $P_1(1)$) is already known. This fact will also be reflected in both the martingale and the stochastic control approach.

i) Solution via Stochastic Control — The Dynamic Programming Approach

The ethos behind the stochastic control approach in the above, discrete-time setting is the dynamic programming principle. It consists of solving the optimisation problem (P), but starting at time t = 1 in either of the two possible states "$P_1(1) = 2$" or "$P_1(1) = 1/2$". After having solved these two particular problems, one is able to determine the optimal strategy at the starting time t = 0 with the help of the already computed optimal strategy $\pi(1)$ for time t = 1. Let us describe this two-step procedure.

Step 1: "t = 1"
Consider first the sub-problem that arises if we find ourselves in the state "$P_1(1) = 2$". Let our current wealth X(1) be equal to the positive number z. If we now choose a particular strategy $\pi(1) = \zeta$ then we can explicitly compute the expected utility from the corresponding terminal wealth $X^{z,\zeta}(2)$ as

$$E\left(\sqrt{X^{z,\zeta}(2)} \mid P_1(1) = 2,\ X(1) = z\right)$$
$$= \tfrac{4}{9}\sqrt{2\zeta z + (1-\zeta)z} + \tfrac{5}{9}\sqrt{(\zeta z)/2 + (1-\zeta)z}$$
$$= \left[\tfrac{4}{9}\sqrt{\zeta + 1} + \tfrac{5}{9}\sqrt{1 - (\zeta/2)}\right]\sqrt{z} =: f(\zeta)\sqrt{z}\,. \qquad (*)$$

To see the first equality, simply note that after an up move of the stock price, the money invested in the stock will double, while it will halve after a down move of the stock price. But the probability for the first event is equal to 4/9, that of the second equals 5/9. Hence, the above relation shows that the optimal strategy can be found by maximising $f(\zeta)$ over $[-1, 2]$ (where the boundaries of this interval are determined by the requirement $X(2) \geq 0$). The unique maximiser ζ^* of $f(\zeta)$ and its value $f(\zeta^*)$ are easily found to be

$$\zeta^* = \tfrac{13}{19}, \qquad f(\zeta^*) = \frac{19}{3\sqrt{38}} =: c^* . \tag{**}$$

The situation in the state "$P_1(1) = 1/2$" is analogous. Again, the money invested in the stock will double after an up move of the stock price and it will halve after a down move. As the probability for these events are the same as in the state "$P_1(1) = 2$", we have the same optimal strategy ζ^* and the same optimal expected utility $f(\zeta^*)\sqrt{z}$. Thus, we have shown, that independent of the state and independent of the wealth at time t = 1, the optimal strategy is given by

$$\pi(1) = \tfrac{13}{19} .$$

Step 2: "t = 0"
After having computed the optimal strategy $\pi(1)$ in every possible state at time t = 1, we are now able to compute the optimal strategy $\pi(0)$ at the starting time. Due to equations (*) and (**), we know the optimal utility, we will receive at time t = 2 if we know the wealth at time t = 1. If we now choose $\pi(0) = \xi$, we can compute

$$E\left(\sqrt{X^{x,(\xi,\pi(1))}(2)}\right)$$
$$= \tfrac{4}{9}\sqrt{2\xi x + (1-\xi)x}\cdot c^* + \tfrac{5}{9}\sqrt{(\xi x)/2 + (1-\xi)x}\cdot c^*$$
$$= f(\zeta)\sqrt{x}\cdot c^* , \tag{***}$$

which will again be maximised by the choice of

$$\zeta^* = \tfrac{13}{19},$$

which then also agrees with the optimal strategy $\pi(0)$. To see the first equality in equation (***), note that the square root terms are the wealth of the investor in the two possible states at time t = 1. But from equations (*) and (**) we know that we simply have to replace $\sqrt{z}$ by the wealth at time t = 1 to get the optimal utility of terminal wealth. Hence, we have determined the optimal portfolio strategy as

$$\pi = (\pi(0), \pi(1)) = (\tfrac{13}{19}, \tfrac{13}{19}).$$

Important point to note :
It should be kept in mind that we had to solve an optimisation problem in **every** state at **all** times before the final time t = 2. This feature is common to all such discrete tree models. In our example, this could be done very efficiently. Due to the homogeneity amongst the sub-problems it was enough to solve the optimisation prob-

lem once only. This also delivered the solution in the other states. In general trees this is, of course, not always the case. If we look at a market with n stocks then the method is in principle the same, although, in all different states we now have to solve an optimisation problem in n unknowns.

b) Solution via the Martingale Method

The main idea behind the martingale approach is a separation between the determination of the optimal terminal wealth and that of a corresponding portfolio process yielding it exactly. This separation principle heavily relies on the completeness of the underlying market model. In particular, in our continuous-time setting it makes intensive use of Theorem 2.7. Although we have not proved it, we will use a discrete-time version of this theorem in our subsequent computations.

Step 1: "Computation of the optimal terminal wealth B"
In Theorem 2.7 the non-negative random variable B has to be F_T-measurable and to satisfy the integrability condition $E(H(T)B) = \tilde{E}(\exp(-rT)B) = x$. In our discrete time setting, $F_T = F_2$ is generated by the four different possible paths of the stock price. Thus, B can attain at most four different values a, b, c, d where each value is associated with a possible such path. We will name the values in such a way that a occurs if the stock price moves up twice, d if it moves down twice, b if it first moves up and then down, and finally c occurs after a down move followed by an up move of the stock price. To check the integrability condition, we need to know the martingale measure $\tilde{P}$ for the stock price in our model. It can easily be checked that the only probability measure $\tilde{P}$, transforming the stock price into a martingale, is the one that always assigns a probability of 1/3 to an up and 2/3 to a down move of the stock price, i.e. the corresponding tree for the stock price under $\tilde{P}$ has the form

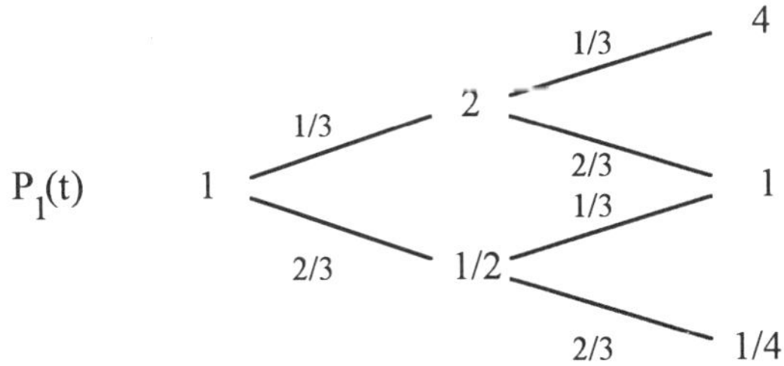

Using the original probabilities for the objective function, the probabilities given by the martingale measure for the integrability condition, and the names a, b, c, d for the four possible values of B, we can rewrite the problem

$$\max_{B} \; E\left(\sqrt{B}\right)$$
$$\text{s.t. } B \; F_T\text{-measurable}, B \geq 0, \; \tilde{E}(\exp(-rT)B) = x$$

as

$$\begin{aligned} &\max_{a,b,c,d} \; \tfrac{16}{81}\sqrt{a} + \tfrac{20}{81}\sqrt{b} + \tfrac{20}{81}\sqrt{c} + \tfrac{25}{81}\sqrt{d} \\ &\text{s.t.} \quad a,b,c,d \geq 0, \quad \tfrac{1}{9}a + \tfrac{2}{9}b + \tfrac{2}{9}c + \tfrac{4}{9}d = x \end{aligned} \tag{O}$$

This non-linear optimisation problem in the four variables a, b, c, d can be solved by the usual Lagrangian methods. Note that due to the specific form of the utility function as a weighted sum of square roots, the non-negativity conditions will not be active in the optimum solution. This solution of (O) is given by

$$a = \left(\tfrac{32}{19}\right)^2 x \approx 2.84\, x\,,$$

$$b = c = \left(\tfrac{20}{19}\right)^2 x = \left(\tfrac{5}{8}\right)^2 a \approx 1.11\, x\,,$$

$$d = \left(\tfrac{25}{38}\right)^2 x = \left(\tfrac{5}{8}\right)^2 b \approx 0.43\, x\,.$$

In particular, note that for the optimal terminal wealth, it does not matter by which path the price $P_2(2) = 1$ is reached.

Step 2: "Computation of a strategy generating the optimal terminal wealth"
Step 1 resulted in the optimal wealth at time t = 2 in every possible state. It is now our task to compute a strategy that will exactly deliver this final wealth. Let us start again in the state "$P_1(1) = 2$" and let $\varphi_0(1)$, $\varphi_1(1)$ be the chosen amounts of money invested in the bond. Depending on the stock price movement from time t = 1 to the final time t = 2, $\varphi_0(1)$, $\varphi_1(1)$ have to satisfy the following two linear equations:

$$\varphi_0(1) + 2\,\varphi_1(1) = \left(\tfrac{32}{19}\right)^2 x \; (= a)$$
$$\varphi_0(1) + \tfrac{1}{2}\varphi_1(1) = \left(\tfrac{20}{19}\right)^2 x \; (= b),$$

i.e. if the stock price goes up, the wealth of the investor has to change to a and if it goes down, it has to change to b. The unique solution of these equations is given by

$$(\varphi_0(1), \varphi_1(1)) = \left(\tfrac{192}{361}x, \tfrac{416}{361}x\right).$$

This also results in

$$\pi(1) = \frac{\varphi_1(1)}{\varphi_0(1)+\varphi_1(1)} = \frac{13}{19}$$

as the optimal portfolio process in the state "$P_1(1) = 2$". However, it is not necessary to compute $\pi(1)$ in addition to $(\varphi_0(1), \varphi_1(1))$, as this pair also describes the action of the investor completely. The corresponding equations for the optimal amounts of money invested in bond and stock $(\psi_0(1), \psi_1(1))$ in the state "$P_1(1) = 1/2$" read

$$\psi_0(1) + 2\,\psi_1(1) = \left(\tfrac{20}{19}\right)^2 x \; (= b)$$

$$\psi_0(1) + \tfrac{1}{2}\psi_1(1) = \left(\tfrac{25}{38}\right)^2 x \; (= d).$$

Due to the fact that the right side of this linear system is simply the right side of the system corresponding to the state "$P_1(1) = 2$" multiplied by $(5/8)^2$, while the left sides of both systems coincide the resulting optimal parameters are given by

$$(\psi_0(1), \psi_1(1)) = \left(\tfrac{5}{8}\right)^2 (\varphi_0(1), \varphi_1(1)) = \left(\tfrac{150}{722}x, \tfrac{325}{722}x\right),$$

$$\pi(1) = \frac{\psi_1(1)}{\psi_0(1)+\psi_1(1)} = \frac{13}{19}.$$

The optimal pairs $(\varphi_0(1), \varphi_1(1))$ and $(\psi_0(1), \psi_1(1))$ now yield the wealth $X(1)$ of the investor at time $t = 1$ in every possible state. We have thus got the right side of the linear system for the optimal amounts $(\zeta_0(0), \zeta_1(0))$ of money invested in bond and stock at time $t = 0$:

$$\zeta_0(0) + 2\,\zeta_1(0) = \varphi_0(1) + \varphi_1(1) = \tfrac{32}{19}x$$

$$\zeta_0(0) + \tfrac{1}{2}\zeta_1(0) = \psi_0(1) + \psi_1(1) = \tfrac{25}{38}x\,.$$

As the right side of this system coincides with the right side of the corresponding system in the state "$P_1(1) = 2$" multiplied by a factor of $(19/32)$, we deduce

$$(\zeta_0(0), \zeta_1(0)) = \tfrac{19}{32}(\varphi_0(1), \varphi_1(1)) = \left(\tfrac{6}{19}x, \tfrac{13}{19}x\right)$$

$$\pi(0) = \frac{\zeta_1(1)}{\zeta_0(1)+\zeta_1(1)} = \frac{13}{19}.$$

As an analogue to the stochastic control solution, the totally symmetric evolution of the stock price simplified the tasks in the different states. In fact, only one linear system had to be solved. It should also be noted that the probability of up or down moves of the stock price did not enter any calculation in Step 2.

Important point to note :
It should be kept in mind that Step 1 consists of the solution of a non-linear optimisation problem, in which the number of unknowns equals the cardinality of F_T. In contrast to the stochastic control method, we only have to solve systems of **linear** equations of dimension n+1 (instead of non-linear optimisation problems with n unknowns) in all states of the world that occur before the final time T. Before we are tempted to make a comparison between the efficiencies of both methods, we should point out that the martingale method is only designed for complete markets. However, we will not do so, in particular, as we have not yet presented the methods for the continuous-time setting.

3.3 The Stochastic Control Method to Solve the Portfolio Problem

The stochastic control method is often called the "Merton approach", reflecting the fact that it was introduced to portfolio optimisation by R.C. Merton in (Merton 1969) and follow up papers such as (Merton 1971). In fact, he applied standard methodology and results from stochastic control theory to the continuous-time portfolio problem. The basic idea in his approach is to look at the stochastic differential equation for the wealth process of an investor as a controlled diffusion process of the form

$$dX_t^u = \mu(t, X_t^u, u)\,dt + \nu(t, X_t^u, u)dW(t),$$

where we have set

$$u = (u_1, u_2) := (\pi, c),$$

$$\mu(t, x, u) := [r(t)(1-u_1'\underline{1}) + b(t)'u_1]\,x - u_2\,,$$

$$\nu(t, x, u) := [u_1'\sigma(t)]\,x$$

and X_t^u is the wealth process corresponding to the strategy (π, c) (readers not familiar with stochastic control methods are recommended to read Appendix C before continuing). To be able to use the technology of the Hamilton-Jacobi-Bellman Equation (for brevity: HJB-Equation, see Appendix C), we have to restrict ourselves to the case of (continuous) deterministic market coefficients r, b, σ, i.e. the market coefficients are deterministic functions in time. To get explicit results in the cases considered below, we will restrict ourselves even further and assume constant market coefficients. This will ensure that then the application of stochastic control theory will be straight forward: set up the HJB-Equation corresponding to the form of the portfolio problem under study and solve it analytically. Such a closed-form solution, though desirable, in practise rarely exists. There are only a few examples, with particular choices of utility functions, such that the HJB-Equation can be solved explicitly. Furthermore, even the numerical tractability of the resulting partial differential equation (for brevity: pde) is very limited.

For our subsequent portfolio problems, we assume that we have a market with one riskless and n risky assets. The market coefficients r, b, σ are constant and $\sigma\sigma'$ is regular (for application of the stochastic control method it is not necessary to have a complete model, i.e. we do **not** assume n = m). We will first start by discussing a special form of the portfolio problem, which cannot be dealt with by the martingale method presented in the next section:

i) Optimal Life-Time Consumption

The problem of optimal life-time consumption consists of maximising utility of consumption with an infinite time horizon, i.e. we consider the problem

$$\max_{(\pi,c)\in A'(x)} E^x\left(\int_0^\infty e^{-\rho t}U(c(t))dt\right), \tag{P}$$

where E^x indicates that the wealth process at the beginning of the time interval is equal to x and where ρ is a positive constant. Further, U should satisfy a polynomial growth condition of the type

$$|U(x)| \le C\left(1+|x|^k\right)$$

for suitable constants $C > 0$ and $k\in\mathbf{N}$. As the requirement $(\pi, c)\in A'(x)$ means that we have to stop the process if the investor's wealth $X^u(t)$ reaches zero, we look at the stochastic control problem given by the value function

$$v(x) = \sup_{u\in U} E^x\left(\int_0^\tau e^{-\rho t}U(u_2(t))dt\right), \tag{7}$$

where where τ denotes the first time when $X^u(t)$ reaches zero and the set $U = U(x)$ consists of all admissible controls (see Definiton C1), i.e. all $\mathbf{R}^{n+1}$-valued, F_t-adapted processes $u(t) = (u_1(t), u_2(t))$ with

$$u_1(t)\in[a, b]^n,\ u_2(t) \ge 0 \quad \forall\, t\in[0,\infty) \text{ a.s.},$$

$$E^x\left(\int_0^T (u_2(t))^m \, dt\right) < \infty \quad \forall\, T > 0,\ m\in\mathbf{N}.$$

Of course, the bound on $u_1(t)$ is very restrictive and **not** required in problem (P). It is introduced only as an artificial constraint to ensure the applicability of stochastic control methods, i.e. the verification theorem C2 of Appendix C. We will see later that this constraint plays no role in our explicit examples, as there, the constants a, b can always be chosen in such a way that the constraint is not active in the optimum solution.

Now, we can formally set up the corresponding HJB-Equation (see Appendix C):

$$0 = \max_{(u_1,u_2)\in[a,b]^n\times[0,\infty[} \left\{ \tfrac{1}{2} u_1'(\sigma\sigma')u_1 x^2 v''(x) + ([r(1-u_1'\underline{1}) + b'u_1]\,x - u_2)v'(x) - \rho v(x) + U(u_2) \right\}$$

(where we will stop the whole process if x reaches zero and add the boundary condition $v(0) = 0$). This equation will now be solved using the customary two steps of the stochastic control approach:

Step 1: "Solve the maximisation problem in the HJB-Equation"
Under the assumption that the value function v(x) is twice continuously differentiable, strictly concave and increasing, and that the unconstrained maxima (u_1^*, u_2^*) are such that they are always inside $[a, b]^n\times[0,\infty)$ we obtain the following representation for these maxima (dependend on the **unknown** value function v(x) and its derivatives !) from the first order conditions for a maximum:

$$u_1^* = -(\sigma\sigma')^{-1}(b-r\underline{1})\frac{v'(x)}{xv''(x)}, \qquad u_2^* = (U')^{-1}(v'(x)).$$

Of course, all the assumptions used to compute (u_1^*, u_2^*) have to be checked after we have found the explicit form of v(x). Note that the first order conditions are also sufficient as we have assumed v(x) and $U(u_2)$ to be strictly concave. Furthermore, it should be noted that the candidate for the optimal portfolio process, u_1^*, is inversely proportional to the product of x and the absolute risk-aversion of the value function. I.e. if the absolute risk-aversion of the value function were be a multiple of 1/x then u_1^* would be constant.

Step 2: "Substitute the maxima into the HJB-Equation and solve the resulting pde"
Substituting (u_1*, u_2*) into the HJB-Equation leads to the following pde:

$$0 = -\tfrac{1}{2}(b-r\underline{1})'(\sigma\sigma')^{-1}(b-r\underline{1})\frac{(v'(x))^2}{v''(x)} + rxv'(x) - (U')^{-1}(v'(x))\,v'(x)$$
$$- \rho v(x) + U((U')^{-1}(v'(x))) . \tag{8}$$

Using Theorem C3, we can state a verification theorem for problem (P):

Proposition 6
Let X*(t) denote the controlled process corresponding to the control u*(t). If there exists a $C^2(0,\infty)$ solution of the HJB-Equation (8), continuous on $[0, \infty)$, such that

$$u_1^*(t) = -(\sigma\sigma')^{-1}(b-r\underline{1})\frac{v'(X^*(t))}{X^*(t)v''(X^*(t))},$$

$$u_2^*(t) = (U')^{-1}(v'(X^*(t)))$$

are admissible strategies, then they are optimal strategies for the problem given by the value function (7). If u_1*(t) is always inside $(a, b)^n$, then u*(t) is also an optimal strategy for the optimal life-time consumption problem (P).

Proof:
We only have to comment on the ultimate statement as the prior ones are a consequence of Theorem C3. If u_1*(t) is always inside $(a, b)^n$, then it is the unconstrained optimum for the optimisation problem inside the HJB-Equation, i.e. we would have the same optimum if the constraint were not present, and thus, the cntrol problem (7) is equivalent to the optimal life-time consumption problem (P).

□

As the pde (8) is highly non-linear, we cannot expect to get an explicit solution for a general utility function U(.). Therefore, we consider the special case of

$$U(x) = \frac{1}{\gamma}x^\gamma, \quad 0 < \gamma < 1,$$

to demonstrate that the above proposition is not formulated on the empty set. This choice of U(x) yields the following form of the pde (8):

$$0 = -\tfrac{1}{2}(b-r\underline{1})'(\sigma\sigma')^{-1}(b-r\underline{1})\,\frac{(v'(x))^2}{v''(x)} + rxv'(x) - \left(\frac{\gamma-1}{\gamma}\right)(v'(x))^{\frac{\gamma}{\gamma-1}} - \rho v(x)\,. \quad (9)$$

To solve it, we conjecture that the value function has the form

$$v(x) = A\,x^{\gamma},$$

where the positive, real constant A has yet to be determined. Inserting this form of v(x) and its derivatives into the pde (9) leads to

$$0 = \left\{-\tfrac{1}{2}(b-r\underline{1})'(\sigma\sigma')^{-1}(b-r\underline{1})A\,\frac{\gamma}{\gamma-1} + rA\gamma - \left(\frac{\gamma-1}{\gamma}\right)(A\gamma)^{\frac{\gamma}{\gamma-1}} - \rho A\right\}x^{\gamma}\,.$$

If we divide this equation by Ax^{γ}, we can reorder it to yield the following equation for A:

$$A^{\frac{1}{\gamma-1}} = \gamma^{\frac{1}{1-\gamma}}\,\{r\gamma + \tfrac{1}{2}(b-r\underline{1})'(\sigma\sigma')^{-1}(b-r\underline{1})\,\frac{\gamma}{1-\gamma} - \rho\,\}/\,(\gamma-1)\,. \quad (10)$$

Hence, there only exists a positive solution to this equation if the nominator of the quotient is negative, i.e. if we have

$$r\gamma + \tfrac{1}{2}(b-r\underline{1})'(\sigma\sigma')^{-1}(b-r\underline{1})\,\frac{\gamma}{1-\gamma} < \rho\,. \quad (11)$$

If this is the case, then then we obtain a unique, positive A. Tracing all our steps backwards, we discover that the conjectured value function $v(x) = Ax^{\gamma}$ satisfies all our assumptions made **before** knowing its explicit form. Thus, all our computations are justified and we have found a solution to the HJB-Equation meeting a polynomial growth condition. Moreover, by using

$$u_1^*(t) = (\sigma\sigma')^{-1}(b-r\underline{1})\frac{1}{1-\gamma}\,, \quad u_2^*(t) = (A/\gamma)^{\frac{1}{1-\gamma}}\,X(t)\,,$$

we can compute the following stochastic differential equation for the wealth process corresponding to the pair $(\pi, c) = (u_1^*, u_2^*)$

$$dX(t) = \left(r + (b-r\underline{1})(\sigma\sigma')^{-1}(b-r\underline{1})/(1-\gamma) - (A/\gamma)^{\frac{1}{1-\gamma}}\right)X(t)\,dt$$
$$+ \left(\frac{1}{1-\gamma}(b-r\underline{1})(\sigma\sigma')^{-1}\sigma dW(t)\right),$$

which yields that the optimal wealth process is a geometric Brownian motion. In particular, it is strictly positive on $[0, \infty)$ if and only if we start with a positive initial endowment of x. We can summarise the result of all our considerations in the following corollary to Proposition 6:

Corollary 7
Under the assumptions of constant market coefficients with regular $\sigma\sigma'$ such that inequality (11) holds, the optimal lifetime consumption problem

$$\max_{(\pi,c)\in A'(x)} E^x\left(\int_0^\infty e^{-\rho t} c(t)^\gamma dt\right)$$

is solved by the strategies

$$\pi^*(t) = \left(\sigma\sigma'\right)^{-1}(b-r\underline{1})\frac{1}{1-\gamma}, \quad c^*(t) = \left(A/\gamma\right)^{\frac{1}{1-\gamma}} X(t)$$

where A is the unique positive solution of equation (10).

Remark 8
a) Note that the optimal portfolio process is constant over time. This, however, does **not** mean that there is no trading necessary after an initial choice of the portfolio. In deed, it means trading at each and every time instant as the components of the security price vector change in different ways, and the necessary corrective action is to rebalance the holdings continuously to maintain the fractions of wealth invested in the different securities at a constant level.
b) Inequality (11) ensures the convergence of the expectation of the integral in our optimisation problem, i.e. it ensures finiteness of the value function.

One can also generalise the foregoing corollary to the case of HARA-functions of Definition 2. By analogy to the computations made in Step 2 of the explicit example (which are of course notationally more involved), we get:

Corollary 7*
Under the assumptions of constant market coefficients with regular $\sigma\sigma'$ such that inequality (11) holds, the optimal lifetime consumption problem

$$\max_{(\pi,c)\in A'(x)} E^x\left(\int_0^\infty e^{-\rho t} U(c(t))dt\right)$$

where U(c) is a HARA-function, is solved by the strategies (π^*, c^*) given by

$$\pi^*(t)X^*(t) = (\sigma\sigma')^{-1}(b - r\underline{1})\left(\frac{1}{1-\gamma}X(t) + \frac{\eta}{\beta r}\right),$$

$$c^*(t) = \frac{1}{\beta}\left(\frac{A}{\beta(1-\gamma)}\right)^{1/\gamma-1} X(t) + \frac{1-\gamma}{\beta}\left(\frac{1}{\beta r}\left(\frac{A}{\beta(1-\gamma)}\right)^{1/\gamma-1} - 1\right)\eta$$

where A is the unique positive constant such that the value function v(x) has the representation

$$v(x) = \frac{A}{\gamma}\left(\frac{1}{1-\gamma}X(t) + \frac{\eta}{\beta r}\right)^{\gamma},$$

i.e. A is uniquely determined by the pde (8) .

The form of the optimal strategy (π^*, c^*) of Corollary 7 is remarkably simple: π^* is just a constant vector, whilst $c^*(t)$ is always proportional to the actual wealth $X^*(t)$. However, this is a rare occasion in the sense of the following proposition :

Proposition 9
Let U(.) be a utility function in the sense of Definition 1 with $U(c) > 0$ for $c > 0$. Given the assumptions on the market model, as presented in this section and assuming that the value function v(x) of the problem

$$\max_{(\pi,c)\in A'(x)} E^x\left(\int_0^\infty e^{-\rho t} U(c(t))dt\right)$$

is a C^2-function, then the problem is solved by a pair (π^*, c^*) of the form

$$\pi^*(t) \equiv \kappa\,, \quad c^*(t) = \delta\, X^*(t) \quad \forall\, t \geq 0,$$

with constants $\kappa \in \mathbf{R}^n$, $\delta > 0$ if and only if we have $U(x) = ax^\gamma + d$ for suitable constants $0 < \gamma < 1$, $\gamma \neq 0$, a, $d > 0$.

Proof:
a) For the "if" part note that we have proved the above claim in the case $0 < \gamma < 1$ (the additional positive constant d does not affect the computations).
b) For the "only if" part note that we have

$$v'(x) = U'(c^*) = U'(\delta x) \quad \forall\, x \geq 0,$$

which leads to

$$v''(x) = \delta\, U''(c^*) \;\; \forall\, x \geq 0,$$

and thus to

$$-\frac{\delta\, U''(c^*)}{U'(c^*)} = \frac{v''(x)}{v'(x)}\,. \tag{12}$$

Equation (12) and the relation

$$\kappa \equiv \pi^*(t) = -(\sigma\sigma')^{-1}(b - r\underline{1})\frac{v'(X^*(t))}{X^*(t)v''(X^*(t))}$$

imply

$$-\frac{U''(x)}{U'(x)} = C\,\frac{1}{x} \quad \forall\, x > 0\,.$$

Solving this differential equation for U(x) leads to the form $U(x) = ax^{\gamma} + d$ for U(x), and making use of the conditions on U(.) stipulated above, we obtain that the constants a, d, γ have the desired forms.

□

There is a similar result in (Merton 1971) which at first sight gives the impression of being more general. However, as pointed out in (Sehti and Taksar 1988), its proof is only valid for the case given above (but with "general" $\gamma < 1$). For more on risk-averse behaviour of the investor and on solutions with more general classes of utility functions in the optimal life-time consumption problem (Presman and Sehti 1991) and (Karatzas, Lehochzky, Sehti and Shreve 1986) make good reading.

ii) Optimal Consumption and Terminal Wealth

We now look at the more complicated situation of solving the portfolio problem with finite time horizon

$$\max_{(\pi,c)\in A'(x)} E^{0,x}\left(\int_0^T e^{-\rho t}U_1(c(t))dt + e^{-\rho T}U_2(X(T))\right). \tag{P}$$

The utility functions $U_1(.)$, $U_2(.)$ are assumed to satisfy the same polynomial growth condition as the utility function in the previous sub-section. Again, if we identify the pair (π, c) with a control strategy (u_1, u_2), then we can set up the corresponding HJB-Equation for the value function v(t, x) given by

$$v(t,x) = \max_{(\pi,c)\in A'(x)} E^{t,x}\left(\int_t^T e^{-\rho(s-t)}U_1(c(s))ds + e^{-\rho(T-t)}U_2(X(T))\right).$$

After having transformed this value function into a form that is amenable to the application of stochastic control methods (see the remarks made in the optimal lifetime consumption example on the additional requirements) the form of the resulting HJB-Equation is given by

$$0 = \max_{(u_1,u_2)\in[a,b]^n \times[0,\infty[} \{\tfrac{1}{2}u_1'(\sigma\sigma')u_1x^2v_{xx}(t, x) + ([r(1-u_1'\underline{1})+b'u_1]\, x-u_2)v_x(t, x)$$
$$+ v_t(t, x)- \rho v(t, x) + U_1(u_2)\}, \tag{13}$$

$$v(T, x) = U_2(x)\,,\; v(t,0) = U_2(0)\,. \tag{14}$$

Again, assuming that v is strictly concave and increasing in x, we can formally solve the maximisation problem in equation (13), leading to

$$u_1^*(t, x) = -(\sigma\sigma')^{-1}(b-r\underline{1})\frac{v_x(t,x)}{xv_{xx}(t,x)}\,,$$

$$u_2^*(t, x) = (U')^{-1}(v_x(t, x))$$

(where as before we assume that the unconstrained maxima lie in the feasible region). As in the previous sub-section, we can now state a verification result:

Proposition 10
Let $X^*(t)$ denote the controlled process corresponding to the control $u^*(t)$. If there exists a $C^{1,2}([0, T)\times(0,\infty))$ solution of the HJB-Equation (13/14) which is continuous on $[0, T]\times[0,\infty)$, such that

$$u_1^*(t) = -(\sigma\sigma')^{-1}(b-r\underline{1})\frac{v_x(t,X^*(t))}{X^*(t)v_{xx}(t,X^*(t))}\,,$$

$$u_2^*(t) = (U')^{-1}(v_x(t, X^*(t)))\,.$$

are admissible strategies and if $u_1^*(t)$ is always inside $(a, b)^n$, then $u^*(t)$ is an optimal strategy for the optimal portfolio and consumption problem (P).

Again we give an explicit solution for the case of utility functions of the form

$$U_1(x) = U_2(x) = \frac{1}{\gamma}x^\gamma$$

with $0<\gamma<1$. As in the optimal lifetime consumption problem, the pde simplifies to

$$0=-\tfrac{1}{2}(b-r\underline{1})'(\sigma\sigma')^{-1}(b-r\underline{1})\frac{(v_x(t,x))^2}{v_{xx}(x)}+rxv_x(t,x)-\left(\frac{\gamma-1}{\gamma}\right)(v_x(t,x))^{\frac{\gamma}{\gamma-1}}$$
$$+v_t(t,x)-\rho v(t,x)\,. \tag{15}$$

However, the additional difficulty lies in the dependence of the value function on time, which is a consequence of the finite time horizon. Therefore, we try a solution of the form

$$v(t,x)=f(t)\,\frac{1}{\gamma}x^{\gamma}.$$

Substituting this form for v(t, x) and its resulting partial derivatives into the pde (15) (and dividing by x^{γ}) leads to the following ordinary differential equation (for brevity: ode) for f(t) :

$$0=f'(t)-f(t)\left(\tfrac{1}{2}(b-r\underline{1})'(\sigma\sigma')^{-1}(b-r\underline{1})\frac{\gamma}{\gamma-1}+r\gamma-\rho\right)-(\gamma-1)(f(t))^{\frac{\gamma}{\gamma-1}}, \tag{16}$$
$$f(T)=1\,.$$

We can rewrite the ode (16) as

$$f'(t)=a_1f(t)+\,a_2(f(t))^{\frac{\gamma}{\gamma-1}},\quad f(T)=1\,.$$

with appropriate abbreviations a_1, a_2. By setting $h(t)=(f(t))^{\frac{1}{1-\gamma}}$, this ode is transformed into a linear one for h(t) (while assuming $a_1\neq 0$)

$$h'(t)=\frac{a_1}{1-\gamma}h(t)+\frac{a_2}{1-\gamma}=\frac{a_1}{1-\gamma}h(t)+1,\quad h(T)=1,$$

which has the unique solution

$$h(t)=\exp\left(-\frac{a_1}{1-\gamma}(T-t)\right)\left(1-\frac{1-\gamma}{a_1}\right)-\frac{1-\gamma}{a_1}>0.$$

Substituting for h(t) and f(t) yields

$$v(t,x)=\left(\exp\left(-\frac{a_1}{1-\gamma}(T-t)\right)\left(1-\frac{1-\gamma}{a_1}\right)-\frac{1-\gamma}{a_1}\right)^{1-\gamma}\frac{1}{\gamma}x^{\gamma}$$

with

$$a_1 = \tfrac{1}{2}(b-r\underline{1})'(\sigma\sigma')^{-1}(b-r\underline{1})\frac{\gamma}{\gamma-1} - r\gamma + \rho\,,\ a_2 = \gamma - 1\ .$$

Thus, we have found the explicit form for a candidate solution to the HJB-Equation. It remains to verify that the wealth process corresponding to the candidate optimal controls $(u_1{}^*, u_2{}^*)$ is non-negative. Using the explicit forms of v_x, v_{xx} leads to

$$\pi(t) := u_1{}^*(t, X(t)) = (\sigma\sigma')^{-1}(b-r\underline{1})\frac{1}{1-\gamma}\,, \tag{17}$$

$$c(t) := u_2{}^*(t, X(t)) = (f(t))^{\frac{1}{\gamma-1}}\, X(t)\,. \tag{18}$$

Inserting these expressions in the sde for the corresponding wealth process leads to

$$dX(t) = X(t)\left(r + (b-r\underline{1})'(\sigma\sigma')^{-1}(b-r\underline{1})\frac{1}{\gamma-1} - (f(t))^{\frac{1}{\gamma-1}}\right)dt$$
$$+ X(t)\left(\frac{1}{1-\gamma}(b-r\underline{1})'\left((\sigma\sigma')^{-1}\right)'\sigma\right)dW(t)$$

which is the equation corresponding to a (generalised) geometric Brownian motion. In particular, $X(t)$ is then strictly positive on $[0, T]$ (of course, if and only if the initial endowment x is positive (see also Section 2.2)). Finally, using Theorem C2 in the Appendix justifies the use of the HJB-Equation technology. Then the fruit of our labour is:

Corollary 11
Under the assumptions of constant market coefficients with regular $\sigma\sigma'$, the optimal portfolio problem

$$\max_{(\pi,c)\in A'(x)} E^{0,x}\left(\int_0^T e^{-\rho t}\frac{1}{\gamma}c(t)^\gamma dt + e^{-\rho T}\frac{1}{\gamma}X(T)^\gamma\right)$$

is solved by the portfolio/consumption pair (π, c) as given in equations (17), (18) where $f(t)$ is the unique solution to the ode (16).

Remark 12
a) Due to the finite horizon, the optimal consumption rate also depends on time. According to whether a_2 is bigger or smaller than a_1 the consumption rate decreases or increases with time, respectively.

b) Comparing Corollary 11 with Corollary 7, it is apparent that condition (11) is not needed in the finite horizon case. In fact, this condition not only ensured positivity of A but also convergence of the (expectation of the) integral in the optimal lifetime consumption problem. Using the explicit form of f(t) and the constants a_1, a_2, it can easily be verified that f(t) is positive, whether relation (11) is satisfied or not.
c) As in the optimum consumption life-time problem, an extension of Corollary 11 to the HARA-case is possible (see (Merton 1971)). We will not present it here, but recommend this extension as a good exercise to practise the HJB-Equation methodology. One could also prove an analogue to Proposition 9 with $U_2(.) \equiv 0$, i.e. the case of a pure consumption problem with finite horizon. As the result and proof have identical form to those of the previous sub-section, we do not present them here.
d) There are not that many examples in portfolio optimisation where we have the existence of an explicit smooth solution to the HJB-Equation. However, our verification theorems always give sufficient conditions for a function being the value function of a portfolio problem. They are far from offering necessary conditions. The introduction of viscosity solutions to the HJB-Equations has revolutionised stochastic control theory as they allow for a substantial weakening of the regularity conditions imposed on the value function. We cannot introduce this interesting subject here but refer the interested reader to the monograph (Fleming and Soner 1993).

3.4 The Martingale Approach to the Continuous-Time Portfolio Problem

As indicated in the discussion of the discrete example given in Section 3.2, the principle idea of the martingale approach is a decomposition of the portfolio problem into a static optimisation problem ("Determination of optimal cash flows (i.e. consumption and/or terminal wealth)") and a representation problem ("Find a strategy that yields an (already determined !) optimal cash flow"). In our methodology of proofs and in our notation we will follow (Karatzas, Lehoczky and Shreve 1987), (Cvitanic and Karatzas 1992) and (Korn and Trautmann 1995). However, in addition we will give some heuristic motivations for the form of the optimal terminal wealth and consumption to show that the somewhat complex and technical methods and results presented below are in fact natural extensions and imitations of methods well-known from deterministic optimisation.

i) The Main Idea

It will be easier to work out the main idea underlying the martingale approach if we restrict ourselves for the moment to the special case of maximising the utility of terminal wealth only. We will return to the general form of the portfolio problem including consumption when we state and prove the final results.

As already said, the main idea of the martingale approach is to **decompose** the **dynamic portfolio problem** (recall that we have to optimise over a set of continuous-time stochastic processes)

$$\max_{\pi \in A'(x)} E\left(U(X^{x,\pi}(T))\right) \tag{P}$$

into a **static optimisation problem**

$$\max_{B \in \mathcal{B}(x)} E(U(B)), \tag{O}$$

$$\mathcal{B}(x) := \{B \mid B \geq 0,\ B\ \ F_T\text{-measurable},\ E(H(T)B) \leq x,\ E(U(B)^-) < \infty\},$$

and a **representation problem**

"Find a portfolio process $\pi^* \in A'(x)$ with

$$X^{x,\pi^*}(T) = B^* \quad \text{a.s. "}, \tag{R}$$

where B^* solves problem (O).

Surely, this decomposition deserves some comments. The essential theoretic result justifying it is Theorem 2.7 ("Completeness of the market model"). Part a) of this theorem states that every wealth process $X^{x,\pi}(t)$ corresponding to a portfolio process $\pi \in A(x)$ satisfies

$$E(H(T)X^{x,\pi}(T)) \leq x .$$

But this fact implies that $\mathcal{B}(x)$, the set of random variables feasible for the optimisation problem (O), contains all the random variables $X^{x,\pi}(T)$ generated by portfolio processes $\pi \in A'(x)$. As it will be shown later , the optimum in problem (O) will be attained by a random variable $B^* \in \mathcal{B}(x)$ with

$$E(H(T)\,B^*) = x .$$

Due to part b) of Theorem 2.7, there exists a portfolio process $\pi^* \in A(x)$ such that the corresponding wealth process $X^{x,\pi^*}(t)$ generates B^*, i.e. we have

$$X^{x,\pi^*}(T) = B^* \quad \text{a.s.}\ .$$

So on one hand, we have that π^* automatically belongs to A'(x). Hence, problems (P) and (O) have the same maximum. On the other hand, the existence of π^* also demonstrates that the representation problem (R) posesses a solution. In other words, completeness of the market model makes the martingale approach work.

Thus, we have transformed our original task of solving the portfolio problem (P) into the two new tasks of solving the problems (O) and (R). The solution of these sub-problems will be the content of the following sub-sections.

ii) Solution of the Optimisation Problem (O)

Before presenting the rigorous and general results, we will first derive the form of the optimal solution heuristically by imitating Lagrangian multiplier methods of deterministic optimisation. To recall these methods, we briefly review the relevant methods and results (see (Luenberger 1969)):

Refresher Course "Lagrangian multipliers and optimisation"

Let f: $\mathbf{R}^n \to \mathbf{R}$, g: $\mathbf{R}^k \to \mathbf{R}^k$ be C^1-functions. Assume further that g is strictly concave and that g is convex. Under these assumptions, solving the optimisation problem

$$\max_x \ f(x) \quad \text{s.t.}\ g(x) = 0$$

is equivalent to finding a zero of the gradient of the corresponding Lagrangian L(x,y) defined as

$$L(x,y) := f(x) - y'g(x)$$

for $(x,y) \in \mathbf{R}^{n+k}$. More precisely, if $(x^*,y^*) \in \mathbf{R}^{n+k}$ solves the (non-linear) system

$$0 = \frac{\partial L}{\partial x_i}(x,y) = \frac{\partial f}{\partial x_i}(x) - \sum_{j=1}^{k} y_j \frac{\partial g_j}{\partial x_i}(x), \quad i = 1,\dots, n$$

$$0 = \frac{\partial L}{\partial y_i}(x,y) = g_i(x)\,, \quad i = 1,\dots, k,$$

then x* is the unique maximiser in the optimisation problem. Moreover, if the vector of Lagrangian multipliers y* consists only of non-negative entries then x* even solves the problem

$$\max_{x} \ f(x)$$
$$\text{s.t. } g(x) \le 0$$

where we have increased the feasible set by changing the equality constraints to inequality constraints (compared to our original problem).

Keeping this short recall in mind, we now give a heuristic motivation for the eventual form of the optimal terminal wealth in the portfolio problem.

Heuristic computation of the optimal terminal wealth
Instead of looking at the problem (O) in its form

$$\max_{B} E(U(B))$$
$$\text{s.t. } B \ge 0, \ E(H(T)B) \le x, \ E\left(U(B)^{-}\right) < \infty \quad , \qquad \text{(O)}$$

we consider the problem

$$\max_{B} \ E(U(B)$$
$$\text{s.t. } E(H(T)B) = x \quad , \qquad (\hat{O})$$

and "solve" it via imitation of the above Lagrangian multiplier method. Therefore, define the analogue of the Lagrangian in our setting as

$$L(B, y) = E(U(B) - y(\ H(T)B - x\ ))$$

for all random variables B, feasible for problem $(\hat{O})$ and $y \in \mathbf{R}$. As above, we seek a zero of the gradient of L(B, y). Although the derivative of L with respect to B is not a derivative in the usual sense (it is a derivative defined on a function space), we treat it as a usual derivative by handling B as if it would be just an ordinary variable. Moreover, we will exchange the derivative with the expectation in L. This whole method can be justified by using suitable functional analytic methods. However, since these computations only serve as a motivation, and also our final method of proof in Theorem 16 will not require them, we omit to present the underlying functional analytic tools. Proceeding in the described formal way leads to the system of equations

$$0 = L_B(B, y) = E\,(U'(B) - yH(T)), \tag{19}$$

$$0 = L_y(B, y) = x - E\,(H(T)B)\,. \tag{20}$$

Hence, every B of the form

$$B = \left(U'\right)^{-1}(yH(T)) =: I(yH(T))$$

solves equation (19) for an arbitrary positive y (recall that H(T) is always positive). Putting this form of B into equation (20) yields

$$x = E(H(T)I(yH(T))) =: \mathcal{X}(y)\,.$$

If we now assume that $\mathcal{Y}(x) := \mathcal{X}^{-1}(x)$ exists, we can solve this last equation which then yields the "optimal" B* as

$$B^* = I(\mathcal{Y}(x)H(T)).$$

If our Lagrangian type considerations can be made rigorous then B* will even solve the problem (O) (if we temporarily ignore the constraint $E(U(B)^-) < \infty$). To see this, note that B* is positive, since $\mathcal{Y}(x)H(T)$ is, and I(z) is also positive for positive z. Further, $\mathcal{Y}(x)$ is the Lagrangian multiplier corresponding to B*. As it is positive, we know that the solution of $(\hat{O})$ is also the solution of the corresponding problem where the equality constraint "E(H(T)B) = x" is replaced by the inequality "$E(H(T)B) \le x$". Note further that for our method we needed that the range of U'(x) is equal to $(0,\infty)$ for solving equation (19) ω-wise, i.e. for making the term under the expectation sign vanish for every $\omega \in \Omega$.

Hence, if we could make our heuristic considerations rigorous, B* would indeed equal the optimal terminal wealth. The following proposition and theorem show that this can be achieved even for the general problem (P) including consumption. Due to the presence of consumption, the notation will be a bit more complex than in the above heuristic computations but will have the similar form. As in (Korn and Trautmann 1995), we will look at a slightly bigger class of utility functions than the one given in Section 1.

Definition 1*

A function $U\colon (0, \infty) \to \mathbf{R}$ such that $U \in C^1$ is strictly concave and satisfies

$$U'(0) := \lim_{x \downarrow 0} U'(x) > 0\ , \quad U'(z) = 0 \tag{1*}$$

for a unique value $z\in(0,\infty]$ will be called a **utility function** .

Remark 14

a) Note first that the utility functions of Definition 1 are of course utility functions in the sense of Definition 1*. Typical examples of functions that are only utility functions in the new sense are $U(x) = a - be^{-cx}$ for positive constants a, b, c and $U(x) = -\frac{1}{2}(x-K)^2$ for a positive constant K.

b) The reason for allowing for utility functions in Definition 1* that are not strictly increasing is that we will now be able to consider the quadratic utility function. Although this function can be critised from many points of view, its use is still very popular in finance, especially in connection with the continuous-time mean variance approach. We will refer to this later in Section 4.3.

c) Similar to Remark 2c), a set of functions $U(t, .)$, $t\in[0, T]$, will also be called a **utility function** if for every fixed $t\in[0, T]$, $U(t, .)$ is a utility function in the second variable (in the sense of Definition 1), $U(., x)$ is a continuous function in the first variable for every fixed positive x, and if there is a $z\in(0,\infty]$ with $U'(t, z) = 0$ for every $t\in[0, T]$ (where we have set $U'(t, z) := \frac{\partial}{\partial t}U(t, z)$).

d) The requirements of Definition 1* imply that U' is strictly decreasing on $[0, z]$ with $U'\colon [0, z]\to[0, U'(0)]$ (where $U'(0)$ and $U'(+\infty)$ are defined as in Definition 1). U' posesses a strictly decreasing, continuous inverse function $I^*\colon [0, U'(0)]\to [0, z]$ which can be extended to all non-negative numbers y by setting

$$I(y) := \begin{cases} I^*(y) & , \quad y\in[0, U'(0)] \\ 0 & , \quad y \geq U'(0) \end{cases}. \tag{21}$$

It will be crucial to note that we have

$$U(I(y)) \geq U(x) + y\,(I(y) - x), \quad y\in(0, \infty), \;\; x \geq 0 \tag{22}$$

where $U(0)$ is defined as

$$U(0) := \lim_{x\downarrow 0} U(x)\,. \tag{23}$$

Relation (22) remains valid for $y = 0$ if we have $z = +\infty$. To see the above inequality, note that for $y\in(0, U'(0)]$ the concavity of U implies

$$U(I(y)) \geq U(x) + U'(I(y))(I(y) - x) = U(x) + y\,(I(y) - x) \;\; \forall\, x \geq 0.$$

In the case of $y > U'(0)$, we have

$$U(I(y)) = U(0) \geq U(x) + U'(0)(0 - x) \geq U(x) + y\,(I(y) - x) \quad \forall\, x \geq 0\,.$$

As we also consider utility functions that are not necessary strictly increasing (such as the quadratic one), we do not always need our whole endowment x to draw the optimal utility from our portfolio problem. We will therefore make a slight change in notation and define the (unconstrained) portfolio problem as the optimisation problem

$$\max_{(\pi, c) \in A'(y),\, y \leq x} J(y\,;\pi, c) \tag{P}$$

Of course, if $U_1(t, x)$, $U_2(x)$ are strictly increasing functions in x then there is no need for the inequality " $y \leq x$ " in the definition of the optimisation problem (this will be a consequence of Theorem 16). In this case, we can confine ourselves to strategies $(\pi, c) \in A'(x)$. As in the heuristic computation of the optimal terminal wealth, we now define the function $X : (0, \infty) \to \mathbf{R}$ for problem (P):

$$X(y) := E\left(H(T) I_2(yH(T)) + \int_0^T H(t) I_1(t, yH(t))\,dt \right) \quad \forall\, y > 0$$

(where $I_1(t, .)$, $I_2(.)$ are the extended inverse functions of $U_1'(t, .)$ and $U_2'(.)$ in the sense of Remark 14 d)). Its main characteristics (which will especially imply the existence of an inverse function $Y(x)$) will be given in the following proposition :

Proposition 15
Assume

$$X(y) < \infty \qquad \forall\, y \in (0, \infty), \tag{24}$$

and in the case

$$U_1'(t, 0) < \infty \quad \forall\, t \in [0, T] \quad \text{and} \quad U_2'(0) < \infty, \tag{25}$$

assume further that $\theta(t)$ is a deterministic function with

$$\int_0^T \|\theta(s)\|^2\, ds > 0\ . \tag{26}$$

Then, X is continuous on $(0, \infty)$ and strictly decreasing with

$$X(\infty) := \lim_{y \to \infty} X(y) = 0\,, \tag{27}$$

$$X(0) := \lim_{y \to 0} X(y) = \begin{cases} \infty, \text{if } \lim_{z\to\infty} U_2'(z) = 0 \text{ or } \lim_{z\to\infty} U_1'(t,z) = 0 \ \forall t \in [0,T] \\ z_1\, E\left(\int_0^T H(t)dt\right) + z_2\, E(H(T)), \qquad \text{else} \end{cases} \tag{28}$$

where z_1, z_2 are the values characterised by the equations $U_1'(t, z_1) = 0 \ \forall\ t \in [0, T]$ and $U_2'(z_2) = 0$, respectively.

Proof :

a) The continuity of X follows by the continuity and monotonicity of $I_1(t, .)$ and $I_2(.)$ together with the help of the dominated convergence theorem.

b) $I_1(t, .)$ is strictly decreasing on $(0, U_1'(t, 0))$. If we can show that we have

$$P(y\, H(t) < U_1'(t, 0) \ \text{ for some } t \in [0, T]) > 0 \tag{29}$$

for every fixed $y \in (0, \infty)$ then it follows that

$$X_1(y) := E\left(\int_0^T H(t) I_1(t, yH(t))dt\right)$$

is strictly decreasing in $y \in (0, \infty)$, because $H(t)I_1(t, yH(t))$ is strictly decreasing in y on the set "$y\, H(t) < U_1'(t, 0)$ for some $t \in [0, T]$" and identically zero on its complement. Since $H(t)$ and $U_1'(t, 0)$ are assumed to be continuous, relation (29) also implies

$$P(\exists\ 0 \le t_1 < t_2 \le T \mid y\, H(t) < U_1'(t, 0) \ \ \forall\ t \in (t_1, t_2)) > 0\,, \tag{30}$$

and we obtain the claimed monotonicity of X_1. Note first that in the case "$U_1'(t, 0) = \infty$" inequality (29) is obvious. If in contrast $U_1'(t, 0)$ is finite then we can use the additional assumption of a deterministic $\theta(t)$ in the following way:

$$\ln(H(t)) = -\int_0^t \theta(s)'dW(s) - \int_0^t \left(r(s) + \tfrac{1}{2}\|\theta(s)\|^2\right)ds$$

is the difference of a bounded random variable (note that r(s) as part of the integrand in the second integral is random but uniformly bounded) and a stochastic integral, which is normally distributed with mean zero and variance equal to

$$\tfrac{1}{2}\int_0^t \|\theta(s)\|^2 ds > 0$$

since $\theta(s)$ is deterministic (compare Proposition B6). Thus, by the characteristics of the normal distribution, we have

$$P(\ln(H(t)) < u \text{ for some } t \in [0, T]) > 0 \quad \forall u > 0 .$$

which implies inequality (29). Analogous considerations for

$$X_2(y) := E(H(T)\, I_2(y\, H(T))$$

imply

$$P(yH(T) < U_2'(0)) > 0 , \tag{31}$$

and X_2 is strictly decreasing in $y \in (0, \infty)$, too. Consequently, $X(y) = X_1(\lambda) + X_2(\lambda)$ is also decreasing in y. Note that relations (29) or (31) are always satisfied if $U_1'(t, 0) = \infty$ for all $t \in [0, T]$ or $U_2'(0) = \infty$. To prove that $X(y)$ is decreasing in the case of a random (but uniformly bounded) process $\theta(t)$, it is therefore enough if either $U_1'(t, 0)$ (for all $t \in [0, T]$) or $U_2'(0)$ are infinite.

c) By the monotone convergence theorem, assumption (24), the monotonicity of I_1 and I_2, and $I_1(t, \infty) = I_2(\infty) = 0$ ($\forall\, t \in [0, T]$) together imply (27).

d) It remains to show (28). First consider the case

$$\lim_{z \to \infty} U_2'(z) = 0. \tag{32}$$

Because I_1 and I_2 are non–negative functions, Fatou's lemma implies

$$\liminf_{y \to 0} X(y) \geq \liminf_{y \to 0} X_2(y) \geq E\,(H(T) \liminf_{y \to 0} I_2(y\, H(T))) = \infty .$$

For the same reasons, in the case

$$\lim_{z \to \infty} U_1'(t,z) = 0 \ \ \forall\, t \in [0, T] \tag{33}$$

we have

$$\liminf_{y \to 0} X(y) \geq \liminf_{y \to 0} X_1(y) \geq E\left(\int_0^T H(t) \liminf_{y \to 0} I_1(t, yH(t))dt \right) = \infty .$$

Conversely, if relations (32) or (33) are not satisfied, then we have

$$\limsup_{y \to 0} X(y) \leq z_1\, E\left(\int_0^T H(t)dt \right) + z_2\, E\,(H(T)),$$

due to the relations $I_1(t, x) \leq z_1$ and $I_2(x) \leq z_2$ for every (t, x) in the regions where I_1 and I_2 are defined. On the other hand, due to Fatou's lemma, we have

$$\liminf_{y \to 0} X(y) \geq E\left(\int_0^T H(t) \liminf_{y \to 0} I_1(t, yH(t))dt + H(T) \liminf_{y \to 0} I_2(yH(T)) \right)$$

$$= z_1 \, E\left(\int_0^T H(t)dt\right) + z_2 \, E\,(\,H(T)\,)\,.$$

Thus, equation (28) is proved.

□

If we now define $X(\infty)$ und $X(0)$ as in equations (27) and (28) and z* by

$$z^* := \begin{cases} z_1 \; E\left(\int_0^T H(t)dt\right) + z_2 \, E(H(T)) \,, & \text{if } z_1 \text{ and } z_2 \text{ are finite} \\ \infty \,, & \text{else} \end{cases}$$

then Proposition 15 yields the existence of a continuous and strictly decreasing inverse function Y: $[0, z^*] \to [0, \infty]$ to X on $[0, \infty]$. This will enable us to give the solution to the unconstrained portfolio problem (a generalisation of Theorem 7.4 in (Cvitanic and Karatzas 1992)) .

Theorem 16

Let $x > 0$. Under assumption (24) and the additional assumption (26) in case of (25), the optimal terminal wealth B* and the optimal consumption process c*(t), $t\in[0, T]$, are given by

$$B^* := \begin{cases} z_2 \,, & \text{if } x \geq z^* \\ I_2(Y(x)H(T)), & \text{else} \end{cases}, \tag{34}$$

$$c^*(t) := \begin{cases} z_1 \,, & \text{if } x \geq z^* \\ I_1(t, Y(x)H(t)), & \text{else} \end{cases}, \tag{35}$$

and there exists an $x^* \in [0, x]$ together with a portfolio process $\pi^*(t)$, $t\in[0, T]$, such that we have

$$(\pi^*, c^*) \in A'(x^*)\,, \qquad X^{x^*,\pi^*,c^*}(T) = B^* \quad \text{a.s.,}$$
$$J(x^*, \pi^*, c^*) = \max_{(\pi,c)\in A'(y),\, y\leq x} J(y\,;\pi, c), \tag{36}$$

i.e. (π^*, c^*) solves the unconstrained portfolio problem. Moreover, if we are not in case (25) then we have $x^* = x$.

Proof :

i) Case "$x \geq z^*$"

Because $U_1(t, .)$ and $U_2(.)$ attain their absolute maxima in $z_1 = I_1(t,0)$ for every t $\in[0, T]$ and in $z_2 = I_2(0)$, respectively, we have

$$\int_0^T U_1(t, I_1(t, 0))dt + U_2(I_2(0)) \geq \int_0^T U_1(t, c(t))dt + U_2(X^{y,\pi,c}(T)) \quad \text{a.s.}$$

for every $(\pi, c)\in A'(y)$ with $y \leq x$. That is, the choices of B^* and $c^*(t)$ according to (34) and (35) are in this case **pathwise optimal** (i.e. yield the optimal utility for almost all $\omega\in\Omega$) and therefore also optimal for the unconstrained portfolio problem. The existence of a portfolio process $\pi^*(t)$ with $(\pi^*, c^*)\in A(z^*)$ and $X^{z^*,\pi,c}(T) = B^*$ a.s. is ensured by Theorem 2.7. Note further that $c^*(t)$ and B^* are now deterministic. Hence, we have

$$E\left(\int_0^T U_1(t, c^*(t))^- dt + U_2\left(X^{z^*,\pi,c}\right)^-\right) = \int_0^T U_1(t, z_1)^- dt + U_2(z_2)^- < \infty ,$$

which implies $(\pi^*, c^*)\in A'(z^*)$. Thus, everything is proved for in this case.
ii) Case "$x < z^*$" (compare (Cvitanic and Karatzas 1992), Theorem 7.4)
By definition of $c^*(t)$, $t \in [0, T]$ and B^* we have

$$E\left(\int_0^T H(t)c^*(t)dt + H(T)B^*\right) = x .$$

The concavity of $U_1(t, .)$ and $U_2(.)$ yields (see also Remark 14 d))

$$U_1(t, c^*(t)) \geq U_1(t, 1) + \mathcal{Y}(x) H(t) (c^*(t) - 1),$$
$$U_2(B^*) \geq U_2(1) + \mathcal{Y}(x) H(T) (B^* - 1).$$

Hence,

$$E\left(\int_0^T U_1(t, c^*(t))^- dt + U_2(B^*)^-\right)$$
$$\leq |U_2(1)| + \int_0^T |U_1(t,1)| dt + \mathcal{Y}(x)\left(x + E(H(T)) + \int_0^T E(H(t))dt\right) < \infty \quad (37)$$

Furthermore,

$$U_1(t, c^*(t)) \geq U_1(t, c(t)) + \mathcal{Y}(x) H(t) (c^*(t) - c(t))$$
$$U_2(B^*) \geq U_2(X^{y,\pi,c}(T)) + \mathcal{Y}(x) H(T) (B^* - X^{y,\pi,c}(T))$$

imply

$$E\left(\int_0^T U_1(t,c^*(t))dt + U_2(B^*)\right)$$
$$\geq J(y, \pi, c) + Y(x)\left(E\left(\int_0^T H(t)(c^*(t) - c(t))dt + H(T)\left(B^* - X^{y,\pi,c}(T)\right)\right)\right)$$
$$= J(y, \pi, c) + Y(x)\left(x - E\left(\int_0^T H(t)c(t)dt + H(T)X^{y,\pi,c}(T)\right)\right) \geq J(y, \pi, c)$$

for every $(\pi, c) \in A(y)$, $y \leq x$, where the last inequality follows from Theorem 2.7 a). The existence of a portfolio process $\pi^*(t)$, $t \in [0, T]$, with $(\pi^*, c^*) \in A(y)$ for a real number $y \leq x$ and

$$X^{y,\pi^*,c^*}(T)) = B^* \quad \text{a.s.}$$

follows from part b) of the same theorem. Moreover, due to relation (37), we have $(\pi^*, c^*) \in A'(x)$ which completes the proof.

□

Remark 17
In particular, if there exist finite values z_1 and z_2 with $U_1'(t, z_1) = 0 \;\; \forall\, t \in [0, T]$ and $U_2'(z_2) = 0$ then Theorem 16 implies that the choice of "optimal consumption" $c(t) = z_1 \; \forall\, t \in [0, T]$ and the choice of "optimal terminal wealth" $B^* = z_2$ are optimal for the unconstrained portfolio problem only if these choices are financiable (by which we mean that the initial capital x of the investor exceeds z^*). If only one of the two values z_1, z_2 is finite, say z_1 (z_2), then Theorem 16 states that it is never optimal to choose this value for the consumption rate (the terminal wealth) at the costs of a smaller terminal wealth (consumption rate). This can be explained by the fact that if the consumption rate is close to z_1, the utility of an additional amount of consumption will converge to zero and will hence be less than the utility gained from the same additional amount of terminal wealth.

The proofs of both the above proposition and theorem show that the pure optimal consumption problem (formally: the choice of $U_2(x) \equiv 0$) and the pure terminal wealth maximisation (formally: the choice of $U_1(t, c) \equiv 0$ for all $t \in [0, T]$) are special cases of the general portfolio problem (P). We formulate this result as a corollary.

Corollary 18
Assume that the conditions of Theorem 16 are satisfied.
a) The optimal consumption process $c^*(t)$, $t \in [0, T]$, for the problem

$$\max_{(\pi,c)\in A'(y),y\le x} E\left(\int_0^T U_1(t,c(t))dt\right)$$

is given by

$$c^*(t) := \begin{cases} z_1, & \text{if } x \ge z_1 \\ I_1(t,\mathcal{Y}(x)H(t)), & \text{else} \end{cases}.$$

Further, there exists an $x^*\in[0, x]$ and a portfolio process $\pi^*(t)$, $t\in[0, T]$, with $(\pi^*, c^*)\in A'(x^*)$ and

$$X^{x^*,\pi^*,c^*}(T) = 0 \quad \text{a.s.}.$$

b) The optimal terminal wealth B^* for the problem

$$\max_{\pi\in A'(y),y\le x} E(U_2(X^{\pi}(T))$$

is given by

$$B^* := \begin{cases} z_2 & , \quad \text{if } x \ge z_2 \\ I_2(\mathcal{Y}(x)H(T)), & \text{else} \end{cases},$$

and there exists an $x^*\in[0, x]$ and a portfolio process $\pi^*\in A'$ with

$$X^{x^*,\pi^*,c^*}(T) = B^* \quad \text{a.s.},$$

i.e. π^* is optimal for the problem of maximising expected utility of terminal wealth.

We will give two explicit examples of portfolio problems, each with a typical form of the utility function, to demonstrate the use of the martingale method .

Example 19: "Logarithmic utility"
Let

$$U_1(t, x) = U_2(x) = \ln(x) \quad \forall\, t\in[0, T].$$

One can easily verify that in this case we have

$$I_1(t, y) = I_2(y) = \frac{1}{y}, \quad \mathcal{X}(y) = \frac{1}{y}(T+1)\ \forall\, y>0, \quad \mathcal{Y}(x) = \frac{1}{x}(T+1) \quad \forall\, x>0.$$

Thus, Theorem16 yields the optimal consumption and terminal wealth as

$$c^*(t) = \frac{x}{T+1}\frac{1}{H(t)} \quad \forall\, t\in[0, T], \qquad B^* = \frac{x}{T+1}\frac{1}{H(T)}.$$

It remains to compute the corresponding optimal portfolio process $\pi^*(t)$. We therefore apply Itô's formula to $X(T) = B^*$, where we have set

$$X(t) := f(t, G(t)),$$

$$G(t) := \int_0^t \left(r(s) + \tfrac{1}{2}\|\theta(s)\|^2\right)ds + \int_0^t \theta(s)'dW(s),$$

$$f(t, z) := x\left(1 - \tfrac{t}{T+1}\right) e^z .$$

With this definiton, we obtain

$$dX(t) = \left(f_t(t,G(t)) + f_z(t,G(t))\left(r(t) + \tfrac{1}{2}\|\theta(t)\|^2\right) + \tfrac{1}{2} f_{zz}(t,G(t))\|\theta(t)\|^2 \right)dt$$
$$+ f_z(t, G(t))\theta(t)'dW(t)$$
$$= X(t)\left(\left(-\frac{1}{T+1-t} + \left(r(t) + \|\theta(t)\|^2\right)\right)dt + \theta(t)'dW(t)\right),$$
$$X(0) = x .$$

Comparing this equation with the general form of the stochastic differential equation for the wealth process X(t) corresponding to the admissible pair (π^*, c^*),

$$dX(t) = [(r(t) + (b(t)-r(t)\underline{1})'\pi^*(t))\, X(t) - c^*(t)]dt + X(t)\pi^*(t)'\sigma(t)dW(t),$$

we obtain

$$\pi^*(t) = (\sigma(t)')^{-1}\theta(t) = (\sigma(t)\sigma(t)')^{-1}(b(t)-r(t)\underline{1}) \quad \forall\, t \in [0, T],$$
$$c^*(t) = \frac{1}{T+1-t}\, X(t)$$

by way of comparison of the diffusion and drift terms of the two representations for X(t). Note that with this choice of (π^*, c^*) the process X(t) satisfies the stochastic differential equation for the wealth process corresponding to (π^*, c^*) and is also positive, because f(t, G(t)) is positive. Further, (π^*, c^*), as given above, satisfies the requirements on an admissible pair. Hence, we have found the optimal portfolio and the optimal wealth process. Note further that, of course, the two representations of $c^*(t)$ coincide, as we have

$$X(t) = x\left(\frac{T+1-t}{T+1}\right)\frac{1}{H(t)} .$$

The second one has the advantage of showing the direct dependence on X(t) (one often refers to that as **feedback form** with respect to X(t)). In the case of constant coefficients and only one stock the form of $\pi^*(t)$ has the following interpretation: The optimal fraction of wealth invested in the stock is given by the difference between the mean rate of stock return and the riskless rate of return divided by the stock's volatility σ^2, a term, which is usually called the (local) relative risk premium for stock investment or the **market price of risk**. Once again, we would like to remind the reader of the fact that although this strategy has an extremely simple form, it requires rebalancing of the portfolio at every time instant t.

Example 20: "Minimal deviation from a target value"
In the case of logarithmic utility, all requirements on a utility function given in Definition 1 were satisfied. Now we will drop two of these requirements. By considering a quadratic utility function, we will no longer have a strict increase in utility with an increasing terminal wealth. Also, the utility function will have a finite derivative in $x = 0$. We will concentrate on a pure terminal wealth maximisation problem. Our goal can be regarded as an approximation problem for a given constant target value (we will look at stochastic target values in Section 6). Let therefore

$$U_1(t,.) \equiv 0 \quad \forall\, t\in[0,T]\,,\quad U_2(x) = -\tfrac{1}{2}\,(x-K)^2 \quad \forall\, x>0\,,$$

where $K > 0$ is a given constant ("**the target value**"). The optimisation problem

$$\min_{(\pi,c)\in A'(y),\; y\le x} \tfrac{1}{2}\, E\left(X^{y,\pi,c}(T)-K\right)^2$$

is equivalent to the unconstrained portfolio problem

$$\max_{(\pi,c)\in A'(y),\; y\le x} E\left(U_2\left(X^{y,\pi,c}(T)\right)\right).$$

We restrict ourselves to constant market coefficients and assume that condition (26) is satisfied which enables us to make use of Corollary 18. To solve this optimisation problem, we have to distinguish between two cases :

a) "$x \ge K\,E(H(T))$"

In this case, Corollary 18 implies that $B^* = K$ is the optimal terminal wealth which can of course be attained by the pure bond strategy of investing $K\exp(-rT)$ in the bond at time $t = 0$ and holding this position until $t = T$ (i.e. we have $\pi^*(t) \equiv 0 \;\forall\, t\in [0, T]$). Of course, this is the uninteresting case.

b) "$x < K\,E(H(T))$"

Here, Corollary 18 implies that

$$B^* = I_2(Y(x)\,H(T)) = (\,K - Y(x)H(T)) \cdot 1_{\left\{H(T) \le \frac{K}{Y(x)}\right\}} = (\,K - Y(x)H(T))^+$$

is the optimal terminal wealth. Unfortunately, the equation

$$\begin{aligned} X(y) &= E\,(H(T)(K - y\,H(T))^+) \\ &= K\,E\big(H(T)1_{A(y)}\big) - y\,E\big(H(T)^2\,1_{A(y)}\big) \end{aligned}$$

(where the set A(y) is defined by $A(y) = \{H(T) \le K/y\}$) cannot be solved explicitly for y to obtain the inverse function Y of X. However, there is a possibility for a numerical solution. In particular, consider the case n = 1. There, we have

$$\begin{aligned} X(y) = {} & K \int_{-\frac{1}{\theta}\left((r+\frac{1}{2}\theta^2)T + \ln(\frac{K}{y})\right)}^{\infty} \frac{1}{\sqrt{2\pi T}} \exp\!\left(-\left(x^2/_{2T}\right) - \left(r + \tfrac{1}{2}\theta^2\right)T - \theta\,x\right)dx \\ & - y \int_{-\frac{1}{\theta}\left((r+\frac{1}{2}\theta^2)T + \ln(\frac{K}{y})\right)}^{\infty} \frac{1}{\sqrt{2\pi T}} \exp\!\left(-\left(x^2/_{2T}\right) - 2\left(r + \tfrac{1}{2}\theta^2\right)T - 2\theta\,x\right)dx \\ = {} & K\exp(-rT)\phi\!\left(\frac{rT - \frac{1}{2}\theta^2 T + \ln(\frac{K}{y})}{\theta\sqrt{T}}\right) - y\,\exp\!\left(\left(-2r + \theta^2\right)T\right)\phi\!\left(\frac{rT - \frac{3}{2}\theta^2 T + \ln(\frac{K}{y})}{\theta\sqrt{T}}\right), \end{aligned}$$

and we get $Y(x)$ by solving the non-linear equation $X(y) = x$ which has a unique solution due to Proposition 15. We will take up this example again in Section 4.3 in which it will help us to solve a continuous-time mean variance problem. There, we also present some numerical examples visualising the above expressions. By a similar comparison method as in the case of the previous example, we can also calculate the corresponding portfolio process π^*. However, its explicit form is of tremendous length and does not give much insight. We therefore omit its presentation here, but again refer to Section 4.3 where we will also make some remarks on (the interpretation of) the optimal portfolio process.

ii) Computation of the Optimal Strategy — the Representation Problem (R)

While the optimisation problem could be solved more or less in a straight forward way, the representation problem will require surprisingly deep methods. Before presenting them, we will first formalise a method indicated in Example 19. It is based on comparing the general form of the wealth process with that obtained from the optimal terminal wealth and consumption as given in Theorem 16. We will then concentrate on markets with constant coefficients r, b, σ and present a special case of a representation result of (Karatzas 1989) for the optimal consumption and portfolio rules (π, c) in feedback form with respect to the current wealth X(t). There, the assumption of constant coefficients enables us to make use of Hamilton-Jacobi-Bellman type methods. To save some notation, we will not consider the slightly more general case of deterministic coefficients.

Method 1: "Comparison of coefficients"

Looking back at Example 19, we see that there, the method used to compute the optimal portfolio process consisted of

- guessing a process X(t) with X(0) = x, X(T) = B* a.s.,
- writing X(t) as a functional of the underlying Brownian motion and the market coefficients,
- application of Itô's formula to this functional and comparison of drift and diffusion terms of the so obtained sde with those in the general form of the sde for a wealth process.

This lead to a candidate process π^* for the admissible portfolio process generating the optimal terminal wealth B* and the optimal consumption c*(t) (as given in Theorem 16). By checking the admissibility of the pair (π^*, c*), we verified that this was indeed the optimal pair and that X(t) was the corresponding wealth process. In fact, using the proof of part b) of Theorem 2.7, we realise that the above "guess" for the process X(t) is nothing else but computing the conditional expectation

$$\frac{1}{H(t)} E\left(\int_t^T H(s)c^*(s)ds + H(T)B^* \middle| F_t \right).$$

This method will be formalised in the following theorem.

Theorem 21

Assume the complete market setting of this section and that we have

$$\frac{1}{H(t)} E\left(\int_t^T H(s)c^*(s)ds + H(T)B^* \middle| F_t \right) = f(t, W_1(t),\dots, W_n(t)) \qquad (39)$$

for a non-negative function $f \in C^{1,2}([0, T) \times \mathbf{R}^n)$ with $f(0,\dots, 0) = x^*$ (where B*, c*, x* are the optimal terminal wealth, consumption and initial wealth for problem (P) as given in Theorem 16). Then the optimal trading strategy $\varphi(t) = (\varphi_0(t),\dots, \varphi_n(t))'$, $t \in [0, T]$, is given by

$$\varphi_i(t) = \frac{1}{P_i(t)}\left(\sigma(t)^{-1}\nabla_x f\left(t, W_1(t),\ldots, W_n(t)\right)\right)_i, \; i = 1,\ldots, n,$$

$$\varphi_0(t) = \left(X(t) - \sum_{i=1}^{n}\varphi_i(t)P_i(t)\right) / P_0(t),$$

where X(t) is the wealth process corresponding to the above trading strategy φ(t) and the consumption process c*(t) of Theorem 16 (provided that φ(t) meets the requirements of Definition 2.3). ∇_x f(.) denotes the gradient of f with respect to the last n variables (i.e. the ones that represent the components of the Brownian motion). The optimal portfolio process π*(t) of Theorem 16 is given by

$$\pi^*(t) = \frac{1}{X(t)}\sigma(t)^{-1}\nabla_x f(t, W_1(t),\ldots, W_n(t)).$$

Proof:

Theorem 2.7 (and its proof) yields the existence of a trading strategy φ(t) such that the wealth process X(t) corresponding to φ(t) and the consumption process c*(t) of Theorem 16 satisfies

$$\frac{1}{H(t)}\, E\left(\int_t^T H(s)c^*(s)ds + H(T)B^* \middle| F_t\right)$$

$$= X(t) = x^* + \sum_{i=0}^{n}\int_0^t \varphi_i(s)dP_i(s) - \int_0^t c(s)ds$$

$$= x^* + \int_0^t\left(\varphi_0(s)P_0(s)r(s) + \sum_{i=1}^{n}\varphi_i(s)P_i(s)b_i(s) - c(s)\right)ds$$

$$+ \sum_{j=1}^{n}\int_0^t \sum_{i=1}^{n}\varphi_i(s)P_i(s)\sigma_{ij}(t)dW(s) \tag{40}$$

On the other hand, by assumption (39) and Itô's formula we have

$$\frac{1}{H(t)}\, E\left(\int_t^T H(s)c^*(s)ds + H(T)B^* \middle| F_t\right)$$

$$= f(0,\ldots, 0) + \int_0^t\left(f_t(s, W_1(s),\ldots, W_n(s)) + \tfrac{1}{2}\sum_{i,j=1}^{n} f_{x_i x_j}(s, W_1(s),\ldots, W_n(s))\right)ds$$

$$+ \sum_{i=1}^{n}\int_0^t f_{x_i}(s, W_1(s),\ldots, W_n(s))dW_i(s). \tag{41}$$

If we note $f(0,\ldots, 0) = x^*$ then both the Itô-processes on the right hand sides of representations (40) and (41), respectively, coincide. By uniqueness of the representation of an Itô-process, the integrands of the Itô-integrals must coincide. By introducing the notation

$$\psi(t) = (\varphi_1(t)P_1(t),\ldots, \varphi_n(t)P_n(t))',$$

we obtain the following linear system

$$\sigma(t)\ \psi(t) = \nabla_x f(t, W_1(t),\ldots, W_n(t)).$$

As $\sigma(t)$ is regular, we can solve for $\psi(t)$ which then yields the desired representation for $\varphi_i(t)$, $i = 1,\ldots, n$. The representation for $\varphi_0(t)$ is obvious and the one for $\pi^*(t)$ is directly implied by the one for $\varphi_i(t)$, $i = 1,\ldots, n$. Further, by assumption, $\varphi(t)$ satisfies the requirements of Definition 2.3. Hence, we have found the optimal trading strategy (and the corresponding portfolio process).

□

Remark 22

a) A representation of the conditional expectation of such simple form as in Equation (39) can only be expected for the case of constant market coefficients. For general market coefficients, we would expect a functional representation of the conditional expectation depending on the whole path of W(s) on [0, t] (and also depending on the values of the market coefficients on [0, t]). In particular, this is the case in Example 19. However, to formulate such a corresponding theorem would require deeper functional analytic methods (see (Occone and Karatzas 1991)).

b) Theorem 21 does not explicitly require that B^* and c^* are the solutions to a portfolio problem. The proof of the theorem shows that we only need both B^* and c^* to be non-negative and to satisfy

$$E\left(\int_0^T H(s)c^*(s)ds + H(T)B^*\right) = x^* \qquad (*)$$

for some positive number x^*. Then application of part b) of Theorem 2.7 is possible. Thus, we can also use Theorem 21 to represent any non-negative B, c satisfying the integrability condition (*). As an example, we look at a market with constant coefficients and $n = 1$. We further choose

$$c(t) \equiv 0\ \forall\ t\in[0, T]\ \text{a.s.},\quad B = \left(W(T) + \theta^2 T\right)^2,$$

$$x^* = E\left(H(T)\left(W(T) + \theta^2 T\right)^2\right) = e^{-rT}T\ .$$

Computing

$$X(t)=\frac{1}{H(t)}E\left(\left(H(T)\left(W(T)+\theta^2T\right)^2\right)\middle|F_t\right)=e^{-r(T-t)}\left(T-t+\left(W(t)+\theta^2t\right)^2\right),$$

and applying Theorem 21 yields the (candidate for the) generating strategy

$$\varphi_1(t)=2\,e^{-r(T-t)}\,\frac{\left(W(t)+\theta^2t\right)^2}{\sigma P_1(t)}\,.$$

It can easily be checked that $\varphi=(\varphi_0,\varphi_1)$ (where we obtain the bond component of this strategy, $\varphi_0(t)$, via $\varphi_0(t)=X(t)-\varphi_1(t)P_1(t)$) meets the requirements of Definition 2.3, and we have thus found the generating strategy.

Method 2: "Feedback representation in the Markovian case"

For developing the next method, we assume that the market coefficients r, b, σ are real constants. Further, we need some assumptions and notations. The utility functions of the investor should have the form

$$U_1(t,c)=\exp(-\mu t)\;U_1{}^*(c)\,,\quad U_2(x)=\exp(-\mu T)\;U_2{}^*(x) \tag{42}$$

where U^*_i are assumed to be utility functions in the sense of Definition 1. Further, they should be twice continuously differentiable and such that for

$$I_1(c)=(\tfrac{\partial}{\partial c}U_1{}^*)^{-1}(c),\quad I_2(x)=(\tfrac{\partial}{\partial x}U_2{}^*)^{-1}(x),$$

we have

$$|\,y\,I_i(y)\,|\le K\left(1+y^{\alpha}+y^{-\alpha}\right)\ \text{ for some } K>0,\ \alpha>0. \tag{43}$$

It is shown in (Karatzas, Lehoczky and Shreve 1987), Lemma 7.2 that relation (43) is valid if the utility functions satisfiy

$$U_i{}^*(0)>-\infty,\quad \lim_{c\downarrow 0}\frac{((U_i{}^*)'(c))^2}{(U_i{}^*)''(c)}\ \text{exists},\quad \lim_{c\to\infty}\frac{((U_i{}^*)'(c))^{\alpha}}{(U_i{}^*)''(c)}=0,\quad i=1,2$$

for some $\alpha>2$ and for a positive constant μ. In our examples, assumption (43) can be checked directly. Therefore, we do not give the above mentioned Lemma 7.2. Let

$$\mathcal{X}(t,y)=E\left[\int_t^T H^t(s)I_1\left(yH^t(s)e^{\mu(s-t)}\right)ds+H^t(T)I_2\left(yH^t(T)e^{\mu(T-t)}\right)\right]$$

with

$$H^t(s) := H(s) / H(t)$$

for $s \geq t$, and assume

$$\mathcal{X}(t, y) < \infty \quad \forall\ (t, y) \in [0, T] \times (0, \infty)\ . \tag{44}$$

Under this assumption, it can be proved for $\theta \neq 0$ that $\mathcal{X}(t, .)$ is a continuous, strictly increasing function on $(0, \infty)$ for all $t \in [0, T]$ and continuously differentiable with respect to the second variable y. Moreover, we have

$$\mathcal{X}(t, 0+) := \lim_{y \downarrow 0} \mathcal{X}(t, y) = \infty\ , \quad \mathcal{X}(t, \infty) := \lim_{y \to \infty} \mathcal{X}(t, y) = 0\ ,$$

and there exists an inverse function $\mathcal{Y}(t, .)$ to $\mathcal{X}(t, .)$ (see (Karatzas 1989)). The proof of these facts is similar to the one of Proposition 15. The required differentiability of $\mathcal{X}(t, y)$ with respect to y is a consequence of an application of the dominated convergence theorem.

After all these assumptions and definitions, we are able to state the following theorem as a special case of the results given in Section 11 of (Karatzas 1989).

Theorem 23
Assume that in the case of constant coefficients r, b, σ, the assumptions (42)–(44) are satisfied. Then, the optimal strategy (π^*, c^*) of Theorem 16 can be given in feedback form of the corresponding optimal wealth process X(t) as

$$c^*(t) = I_1(\mathcal{Y}(t, X(t)))$$

$$\pi^*(t) = -(\sigma\sigma')^{-1}[b - r\underline{1}]\ \frac{\mathcal{Y}(t, X(t))}{\mathcal{Y}_x(t, X(t))X(t)}$$

where $\mathcal{Y}_x(t, x)$ is the partial derivative of $\mathcal{Y}(t, x)$ with respect to x.

Before proving the theorem, we will show how it can be used by giving an example where we choose the same utility function as in our explicit examples in Section 3.

Example 24 "Representation problem and power functions"
In this example, we consider utility functions U_i^* of the form

$$U_1^*(x) = U_2^*(x) = \gamma\, x^\gamma\ , \quad 0 < \gamma < 1\ .$$

We then obtain

$$X(t, y) = E\left[\int_t^T y^{1/\gamma-1}(H^t(s))^{\gamma/\gamma-1} e^{\mu(s-t)/\gamma-1} ds + y^{1/\gamma-1}(H^t(T))^{\gamma/\gamma-1} e^{\mu(T-t)/\gamma-1}\right]$$
$$= y^{1/\gamma-1}\, p(t)$$

with

$$p(t) = \begin{cases} \frac{1}{K}(1-e^{-K(T-t)}) + e^{-KT} & , K \neq 0 \\ 1+T-t & , \text{else} \end{cases}, \quad K = \frac{1}{1-\gamma}\left(\mu - r\gamma - \frac{\gamma\|\theta\|^2}{2(1-\gamma)}\right).$$

We can now directly compute

$$Y(t, x) = \left(\frac{x}{p(t)}\right)^{\gamma-1}, \quad Y_x(t, x) = \frac{\gamma-1}{p(t)}\left(\frac{x}{p(t)}\right)^{\gamma-2},$$

and from Theorem 23, we obtain

$$c^*(t) = \frac{1}{p(t)} X(t), \quad \pi^*(t) = (\sigma\sigma')^{-1}\frac{b - r\underline{1}}{1-\gamma}.$$

Proof (of Theorem 23):
The proof will be split into four parts.
i) The wealth process $X^{(t,x)}(s)$, corresponding to the optimal strategy for the problem starting at time t with initial wealth x and with the value function

$$v(t, x) = \sup_{(\pi,c)\in A'(t,x)} E\left[\int_t^T e^{-\mu s} U_1(c(s))ds + e^{-\mu T} U_2(X(T))\right],$$

(where A'(t, x) is just the set of all members of A'(x) restricted to [t, T]) is given by

$$X^{(t,x)}(s) = X(s, Y(t, x)\exp(\mu(s-t))H^t(s)), \quad X^{(t,x)}(t) = x, \; s\in[t, T]. \tag{45}$$

To see this, note that in analogy with the proof of part b) of Theorem 2.7, we have

$$X^{(t,x)}(s) =$$
$$E\left[\int_s^T H^s(u) I_1\left(Y(t,x)H^t(u)e^{\mu(u-t)}\right)ds + H^s(T) I_2\left(Y(t,x)H^t(T)e^{\mu(T-t)}\right)\middle| F_s\right]$$
$$= E\left[\int_s^T H^s(u) I_1\left(Y(t,x)H^t(s)e^{\mu(s-t)}H^s(u)e^{\mu(u-s)}\right)ds \right.$$
$$\left. - H^s(T) I_2\left(Y(t,x)H^t(s)e^{\mu(s-t)}H^s(T)e^{\mu(T-s)}\right)\middle| F_s\right]$$
$$= X(s, Y(t, x)e^{\mu(s-t)}H^t(s)).$$

ii) Application of Itô's formula to the process

$$\eta^{(t,x)}(s) := Y(t, x)e^{\mu(s-t)}H^t(s), \quad s\in[t, T]$$

yields the stochastic differential equation

$$d\eta^{(t,x)}(s) = \eta^{(t,x)}(s)[(\mu-r)ds - \theta' dW(s)], \quad \eta^{(t,x)}(t) = Y(t, x). \qquad (46)$$

Of course, if we replace $Y(t, x)$ by y in the definiton of $\eta^{(t,x)}$, this would only affect the initial condition in equation (46).

iii) With the help of the Feynman-Kac type theorem, Theorem B20, we can verify that $S(t, y) := yX(t, y)$ solves the linear (parabolic) partial differential equation (for brevity: pde)

$$S_t + \tfrac{1}{2}\|\theta\|^2 y^2 S_{yy} + (\mu-r)yS_y + \mu S + yI_1(y) = 0$$

which yields the following pde for $X(t, y)$:

$$X_t + \tfrac{1}{2}\|\theta\|^2 y^2 X_{yy} + \left(\mu - r - \|\theta\|^2\right)yX_y - rX + I_1(y) = 0, \quad 0 \le t < T, \quad y > 0. \qquad (47)$$

iv) Application of Itô's formula to representation (45) of $X^{(t,x)}(s)$ and the use of the equations (46) and (47) yield

$$dX^{(t,x)}(s) = dX(s, \eta^{(t,x)}(s))$$

$$= \left[X_t + \tfrac{1}{2}\left(\eta^{(t,x)}(s)\|\theta\|\right)^2 X_{yy} + \eta^{(t,x)}(s)(\mu - r)X_y\right]ds - \eta^{(t,x)}(s)X_y\,\theta' dW(s)$$

$$= \left[-I_1\left(\eta^{(t,x)}(s)\right) + \|\theta\|^2\eta^{(t,x)}(s)X_y + rX\right]ds - \eta^{(t,x)}(s)X_y\,\theta' dW(s). \qquad (48)$$

On the other hand, the general form of the equation for the wealth process $X^{(t,x)}(s)$ corresponding to the admissible pair (π, c) is given by

$$dX^{(t,x)}(s) = \left((r + (b - r\underline{1})'\pi(s))X^{(t,x)}(s) - c(s)\right)ds + X^{(t,x)}(s)\pi(s)'\sigma dW(s).$$

Comparison of drift and diffusion coefficients of this equation with those of equation (48) and multiple use of the identity $z = Y(t, X(t, z))$ and of relation (45) lead to the representations

$$c^*(s) = I_1\left(s, \eta^{(t,x)}(s)\right) = I_1\left(s, Y\left(s, X\left(s, \eta^{(t,x)}(s)\right)\right)\right) = I_1(s, Y(s, X(s))),$$

$$\pi^*(s) = -(\sigma\sigma')^{-1}[b - r\underline{1}]\,\frac{X_y(s,\eta^{(t,x)}(s))\eta^{(t,x)}(s)}{X^{(t,x)}(s)}$$

$$= -(\sigma\sigma')^{-1}[b - r\underline{1}]\,\frac{Y(s, X(s,\eta^{(t,x)}(s)))}{Y_x(s, X(s,\eta^{(t,x)}(s)))X^{(t,x)}(s)}$$

$$= -(\sigma\sigma')^{-1}[b - r\underline{1}]\,\frac{Y(s, X^{(t,x)}(s))}{Y_x(s, X^{(t,x)}(s))X^{(t,x)}(s)}\,.$$

Finally, we get the desired representations for (π^*, c^*) with $(t, x) = (0, x)$ and by replacing s by t in the above representations for $(\pi^*(s), c^*(s))$.

□

Generalisations of this result can be found in (Occone and Karatzas 1991) where the authors apply Malliavin–Calculus techniques to obtain similar representations for the optimal portfolio and consumption process in complete market setting with random coefficients.

3.5 The Martingale Method Revisited — Pliska's Version

As the headline already says, this approach is another variant of the martingale method. It is given in (Pliska 1986) and differs from the martingale method, as presented in the previous section, by the method of solution of the static optimisation problem (and also in some of the assumptions on the utility function). The methods used are of functional analytic nature. They resemble the method of optimising functionals with the help of the Fenchel-Duality-Theorem (Readers not familiar with this theory are recommended to read Part E of the Appendix). Although some of the results in (Pliska 1986) are formulated in a market setting which is more general than the one introduced in Chapter 2, we will restrict our presentation to the complete market setting as in Section 3.4. In contrast to the preceding sections, Pliska's approach does not consider the possibility of consumption and the investor's activities are described by the trading strategy and not by the portfolio process used, i.e. by absolute numbers of shares held and not by fractions of wealth invested in the different securities. It will also be convenient to look at discounted prices and terminal wealth. We therefore introduce some notations and definitions (which are in some cases adaptations of previous definitions to our current situation).

Let

$$Z(t) = (Z_1(t),\dots, Z_n(t))' := \left(\frac{P_1(t)}{P_0(t)},\dots,\frac{P_n(t)}{P_0(t)}\right)'$$

be the vector of discounted security prices.

Definition 25

An **admissible trading strategy** is an F_t-adapted, $\mathbf{R}^{n+1}$-valued process $\varphi(t) = (\varphi_0(t), \dots, \varphi_n(t))'$ with

$$\int_0^T (\varphi_i(t)P_i(t))^2 \sum_{j=1}^{n} \sigma_{ij}^2(t)dt < \infty \quad \text{a.s.} \quad \forall\, i = 1,\dots, n$$

such that the **discounted wealth process**

$$\hat{V}_t(\varphi) = \varphi_0(t) + \sum_{i=1}^{n} \varphi_i(t) Z_i(t)$$

is given by

$$V_t(\varphi) = \hat{V}_0(\varphi) + \sum_{i=1}^{n} \int_0^t \varphi_i(s)dZ_i(s), \tag{49}$$

and such that $H(t)V_t(\varphi)$ is a martingale (where $V_t(\varphi) = \varphi(t)'P(t)$ is the wealth process corresponding to the strategy φ, $H(t)$ as defined in the previous section). We will denote by $S(x)$ the set of admissible trading strategies with $\hat{V}_0(\varphi) = V_0(\varphi) = x$.

Remark 26

a) Condition (49) is the usual self-financing condition formulated in discounted terms. Using Itô's formula, it is easy to show that condition (49) is equivalent to the (non-discounted) formulation as given in Chapter 2.

b) With the help of the equivalent martingale measure $\tilde{P}$ introduced in Section 2.4, one can rephrase the martingale requirement in Definition 25 above as "$\hat{V}_t(\varphi)$ is a martingale with respect to $\tilde{P}$". In the following, it will often be (notationally) convenient to compute expectations with respect to this equivalent martingale measure. For our use of $\tilde{P}$, it is not necessary to have a knowledge of the general theory on equivalent martingale measures. It is enough to have the defining relation

$$\tilde{P}(A) = E(H(T)P_0(T)1_A) \quad \forall\, A \in F_T \tag{50}$$

in mind. Denoting by $\tilde{E}(.)$ the expectation with respect to $\tilde{P}$ we have

$$\tilde{E}(B) = E(H(T)P_0(T)B)$$

for all F_T-measurable random variables B such that the expectation on the right side of this relation exists. Note that due to the boundedness of $P_0(T)$, the requirement of a finite expectation with respect to $\tilde{P}$ is equivalent to the familiar condition of

$$E(H(T)B) < \infty$$

for a possible terminal wealth B.

c) Comparing the requirements on an admissible trading strategy φ in Definition 25 with that of the previous section (i.e. the requirements on φ that are implied by $(\pi,0) \in A'(x)$), we observe that on one hand, in Definition 25 we have not required a non-negative wealth process, but on the other hand, the martingale property required to hold for $H(t)V_t(\varphi)$ implies

$$E(H(t)V_t(\varphi)) = x \quad \forall\, t \in [0, T].$$

However, if in the last section we were in the case of an increasing utility function with $U'(0) = +\infty$, the non-negativity constraint was not active in the optimal solution **and** the optimal solution automatically satisfied

$$E(H(T)X^*(T)) = x$$

where $X^*(T)$ was the optimal terminal wealth. Thus, we would have ended up with exactly the same solution if we would have used Definition 25 to define an admissible trading strategy rather than the constraint $(\pi, 0) \in A'(x)$. Hence, in this case, Pliska's variant will lead to the same solution as the martingale method of Section 4. Contrary, the two methods lead to different optimal solutions (due to the different sets of feasible trading strategies) if we have $U'(0) < +\infty$. In this case, there will exist a set of positive probability on which the optimal terminal wealth of Section 4 will be zero. Hence, the non-negativity constraint is binding, and removing it will lead to an optimal terminal wealth which will be negative with positive probability if the setting of this section will be used. We would also like to point out that in the martingale approach, as presented in Section 4, the non-negativity of the wealth process was needed mainly for two reasons. On one hand, it ensured that one could use utility function such as the logarithm or the square root which are not defined for negative numbers. On the other hand, in the proof of Theorem 16, part a) of Theorem 2.7 was used. But part a) of Theorem 2.7 required a non-negative wealth process (more precisely, one that is bounded from below). Thus, if we do not rely on it, we do not need the (of course economically reasonable) requirement of a positive wealth process.

We will define the analogue to the class B(x) of possible terminal wealths given in Section 2.2, but we will leave aside the non-negativity requirement and will formulate the defining relation in discounted terms. Let therefore be

$$G := \{X \mid X \ F_T\text{-measurable}, \ \tilde{E}(X) < \infty \}.$$

$$\tilde{C}(x) := \left\{ X \in G \,\middle|\, X = x + \sum_{i=1}^{n} \int_0^T \varphi_i(s) dZ_i(s) \ \text{a.s. for some } \varphi \in S(x) \right\}$$

will be called the set of **attainable terminal wealths** (however, the actual terminal wealth will be $X \cdot P_0(T)$, due to the fact that we have used discounted terms in the defining equation of an attainable terminal wealth). Our main object will be the following problem of terminal wealth maximisation,

$$\max_{\varphi \in S(x)} E\left(\tilde{U}(V_T(\varphi))\right), \tag{P}$$

where the utility function $\tilde{U}(x)$ is assumed to be continuous, strictly increasing and concave in $x \in \mathbf{R}$. As in Section 4, this problem will be decomposed into the sub-problems of finding the optimal attainable wealth,

$$\max_{X \in \tilde{C}(x)} \tilde{J}(X) = \max_{X \in \tilde{C}(x)} E\left(\tilde{U}(X \cdot P_0(T))\right), \tag{O}$$

and the resulting representation problem

"Find an admissible strategy generating an optimal solution of (O)". (R)

We will not give any new method for solving the representation problem (R). Our computations in the example below will closely resemble Method 1 of Section 4. Instead, we will concentrate on the solution method for problem (O). In a first step, we transform the feasible set for the optimisation problem. Define

$$C := \tilde{C}(x) - x := \tilde{C}(0),$$

$$U(z) := U(z, \omega) := \tilde{U}\left((z + x) P_0(T, \omega)\right) F(T, \omega)$$

with $F(T) = 1/(H(T)P_0(T))$ (where the notation $U(z, \omega)$ should indicate the fact that $U(z)$ is a random function for every z). Note that relation (50) implies that $1/F(T)$ is the Radon- Nikodym derivative of $\tilde{P}$ with respect to P. Due to the relation

$$\tilde{E}(U(X)) = E(U(X)/F(T)) = E(\tilde{U}((X+x)P_0(T))),$$

problem (O) is equivalent to the following problem which is (formally) independent of the initial wealth x (or more precisely, which assumes an initial wealth of zero),

$$\max_{X \in C} J(X) := \max_{X \in C} \tilde{E}(U(X)). \tag{O*}$$

From now on, we will always assume that there exists at least one X with

$$\tilde{E}(U(X)) > -\infty, \tag{51}$$

and we introduce the **orthogonal subspace** $C^{\perp}$ of C in G , i.e.

$$C^{\perp} := \{Y \mid Y \text{ bounded, } F_T\text{-measurable with } \tilde{E}(XY) = 0 \ \forall\, X \in C\}.$$

The term "orthogonal" is formulated with respect to the equivalent martingale measure $\tilde{P}$. It can, of course, also be formulated with respect to the original measure P. Then the expectation requirement in the definition of $C^{\perp}$ reads "E(XY/F(T)) = 0".

Definition 27
For all bounded random variables Y define the **concave conjugate functional** of J by

$$J^*(Y) := \inf_{X \in L} \{\tilde{E}(XY) - J(X)\}$$

and the **concave conjugate functional** of U by

$$j^*(y) := j^*(y, \omega) := \inf_{x \in R} \{xy - U(x)\}.$$

A sufficient condition for optimality of $X_0 \in C$ for problem (O*) is given in

Theorem 28
Assume that for $X_0 \in C$, $Y_0 \in C^{\perp}$ we have

$$j^*(Y_0(\omega), \omega) = X_0(\omega)Y_0(\omega) - U(X_0(\omega), \omega) \tag{52}$$

for almost all ω. Then we have

$$J(X) \leq J(X_0) = -J^*(Y_0) \leq -J^*(Y) \quad \forall\, X \in C, Y \in C^{\perp}, \tag{53}$$

i.e. X_0 solves the (primal) problem (O*), and Y_0 solves the **dual problem** corresponding to (O*),

$$\min_{Y \in \mathcal{C}^{\perp}} J^*(Y).$$

Proof :
Theorem 3.C in (Rockafellar 1976) implies that under Assumption (51) J* is given by $J^*(Y) = \tilde{E}(j^*(Y))$. This representation together with condition (52) implies

$$J^*(Y_0) = \tilde{E}(X_0 Y_0) - J(X_0) = -J(X_0) \tag{54}$$

as we have $X_0 \in \mathcal{C}$, $Y_0 \in \mathcal{C}^{\perp}$. Further, the defining relation for J* yields

$$J^*(Y_0) \leq \tilde{E}(XY_0) - J(X) \quad \forall\, X \in G. \tag{55}$$

Because we have $\tilde{E}(XY_0) = 0$ for $X \in \mathcal{C}$, relation (55) implies

$$J^*(Y_0) \leq -J(X). \tag{56}$$

Further, for arbitrary $Y \in \mathcal{C}^{\perp}$ we have

$$J^*(Y) \leq \tilde{E}(X_0 Y) - J(X_0) = -J(X_0)$$

Finally, combining the relations (54), (55), and (56) yields the desired result (53).

□

The duality relation between the solutions of the primal and the dual problem, $J(X_0) = -J^*(Y_0)$, will in the following be used to solve the optimisation (O*). To prove Theorem 28, we have in fact not used the completeness of our market model which means that it is also valid in a more general setting. However, completeness of the market will greatly simplify the solution of the dual problem. Note first that due to the completeness of the market we have

$$\mathcal{C} = \{X \in G \mid \tilde{E}(X) = 0\}.$$

By using this characterisation of $\mathcal{C}$, we can show that $Y_0 \in \mathcal{C}^{\perp}$ satisfying condition (52) in Theorem 28 must have an extremely simple form.

Proposition 29
Let $X_0 \in \mathcal{C}$, $Y_0 \in \mathcal{C}^{\perp}$ satisfy equation (52). Then Y_0 is a positive constant.

Proof:
Step 1 "$Y_0 \geq 0$"
Let $A := \{\omega \in \Omega \mid Y_0(\omega) < 0\}$, $X := X_0 + 1_A$. Clearly, $X \in G$ and $J(X) \geq J(X_0)$ due to the monotonicity of U(.). Relations (53) and (55) imply

$$-J(X_0) = J^*(Y_0) \le \tilde{E}(XY_0) - J(X),$$

hence

$$0 \le \tilde{E}(XY_0) = \tilde{E}(X_0Y_0) + \tilde{E}(1_A Y_0) = \tilde{E}(1_A Y_0).$$

By definition of A, this implies $1_A Y_0 = 0$ a.s. and also $P(A) = 0$.

Step 2 "$Y_0 = K > 0$ for a constant K"

Set $K := \tilde{E}(Y_0)$, $Z := Y_0 - K$. Hence, $\tilde{E}(Z) = 0$ and thus $Z \in C$. By noting that both Y_0 and K are in $C^{\perp}$, we also have $Z \in C^{\perp}$. Consequently, $Z = 0$, i.e. $Y_0 = K$. Due to Step 1, K is non-negative. If we assume $K = 0$ then condition (52) takes the form

$$j^*(0, \omega) = -U(X_0(\omega), \omega) = \inf_{x \in \mathbf{R}} \{-U(x, \omega)\}$$

which contradicts the assumption of a strictly increasing function U(x). Thus, K must be a positive constant.

□

Under some additional assumptions on the (transformed) utility function U, we can prove existence of a solution to the problem (O*) (and thus also the existence of a solution to (O)).

Theorem 30 "Existence of an optimal terminal wealth"

Assume that the function $y \to U(y, \omega)$ is continuously differentiable and that we have

$$\lim_{y \to \infty} \frac{\partial U(y,\omega)}{\partial y} = 0 \quad \text{a.s.}, \quad \lim_{y \to -\infty} \frac{\partial U(y,\omega)}{\partial y} = \infty \quad \text{a.s.},$$

$$\beta \ge \frac{\partial U(0,\omega)}{\partial y} \ge \varepsilon > 0 \quad \text{a.s.} \tag{57}$$

for suitable constants β and ε. Then there exists a solution X* to problem (O).

Proof :

a) We will first solve the dual problem to (O*). Due to Proposition 29, it is enough to solve the much simpler (dual) problem

$$\min_{K \in (0,\infty)} J^*(K).$$

For each positive K and almost every ω there exists a minimiser X(K, ω) of the function $x \to Kx - U(x, \omega)$ due to the requirements on U, i.e. we have the representations

$$j^*(x, \omega) = KX(K, \omega) - U(X(K, \omega), \omega) \qquad (K > 0)$$

$$J^*(K) = \tilde{E}(j^*(K, \omega)). \tag{58}$$

We will show that $J^*(K)$ has a minimiser in $(0, \infty)$. Note therefore that we have

$$\frac{\partial}{\partial K} j^*(K, \omega) = X(K, \omega)$$

which follows from the fact that, by definition of $X(K, \omega)$, we must have

$$\frac{\partial}{\partial x} U(X(K, \omega), \omega) = K .$$

Application of the dominated convergence theorem results in

$$(J^*)'(K) = \tilde{E}\left(\frac{\partial}{\partial K} j^*(K, \omega)\right) = \tilde{E}(X(K, \omega)).$$

Further, for $K < \varepsilon/2$, we must have $X(K, \omega) > 0$ a.s. . To see this, note that we have

$$K = \frac{\partial}{\partial x} U(X(K, \omega), \omega) < \frac{\partial}{\partial x} U(0, \omega)$$

by condition (57). Positivity of $X(K, \omega)$ then follows by the concavity of $U(., \omega)$ which implies that its derivative with respect to x is decreasing in x. Thus, we have proved that $(J^*)'(K)$ is positive for small values of K. By a similar argument, for very large K, say $K > 2\beta$, we must have $X(K,\omega) < 0$ and thus $(J^*)'(K) < 0$. We have therefore not only proved the existence of a strictly positive zero of $(J^*)'$, but also that there exists such a zero $\hat{K}$ that is a minimiser of J^* on $(0, \infty)$.

b) Next, we set

$$X_0(\omega) := X(\hat{K}, \omega)$$

with $\hat{K}$ as in part a). Then $X_0(\omega)$ and $Y_0(\omega) := \hat{K}$ satisfy condition (52). By applying Theorem 28, we can conclude that X_0 is a solution to subproblem (O^*). Hence, $X^* = X_0 + x$ solves Problem (O).

□

The proof of Theorem 30 can be written in the form of an algorithm for obtaining X_0 and thus the optimal terminal wealth $X^* = (X_0 + x)P_0(T)$. A similar algorithm will appear in Section 4.4 where a constrained portfolio problem will be solved by a dual approach.

Algorithm: "Computing the optimal terminal wealth: Pliska's method"

1. Compute $j^*(K,\omega)$ and its minimiser $X(K,\omega)$
2. Compute $J^*(K)$ via representation (58)
3. Compute $\hat{K}$, the maximiser of $J^*(K)$
4. Substitute $\hat{K}$ in representation (58) to obtain $J(\hat{K})$
5. $X_0(\omega) = X(\hat{K}, \omega)$, $X^* = (X_0 + x)P_0(T)$

It remains to solve the representation problem. We will do this by a comparison method similar to the one presented in Section 4. Here, comparison of two different representations of the discounted wealth process will yield the optimal trading strategy. This method will be presented in the following example.

Example 31 "Exponential utility"
We assume constant market coefficients r, b, σ and also n = 1. The case of a market with an arbitrary number of risky securities is only notationally more complicated (this extension is recommended as an exercise to the reader). We consider a utility function of the form

$$\tilde{U}(x) = a - \frac{d}{c} e^{-cx}$$

where a, c, d are constants with c, d > 0. Further, we assume an initial capital of x = 0. Hence, we have the following form of the "auxiliary utility function" U(x):

$$U(x) = \tilde{U}(xe^{rT})\,F(T) = \left(a - \frac{d}{c} e^{-ce^{rT}x}\right) e^{\frac{1}{2}\theta^2 T + \theta W(T)}.$$

Of all assumptions of Theorem 30 only assumption (57) is violated. Condition (57) was only necessary to ensure the existence of a stationary point of $J^*(K)$. As condition (57) is not satisfied with our choice of utility function, we have to show this existence directly. The reason that U(x) does not meet requirement (57) is that U(x) still depends on F(T) which is positive but not bounded from below or above.

Let us start with the algorithm to compute the optimal terminal wealth. A straight forward computation results in the following forms for $j^*(K,\omega)$ and $X(K,\omega)$:

$$X(K,\omega) = \frac{e^{-rT}}{c} \ln\left(\frac{de^{rT}F(T)}{K}\right),$$

$$j^*(K,\omega) = \begin{cases} \frac{Ke^{-rT}}{c}\ln\left(\frac{de^{rT}F(T)}{K}\right) - aF(T) + \frac{Ke^{-rT}}{c} & ,K>0 \\ -aF(T) & ,K=0 \\ -\infty & ,K<0 \end{cases}.$$

As a consequence of representation (58), we have

$$J^*(K) = \frac{Ke^{-rT}}{c}\left\{\ln(d) + \tilde{E}(\ln(F(T))) - \ln(K) + rT + 1\right\} - a \quad , \quad K > 0 .$$

By differentiating J*(K) with respect to K and additionally noting

$$\tilde{E}(\ln(F(T))) = E(H(T)P_0(T)\ln(F(T))) = -\tfrac{1}{2}\theta^2 T ,$$

we obtain the unique maximising value for J*(K) as

$$\hat{K} = de^{rT}\exp(\tilde{E}(\ln(F(T)))) = de^{(r-\frac{1}{2}\theta^2)T}.$$

This results in the following representations for the optimal utility, attainable and terminal wealth:

$$J(\hat{K}) = a - \tfrac{d}{c}\exp(\tilde{E}(\ln(F(T)))) = a - \tfrac{d}{c}e^{-\frac{1}{2}\theta^2 T},$$

$$X_0 = \tfrac{e^{-rT}}{c}\left\{\ln(F(T)) - \tilde{E}(\ln(F(T)))\right\} = \tfrac{e^{-rT}}{c}\left\{\theta^2 T + \theta W(T)\right\},$$

$$X^* = \tfrac{1}{c}\left\{\ln(F(T)) - \tilde{E}(\ln(F(T)))\right\} = \tfrac{1}{c}\left\{\theta^2 T + \theta W(T)\right\}.$$

Note that the optimal terminal wealth X* is normally distributed (with a positive mean), i.e. there is a positive probability to end up with a negative total wealth.

It remains to compute the optimal strategy, i.e. the strategy φ generating X*. By requiring

$$X^* = V_T(\varphi) \text{ a.s.}$$

and using the martingale property of $H(t)V_t(\varphi)$, we get

$$V_t(\varphi) = E\left(H(T)X^*\middle|F_t\right) = \tfrac{e^{-r(T-t)}}{c}\left\{\theta W(t) + \theta^2 t\right\},$$

$$\hat{V}_t(\varphi) = \tfrac{e^{-rT}}{c}\left\{\theta W(t) + \theta^2 t\right\}. \tag{59}$$

On the other hand, by using the self-financing condition (49), we obtain

$$d\hat{V}_t(\varphi) = \varphi_1(t)dZ_1(t) = \varphi_1(t)Z_1(t)[(b-r)\,dt + \sigma dW(t)]\ . \tag{60}$$

Equating the drift and diffusion parts of the two different representations (59) and (60) for the discounted wealth process $\hat{V}_t(\varphi)$ yields

$$\varphi_1(t) = \frac{(b-r)e^{-rT}}{c\sigma^2 Z_1(t)} = \frac{(b-r)e^{-r(T-t)}}{c\sigma^2 P_1(t)}\ .$$

As the rest of the wealth must be invested into the bond, we obtain

$$\varphi_0(t) = (\hat{V}_t(\varphi) - \varphi_1(t)P_1(t))e^{-rt} = \frac{e^{-rT}}{c}\left\{\theta W(t) + \theta^2 t - \frac{b-r}{\sigma^2}\right\}$$

Obviously, $(\varphi_0(t), \varphi_1(t))$ meet all the requirements of Definition 25 which means that they constitute an admissible trading strategy. For the case of "r = 0" (or in discounted terms), we observe that the amount of money $\varphi_1(t)P_1(t)$ invested into the risky security is constant over time, only the value of the bond holdings is varying. This type of optimal strategy demonstrates the substantially different properties of the exponential utility function compared to the power functions or the logarithm. This is also reflected by the fact that for the exponential function possesses the measure of absolute risk aversion is constant as a function of x. In the case of the power functions or the logarithm, the optimal fraction of the wealth invested in the risky asset was constant. Here, the optimal fraction invested in the risky asset is inversely proportional to the actual wealth of the investor $\hat{V}_t(\varphi)$, i.e. he behaves in a more risky way if his wealth is small than if his wealth is big.

3.6 An Application: "Minimising the Difference to the Terminal Wealth of a Richer Investor"

In this section we will give an application of the foregoing theoretical results . We consider the (very relevant !) situation that there exists an investor, maximising his expected utility from terminal wealth, who possesses an initial endowment that exceeds our initial capital. However, we assume that we **exactly know** the portfolio process that he will follow. Our goal will be to invest in such a way that our terminal wealth will be as close as possible to that of the richer investor (see (Korn 1997)).

We start by stating the problem in a more general, abstract way: Given a non-negative random variable B (with existing variance) and an initial capital of $x > 0$, we consider the following problem :

$$\min_{\pi \in A'(x)} E\left(B - X^{\pi}(T)\right)^2 \tag{61}$$

where $X^{\pi}(t)$, $t \in [0,T]$, is the wealth process corresponding to the portfolio process $\pi(t)$ in the complete market setting of Section 4 and where we do not allow for consumption. If we define the (random) utility function

$$U(x) = -\tfrac{1}{2}(B - x)^2, \quad x \in \mathbf{R}, \tag{62}$$

then problem (61) is similar to the portfolio problems treated in Section 4 if we transform the minimisation problem (61) into the equivalent maximisation problem

$$\max_{\pi \in A'(x)} E\left(U\left(X^{\pi}(T)\right)\right) . \tag{63}$$

Define the function I(y) (for each $B(\omega)$) by

$$I(y) = I(y, \omega) = \begin{cases} B(\omega) - y \ , & \text{if } y \le B(\omega) \\ 0 \ , & \text{else} \end{cases} ,$$

i.e. I(y) is the inverse function of U'(x) on $(-\infty, B]$ and vanishes on (B, ∞). With the notations of Section 4, we write

$$\mathcal{X}(y) = E(H(T)I(y)) = E\left(H(T)(B - yH(T))^+\right), \qquad y > 0.$$

As in Section 4, we can show the existence of an inverse function $\mathcal{Y}(x)$ (defined on $(0, \mathcal{X}(0+))$) to the strictly decreasing function $\mathcal{X}(y)$ (defined on $(0,\infty)$) if we make the assumption of a (non-vanishing) deterministic mean-variance trade-off, i.e. that $\theta(t)$ is a deterministic function satisfying condition (26). Under these assumptions we can give the key result of this section:

Proposition 32
Let $x > 0$ and assume a deterministic mean-variance trade-off satisfying condition (26). Then, there exists a portfolio process $\pi(t) \in A'(x)$ attaining the minimum in the optimisation problem (61). The corresponding optimal terminal wealth $X^{\pi}(T)$ is given by

$$X^{\pi}(T) = \begin{cases} B & , \text{ if } x \geq E(H(T)B) \\ (B - Y(x)H(T))^{+}, & \text{else} \end{cases} \tag{64}$$

Proof:
If we can show that the inequality

$$U(I(y, \omega)) \geq U(x) + y(I(y, \omega) - x), \quad y, x \in [0, \infty) \tag{65}$$

is valid with our choice of the random utility function U(x) given in (62) (for almost every $\omega \in \Omega$), then the proof of Theorem 16 also applies in the situation of problem (63). This will then yield the assertion for problem (61). In the case "$0 < y \leq B(\omega)$", the concavity of U implies

$$U(I(y, \omega)) \geq U(x) + U'(I(y, \omega))(I(y, \omega) - x) = U(x) + y(I(y, \omega) - x) \tag{66}$$

for $x \in [0, \infty)$ while in the case "$y > B(\omega)$", inequality (66), the concavity of U and the definition of $I(y, \omega)$ yield

$$U(I(y, \omega)) = U(0) \geq U(x) + U'(0)(0 - x) \geq U(x) + y(I(y, \omega) - x)$$

which completes the proof.

□

Remark 33
This result can also be used for applications in option hedging if an investor cannot afford to fully hedge his position (see (Korn 1997b)).

We are now ready to consider our original goal of "Minimising the difference to the terminal wealth of a richer investor". For convenience, we will restrict ourselves to the situation of a market consisting of a bond and a single stock. We assume constant market coefficients b, σ and a zero riskless interest rate, i.e. $r = 0$. We further assume that there exists an investor with an initial endowment of x_B who maximises his expected utility from terminal wealth with time horizon $t = T$. Even more, we assume to know that he uses the constant portfolio process $\pi(t) \equiv \pi^* \in \mathbf{R}$ (and does not consume parts of his wealth on [0, T]). As we have seen in the preceding section, such a portfolio process will be an optimal one if for example the investor has a utility function of the form

$$U(x) = x^{\alpha} \quad \text{or} \quad U(x) = \ln(x)$$

for $0 < \alpha < 1$. Therefore, the assumption of a constant portfolio strategy is not an artificial one in our setting. Although, the following considerations can also be made for the situation when the investor uses a non-constant portfolio strategy (as long as we know its exact form), the relevant computations will not be as explicit as in the constant case. In the constant case, his corresponding wealth process $X_B(t)$ is given as

$$X_B(t) = x_B e^{(\pi^* b - \frac{1}{2}\pi^{*2}\sigma^2)t + \pi^*\sigma W(t)}.$$

The initial capital x_B of this investor is assumed to exceed our initial endowment of x which justifies to talk of this investor as the richer one. As already said, our goal is to achieve a terminal wealth that is as close as possible to that of the richer one. We therefore try to minimise the mean squared distance to the terminal wealth of the richer investor, i.e. we try to trade in such a way that we minimise

$$E\left(\left(X^\pi(T) - X_B(T)\right)^2\right)$$

over all admissible portfolio strategies π delivering a non-negative terminal wealth.

By setting $B = X_B(T)$ in Proposition 32 our optimal terminal wealth is given by

$$X^\pi(T) = \left(X_B(T) - Y(x)H(T)\right)^+$$

$$= \left(x_B \exp\left(\left(\pi^* b - \tfrac{1}{2}(\pi^*\sigma)^2\right)T + \pi^*\sigma W(T)\right) - Y(x)\exp\left(-\tfrac{b}{\sigma}W(T) - \tfrac{1}{2}\left(\tfrac{b}{\sigma}\right)^2 T\right)\right)^+$$

From the form of $X^\pi(T)$, it is obvious that the terminal wealth is non-negative. Further, due to the positivity of H(T), it is also clear that our terminal wealth will never exceed the one of the richer investor. We still have to compute the real number $Y(x)$ to specify our terminal wealth. Note therefore that $X(y)$ is determined by

$$X(y) = E\left(H(T)\left(X_B(T) - yH(T)\right)^+\right)$$

$$= x_B \Phi\left(\frac{\left(\pi^*\sigma - b/\sigma\right)T + K(y)}{\sqrt{T}}\right) - y\, e^{(b/\sigma)^2 T}\, \Phi\left(\frac{K(y) - 2b/\sigma\, T}{\sqrt{T}}\right)$$

with

$$K(y) = \frac{1}{\sigma\pi^* + b/\sigma}\left(\ln\left(x_B/y\right) + \left(\pi^* b - \tfrac{1}{2}\pi^{*2}\sigma^2 + \tfrac{1}{2}\left(b/\sigma\right)^2\right)T\right)$$

where $\Phi(x)$ is the distribution function of the standard normal distribution. Because $X(y)$ is strictly decreasing and continuously differentiable on $(0, \infty)$, we can use the Newton-Raphson method to solve the non-linear equation $X(y) = x$ to obtain its unique positive solution $Y(x)$ (Note that we have $X(0) = x_B > x$ and $X(\infty) = 0$).

Hence, we are also able to compute the minimal mean squared distance to the terminal wealth of the richer investor delivered by an admissible strategy as

$$E\left(\left(X^{\pi}(T)-X_B(T)\right)^2\right) =$$

$$x_B{}^2 e^{(2\pi^* b+\pi^{*2}\sigma^2)T}\,\Phi\left(\frac{-K(Y(x))-2\sigma\pi^* T}{\sqrt{T}}\right)+Y(x)^2 e^{\left(b/\sigma\right)^2 T}\,\Phi\left(\frac{-K(Y(x))+2\,b/\sigma\,T}{\sqrt{T}}\right)$$

We will now give some numerical examples to judge the performance of our strategy. For this purpose, we will not only compare it to the terminal wealth of the richer investor but also to that of an investor with an initial endowment of x (which is the same amount as we are initially endowed with) using the same strategy π^* as the first investor, i.e. we will examine if our strategy performs better than the one that consists simply of imitating the richer investor. Let $X^*(t)$ be the wealth process of the investor with initial capital x and portfolio strategy π^*. The mean squared distance of the terminal wealth of this investor to the richer one is given by

$$E\left((X^*(T)-X_B(T))^2\right) = (x_B - x)^2 e^{(2\pi^* b+\pi^{*2}\sigma^2)T}$$

(note that we have $X^*(T) = x\,(X_B(T)\,/\,x_B)$). The first desirable feature of our proposed strategy is that it has a higher expected terminal wealth then that achieved by following the strategy π^* (under an additional requirement on the market coefficients), i.e. we have

Proposition 34

Under the above assumptions, we have $E\left(X^{\pi}(T)\right) > E(X^*(T))$ if $\pi^* b > -(b/\sigma)^2$.

Proof :

To prove this result we compute the two expected values as

$$E(X^*(T)) = x e^{\pi^* bT},$$

$$E\left(X^{\pi}(T)\right) = x_B e^{\pi^* bT}\,\Phi\left(\frac{\sigma\pi^* T+K(Y(x))}{\sqrt{T}}\right)-Y(x)\,e^{\left(b/\sigma\right)^2 T}\,\Phi\left(\frac{K(Y(x))-b/\sigma\,T}{\sqrt{T}}\right)$$

$$= x e^{\pi^* bT}+e^{\pi^* bT}\left\{x_B\Phi\left(\frac{\sigma\pi^* T+K(Y(x))}{\sqrt{T}}\right)-Y(x)e^{-b\pi^* T}\,\Phi\left(\frac{K(Y(x))-b/\sigma\,T}{\sqrt{T}}\right)-x\right\}$$

$$= xe^{\pi^* bT} + e^{\pi^* bT}\left\{ X\left(Y(x)\exp\left(-\left(\left(\tfrac{b}{\sigma} \right)^2 + \pi^* b \right) T \right) \right) - x \right\}.$$

Comparison of both expectations, using the assumption "$\pi^* b > -(b/\sigma)^2 > 0$", and the fact that $X(y)$ is strictly decreasing in y proves the proposition.

□

Remark 35

The requirement "$\pi^* b > -(b/\sigma)^2$" in Proposition 34 is in deed not a very strong one. In particular, if b is positive (which is the natural case), it is satisfied if the investor is not selling the risky security short which seems to be a bad strategy in this case. As our terminals wealths depends on the portfolio process used by the richer investor, we could therefore say that, as long as he behaves in a reasonable way, our strategy yields a higher expected terminal wealth as if we would choose π^*, too.

Under the conditions of Proposition 34, our optimal terminal wealth $X^{\pi}(T)$ has a higher expectation than $X^*(T)$ **and** a smaller L^2-distance to the terminal wealth $X_B(T)$ of the richer investor (as being the solution of optimisation problem (63)). Hence, it is better for us not to use the same strategy as the richer investor. The typical results of the different strategies will be demonstrated in the following examples.

Let always A (resp. A*) be the square root of the L^2-distance of the optimal terminal wealth $X^{\pi}(T)$ (resp. of the terminal wealth $X^*(T)$) to $X_B(T)$. In Table 1 we look at the results for various values x of our initial capital. Further, we have chosen the following parameters: $x_B = 100$, $b = 0.05$, $\sigma = 0.25$, $T = 1$, $\pi^* = b/\sigma^2 = 0.8$.

	$E(X^{\pi}(T))$	A	$E(X^*(T))$	A*
99	103.12	0.98	103.04	1.06
90	94.47	9.80	93.67	10.62
75	80.06	24.51	78.06	26.55
50	55.86	49.07	52.04	53.09
25	30.30	74.45	26.02	79.64
10	13.32	91.37	10.41	95.57
1	1.61	104.04	1.04	105.12

Table 1: Comparison between different methods to approximate the terminal wealth of a richer investor (varying initial capital)

We can easily calculate that the expected terminal wealth of the richer investor is given by $E(X_B(T)) = 104.08$. With the exception of the cases when the initial capital is very small (i.e. $x = 1, 10$), the use of our method also leads to a square root of L^2-distance that is smaller than the difference between the initial capitals. This means that despite the fact of having a smaller amount of initial capital, the poorer investor x (i.e. we) performs better in absolute values (!) than the richer one that uses the strategy π^* (which is optimal for a suitable choice of the utility function; in our example in Table 1, it would be optimal for the choice of $\ln(x)$ as utility function).

In Table 2, we compare the performance of our method to different constant portfolio strategies used by the richer investor. We use the following data: $x_B = 100$, $x = 80$, $b = 0.05$, $\sigma = 0.25$, $T = 1$. The entries of Table 2 again show the superior behaviour of our method. Over the whole range of π it performs better than the constant strategies. On one hand it outperforms the poorer investor using a constant strategy. On the other hand it shortens the initial distance to the richer investor from 20 to 19.60.

π^*	$E(X_B(T))$	$E(X^{\pi}(T))$	A	$E(X^*(T))$	A*
0.2	101.01	81.79	19.60	80.80	20.23
0.5	102.53	83.32	19.60	82.03	20.66
0.8	104.08	84.87	19.60	83.26	21.23
1	105.13	85.91	19.60	84.10	21.69
1.5	107.79	88.56	19.61	86.23	23.13

Table 2: Comparison between different methods to approximate the terminal wealth of a richer investor (different behaviour of the richer one)

It is remarkable that the L^2-distance seems to be independent of the strategy used by the rich investor. However, for more extreme values of π^* the L^2-distance increases with increasing π^* (but stays below 20). This is no serious disadvantage, because such extreme values for π^* are irrelevant for practical purposes.

We also examine the performance respect to the drift parameter of the stock. The results are summarised in Table 3. The data used for these examples are $x_B = 100$, $x = 80$, $\sigma = 0.25$, $T = 1$, $\pi^* = 0.8$. The remarkable fact that can be read off from Table 3 is that the L^2-distance between the terminal wealth of the richer investor and our terminal wealth decreases with increasing b. This is exactly the opposite to the behaviour of the corresponding L^2-distance of the poorer investor using the same strategy as the richer one.

b	$E(X_B(T))$	$E(X^{\pi}(T))$	A	$E(X^*(T))$	AP
0.01	100.80	80.84	19.98	80.64	20.57
0.05	104.08	84.87	19.60	83.26	21.23
0.10	108.33	91.27	18.47	86.66	22.10

Table 3 : Comparison between different methods to approximate the terminal wealth of a richer investor (different drift parameters)

For some remarks on the trading strategy corresponding to our method, we refer to Section 4.3. It can be computed with the help of Theorem 21 and is done in (Laue 1998), but its tremendous length makes it hard to gain a lot of insight.

CHAPTER 4

CONSTRAINED CONTINUOUS-TIME PROBLEMS

4.1 Portfolio Problems with Constraints

Of course, it is clear that from a scientific point of view and for practical purposes, it would be nice to get rid of some of the simplifying assumptions of the continuous-time market model and the portfolio problem presented in the previous two chapters. The way of bringing the theory and the model closer to practical situations, given in this chapter, lies in the introduction of constraints. These could be constraints on the trading strategies or the portfolio processes such as short-selling constraints or leverage constraints. One could also imagine about interval constraints, such as "invest at least 10 % and at most 40 % of your wealth in shares of stock i " or prohibition to invest in a certain stock. Such types of constraints will be dealt with in Section 4. As an application of the theory developed in this section, we will also look at some explicit examples of constrained problems in Section 5. Constraints on the terminal wealth, such as certain liquidity demands or a minimum expectation of terminal wealth, are also natural demands. In particular, one can think of a continuous-time analogue to the Markowitz mean-variance approach. Such constrained portfolio problems are the subject of Sections 2 and 3 below. A common feature of the introduction of all the above mentioned constraints is that our market model will no longer be complete. Consequently, the optimal strategies computed in the previous chapter need no longer be feasible ones. Therefore, we have to introduce new methods of solving these problems. As the proofs of some theorems in Sections 2 and 4 are rather technical, it is recommended to skip them at a first reading. One should instead concentrate on the main results and in particular on the resulting algorithms to solve the constrained problems.

4.2 A Dual Method to solve Portfolio Problems with Constraints on the Terminal Wealth

Looking at the decomposition of the portfolio problem (P) in Section 3.4 into the static optimisation problem (O) and the representation problem (R), one can immediately imagine about two kinds of generalisation of the portfolio problem. One way is to put constraints on the generating portfolio process π^* in the representation problem which is done in (Cvitanic and Karatzas 1992), (He and Pearson 1991) or (Karatzas, Lehoczky, Shreve and Xu 1991). This will be the subject of Sections 4 and 5. Another possibility is to put constraints on the terminal wealth B in the optimisation problem (O). This generalisation is the main subject of (Korn and Trautmann 1995) and will be dealt with in this section. More precisely, we will consider problems of the following type :

$$\max_{(\pi,c)\in A_T'(y),\, y\le x} J(y;\pi,c) \tag{PT}$$
$$\text{s.t.}\quad E\left(G_i\left(X^{y,\pi,c}(T)\right)\right)\left(G_i\left(X^{y,\pi,c}(T)\right)\right) \le 0 \;,\; i=1,..,k$$

where $J(y;\pi,c)$ is given as in Chapter 3 and $G(u) = (G_1(u),\ldots, G_k(u))'$ is an $\mathbf{R}^k$-valued function such that

$$(U_2 - d'G)(x) \quad \text{is a utility function for all } d\in[0,\infty)^k,\ x\ge 0\,. \tag{1}$$

The set $A_T'(y)$ contains the admissible strategies for the constrained portfolio problem (PT) (not necessarily the feasible ones !) and is defined as

$$A_T'(y) = A'(y) \setminus \left\{(\pi,c)\in A'(y) \middle| E\left(G_i\left(X^{y,\pi,c}(T)\right)^+\right) = \infty \text{ for some } i\in\{1,\ldots,k\}\right\} \tag{2}$$

Remark 1

a) The introduction of the set of admissible strategies for the constrained optimisation problem (PT) has the same reason as the introduction of the set of admissible strategies $A'(y)$ for the unconstrained portfolio problem in the previous chapter: Strategies (π, c) having a terminal wealth $X^{y,\pi,c}(T)$ with

$$E\left(G_i\left(X^{y,\pi,c}(T)\right)\right) = -\infty$$

for some $i\in\{1,\ldots, k\}$ should also be called admissible, if they do not satisfy

$$E\left(G_i\left(X^{y,\pi,c}(T)\right)^+\right) = +\infty$$

for some $i\in\{1,\ldots, k\}$.

b) Condition (1), which is imposed on the constraint function G, is always satisfied if for all component functions G_i, i = 1,..., k, the functions $(-G_i)$ are utility functions in the sense of Definition 3.1* .
c) A typical example of problems of the form (PT) is given by the linear-quadratic one which will prove to be equivalent to a mean-variance problem (with an appropriate choice of the constants K_1, K_2):

$$\begin{aligned} &\min_{\pi \in A_T'(y),\, y \le x} \tfrac{1}{2} E\left(X^{y,\pi}(T) - K_1\right)^2 \\ &\text{s.t.} \quad E\left(X^{y,\pi}(T)\right) \ge K_2 \end{aligned} \tag{LQ}$$

The solution method for problem (PT), given below, will be based on a modification of the corresponding method of deterministic optimisation which relies on the saddle-point theorem. Therefore, we first sketch the deterministic method (see Section 9.5 in (Fletcher 1981)).

Short Survey: "Lagrangian method of constrained optimisation"
Consider the optimisation problem

$$\begin{aligned} &\max_{x \in R} f(x) \ , \\ &\text{s.t.} \ g(x) \le 0 \end{aligned} \tag{3}$$

where f is strictly concave, $g(x) = (g_1(x),\ldots, g_k(x))'$, $k \in \mathbf{N}$, and the component functions g_i: $\mathbf{R} \to \mathbf{R}$, i = 1,..., k, are to be convex. Define the Lagrangian L: $\mathbf{R} \times \mathbf{R}^k \to \mathbf{R}$ by

$$L(x, d) = f(x) - d'g(x) , \tag{4}$$

where $d \in \mathbf{R}^k$ is the vector of Lagrangian multipliers. Then, we have the following identity between three optimisation problems

$$\max_{\{x : g(x) \le 0\}} f(x) = \max_{x \in \mathbf{R}} \min_{d \ge 0} L(x, d) = \min_{d \ge 0} \max_{x \in \mathbf{R}} L(x, d) \tag{5}$$

in the sense that not only are the optimal values of the objective functions equal, but also the optimal arguments (x*, d*) of the last two problems coincide, and x* is the maximiser of the first problem. The advantage gained by this equality is that we can now solve the original optimisation problem (3) in three different ways. One way which is often very useful is to solve the problem on the right side of equation (3) by the following two step procedure :

Step 1: Solve the unconstrained optimisation problem

$$\max_{x \in \mathbf{R}} L(x, d) \tag{6}$$

for fixed (but arbitrary) $d \in [0, \infty)^k$.

Step 2: Solve the "dual" problem

$$\min_{d \geq 0} \varphi(d) . \tag{7}$$

Here, the "dual objective function" $\varphi(d)$ is defined as

$$\varphi(d) := \sup_{x \in \mathbf{R}} L(x, d) \quad , d \in [0, \infty)^k.$$

Hence, we have decomposed the constrained problem (3) into the (set of) unconstrained problem(s) (6) and the problem (7) with non-negativity constraints.

Keeping this short survey in mind, we can adjust the foregoing notations and definitions to solve the (non-deterministic) constrained optimisation problem (PT). Let again x be the fixed initial wealth of the investor. In analogy to the Lagrangian in equation (4) we define

$$L((\pi, c),d) := E\left(\int_0^T U_1(t,c(t))dt + (U_2 - d'G)\left(X^{y,\pi,c}(T)\right)\right) \tag{8}$$

for $(\pi, c) \in A_T'(y)$, $y \leq x$, $d \in [0, \infty)^k$. Then we have

$$\min_{d \geq 0} L((\pi, c),d) = \begin{cases} -\infty \quad , & \text{if } \; EG(X^{y,\pi,c}(T)) \not\leq 0 \\ J(y;\pi,c), & \text{else} \end{cases} , \tag{9}$$

i.e. the optimisation problem (PT) is equivalent to the problem

$$\max_{(\pi,c) \in A_T'(y), y \leq x} \; \min_{d \geq 0} L((\pi, c),d) \quad . \tag{10}$$

The obvious relation

$$\max_{(\pi,c)\in A_T'(y),\, y\le x}\ \min_{d\ge 0} L((\pi, c),d) \le \max_{(\pi,c)\in A_T'(y),\, y\le x} L((\pi, c),d^*)$$

for every $d^*\in[0, \infty)^k$ implies

$$\max_{(\pi,c)\in A_T'(y),\, y\le x}\ \min_{d\ge 0} L((\pi, c),d) \le \min_{d\ge 0}\ \max_{(\pi,c)\in A_T'(y),\, y\le x} L((\pi, c),d). \quad (11)$$

Referring to the notations

$$\varphi(d) := \sup_{(\pi,c)\in A_T'(y),\, y\le x} L((\pi, c),d), \quad d\in[0, \infty)^k,$$

$$\psi(\pi, c) := \inf_{d\ge 0} L((\pi, c),d), \quad (\pi, c)\in A_T'(y),\ y \le x,$$

we only have to find a pair $((\pi^*, c^*), d^*)\in A_T'(y)\times[0, \infty)^k$ for some $y \le x$ with

$$\varphi(d^*) = \psi(\pi^*, c^*) \quad (12)$$

to prove equality in relation (11). The existence of such a pair will be shown in the proof of the following theorem which is the main result of this section.

Theorem 2 "Solution of the constrained problem"
Let G_i: $[0, \infty) \to \mathbf{R}$ be convex functions for all $i =1,\ldots, k$, and let the assumptions of Theorem 3.16 be satisfied for all $d\in[0,\infty)^k$ below.
a) Assume that there exists a strategy $(\pi_0, c_0)\in A_T'(y)$, $y \le x$ with

$$E\left(G\left(X^{y,\pi_0,c_0}(T)\right)\right) < 0 \quad , \quad (13)$$

and that problem (PT) has a finite optimal solution. Then we have

$$\max_{(\pi,c)\in A_T'(y),\, y\le x}\ \min_{d\ge 0} L((\pi, c),d) = \min_{d\ge 0}\ \max_{(\pi,c)\in A_T'(y),\, y\le x} L((\pi, c),d), \quad (14)$$

and there exists a pair $((\pi^*, c^*), d^*)\in A_T'(y)\times[0,\infty)^k$ for some $y \le x$ with

$$\varphi(d^*) = \psi(\pi^*, c^*) \quad (15)$$

and

$$0 = (d^*)'\, E\left(G\left(X^{y,\pi^*,c^*}(T)\right)\right) . \quad (16)$$

b) If we have

$$\min_{d \geq 0} \varphi(d) = -\infty \tag{17}$$

(i.e. the dual problem has no finite optimal solution), then there exists no feasible solution for the constrained portfolio problem(PT) (i.e. no pair $(\pi, c) \in A_T'(y)$ with $E(G(X^{y,\pi,c}(T))) \leq 0$ for any $y \leq x$).

c) If we have

$$\max_{(\pi,c) \in A_T'(y), y \leq x} \psi(\pi, c) = +\infty \tag{18}$$

(i.e. the primal problem has no finite optimal solution), then there exists no feasible solution for the dual problem, i.e. there does not exist any $d \in [0,\infty)^k$, such that we have

$$\varphi(d) < +\infty . \tag{19}$$

Remark 3

Before proving this theorem, we would like to point out that it offers a decomposition of the solution of the constrained portfolio problem (PT) into two steps which parallels the decomposition of the method in the deterministic case, presented in the short survey preceding the statement of the theorem (due to this similarity, we call this new method a dual method, too) :

Step 1: Solve the unconstrained portfolio problem

$$\max_{(\pi,c) \in A_T'(y), y \leq x} L((\pi, c),d) \tag{20}$$

for $d \in [0,\infty)^k$ arbitrary but fixed (with one of the solution methods presented in Chapter 3 or with any other method).

Step 2: Minimise the function

$$L((\pi^*(d), c^*(d)),d)$$

with respect to $d \in [0,\infty)^k$ where $(\pi^*(d), c^*(d))$ is an optimal pair for the maximisation problem of Step 1.

Note that the optimisation problem in Step 2 is a static, deterministic one. The form $(\pi^*(d), c^*(d))$ of the optimal pairs (π^*, c^*), obtained for a fixed d in Step 1, should indicate the dependence on d. Before we give a particular application of this method in Section 3, we will prove Theorem 2:

Proof (of Theorem 2):
a) As already noted, we only have to show the existence of the pair $((\pi^*, c^*), d^*)$. Then, equation (14) follows with the help of inequality (11), and equality in (16) will be implied by equation (9). Therefore, let

$$K_1=\{(u,z)\in \mathbf{R}^{1+k} \mid u\le J(y;\pi,c),\ E(G(X^{y,\pi,c}(T))) \le z \text{ for some } (\pi,c)\in A_T'(y),\ y\le x\},$$

$$K_2 = \{(u, z)\in \mathbf{R}^{1+k} \mid u \ge \max_{(\pi,c)\in A_T'(y),\, y\le x} J(y;\pi,c),\ z\le 0\}.$$

The assumptions of the theorem imply that K_2 is a non-empty convex set. We also show that

$$K_1 \text{ is a non-empty, convex set.} \tag{21}$$

It follows directly from the assumptions made in part a) that the set K_1 is non-empty. It remains to prove the convexity of K_1. Let $(a, b), (c, d)\in K_1$. Then there exist pairs $(\pi_1, c_1)\in A_T'(y_1)$, $(\pi_2, c_2)\in A_T'(y_2)$ for some $y_1, y_2 \le x$ with

$$a \le J(y_1;\pi_1,c_1),\ E\Big(\Big(G\Big(X^{y_1,\pi_1,c_1}(T)\Big)\Big)\Big) \le b, \tag{22}$$

$$c \le J(y_2;\pi_2,c_2),\ E\Big(\Big(G\Big(X^{y_2,\pi_2,c_2}(T)\Big)\Big)\Big) \le d. \tag{23}$$

With the definitions of

$$Y := \lambda\, X^{y_1,\pi_1,c_1}(T) + (1-\lambda)\, X^{y_2,\pi_2,c_2}(T), \tag{24}$$

$$c(t) := \lambda\, c_1(t) + (1-\lambda)\, c_2(t) \tag{25}$$

for $\lambda\in[0, 1]$ we obtain

$$\begin{aligned} x^* &= E\left(\int_0^T H(t)c(t)dt + H(T)Y\right) \\ &= \lambda\, E\left(\int_0^T H(t)c_1(t)dt + H(T)X^{y_1,\pi_1,c_1}(T)\right) \\ &\quad + (1-\lambda)\, E\left(\int_0^T H(t)c_2(t)dt + H(T)X^{y_2,\pi_2,c_2}(T)\right) \\ &\le \lambda y_1 + (1-\lambda) y_2 \le x, \end{aligned}$$

where the first inequality follows from the relations (22) and (23). Part b) of Theorem 2.7 ("Completeness of the market") yields the existence of a portfolio process π with $(\pi, c) \in A(x^*)$ and

$$Y = X^{x^*,\pi,c}(T) \ \text{ a.s.}\,. \tag{26}$$

Equation (26), the convexity of G_i, the concavity of U_1, U_2, and $U_2 - d'G$ imply that we have $(\pi, c) \in A_T'(x^*)$. Further, note

$$E\left(G\left(X^{x^*,\pi,c}(T)\right)\right) = E(G(Y))$$

$$\leq \lambda E\left(G\left(X^{y_1,\pi_1,c_1}(T)\right)\right) + (1-\lambda) E\left(G\left(X^{y_2,\pi_2,c_2}(T)\right)\right)$$

$$\leq \lambda\, b + (1-\lambda)\, d\,. \tag{27}$$

Moreover, the concavity of U_1, U_2 and the definitions (24), (25) together imply

$$J(x^*; \pi, c) \geq \lambda\, J(y_1, \pi_1, c_1) + (1-\lambda)\, J(y_2, \pi_2, c_2) \geq \lambda\, a + (1-\lambda)\, c\,. \tag{28}$$

As a consequence of relations (27) and (28), we have

$$\lambda(a, b) + (1-\lambda)(c, d) \in K_1 \quad \forall\, \lambda \in [0, 1]\,,$$

i.e. we have proved the convexity of K_1. The assumption of a finite optimal solution to the constrained problem (PT) implies that we must have

$$\text{int}(K_2) \neq \varnothing$$

(where $\text{int}(K_2)$ is the interior of K_2), and the definitions of K_1 and K_2 yield

$$K_1 \cap \text{int}(K_2) = \varnothing.$$

The separation theorem for convex sets (see e.g. (Ioffe and Tichomiroff 1979)) implies the existence of a functional $w^* \neq 0$ with $w^* = (w_1^*, (w_2^*)')' \in \mathbf{R}^{1+k}$ and

$$(w^*)'x \leq (w^*)'y \quad \forall\, x \in K_1,\, y \in K_2\,,$$

and hence

$$\sup_{x \in K_1} (w^*)'x \leq \inf_{y \in K_2} (w^*)'y\,. \tag{29}$$

As a consequence of relation (29) and of the form of K_2, w^* has to satisfy

$$w_1^* \in [0, \infty)\,, \quad w_2^* \le 0 \quad \text{(component wise)}\,.$$

To prove this, choose y^* as the maximum of the objective function of the constrained portfolio problem (PT) (the existence of this maximum is assumed in part a)). Then $(y^*, z) \in K_2$ satisfies $z \le 0$, but there also exist pairs (y^*, z) in K_1 with strictly positive z. Hence, relation (29) implies

$$w_2^* \le 0\,. \tag{30}$$

Furthermore, K_1 is bounded from below in y (by y^*), which is not the case for K_2. Thus, relation (29) and the fact that there exist pairs of the form (u, 0) which are elements of both K_1 and K_2 result in

$$w_1^* \ge 0\,.$$

Even more, we can show

$$w_1^* > 0\,, \tag{31}$$

because the assumption $w_1^* = 0$ in connection with (29) and the fact that, under the assumptions of a), we have $(y^*, 0) \in K_2$ together imply

$$(w^*)' \begin{pmatrix} \pi \\ z \end{pmatrix} = w_2^* z \le 0 \quad \forall\ (y, z') \in K_1\,. \tag{32}$$

But this is a contradiction to the existence of the pair (π_0, c_0) with property (13), because the assumption $w_1^* = 0$ and relation (30) imply the existence of some negative components of w_2^* (remember $w^* \ne 0$). Hence, the pair (π_0, c_0) would not satisfy condition (32). Consequently, we have proved inequality (31). Due to relation (31), we can now w.l.o.g. assume that we have

$$w_1^* = 1$$

as the component wise multiplication of w^* with a positive constant is irrelevant for the separation property of w^*. If again we choose y^* as the maximum of the objective function of the constrained portfolio problem (PT) then this implies

$$y^* = \max_{\substack{(\pi,c)\in A_T'(y),\, y\le x \\ E\left(G\left(X^{y,\pi,c}(T)\right)\right)\le 0}} J(y;\pi,c) = \max_{(\pi,c)\in A_T'(y),\, y\le x}\ \min_{d\ge 0} L((\pi,c),d)$$

and

$$(y^*, 0) \in K_1 \cap K_2\,. \tag{33}$$

Thus, due to relations (29) and (33), we have

$$\sup_{(x,z)\in K_1} (w^*)'\begin{pmatrix} x \\ z \end{pmatrix} = y^* = \inf_{(x,z)\in K_2} (w^*)'\begin{pmatrix} x \\ z \end{pmatrix}$$

Furthermore,

$$\begin{aligned} y^* &= \sup_{(u,z)\in K_1} (w_1^*, (w_2^*)')'\begin{pmatrix} u \\ z \end{pmatrix} = \sup_{(u,z)\in K_1} (u + (w_2^*)'z) \\ &\geq \max_{(\pi,c)\in A_T'(y),\, y\leq x} \left(J(y;\pi,c) + (w_2{}^*)'E\left(G\left(X^{y,\pi,c}(T)\right)\right)\right) \\ &\geq \max_{\substack{(\pi,c)\in A_T'(y),\, y\leq x \\ E\left(G\left(X^{y,\pi,c}(T)\right)\right)\leq 0}} J(y;\pi,c) \quad = y^* , \end{aligned} \tag{34}$$

where the first inequality follows from the definition of K_1. Relation (34), the definition of $\varphi(d)$ and the notation $d^* = -w_2^*$ imply

$$\begin{aligned} \min_{d\geq 0}\ \max_{(\pi,c)\in A_T'(y),\, y\leq x} L((\pi,c),d) &\leq \varphi(d^*) = \\ &= y^* = \max_{(\pi,c)\in A_T'(y),\, y\leq x}\ \min_{d\geq 0} L((\pi,c),d). \end{aligned} \tag{35}$$

Hence, equality (14) is proved (recall that the opposite inequality to (35) is always valid). The existence of the strategy (π^*, c^*) satisfying equation (15) follows from Theorem 16 as solution of the unconstrained portfolio problem

$$\max_{(\pi,c)\in A_T'(y),\, y\leq x} L((\pi, c),d^*) .$$

b) Relations (17) and (11) imply

$$-\infty = \min_{d\geq 0} \varphi(d) \geq \max_{(\pi,c)\in A_T'(y),\, y\leq x}\ \min_{d\geq 0} L((\pi, c),d) \geq \min_{d\geq 0} L((\pi, c),d)$$

for every $(\pi, c)\in A_T'(y)$, $y \leq x$. By combining this relation with equation (9), we can conclude that there does not exist any strategy $(\pi, c)\in A_T'(y)$, $y \leq x$, with

$$E\left(G\left(X^{y,\pi,c}(T)\right)\right) \leq 0 .$$

c) Relations (11) and (18) imply the assertions of part c) in a similar way as relations (11) and (17) did in the proof of part b).

□

4.3 A Continuous-Time Mean-Variance Problem

In this section, we formulate and solve a continuous-time analogue of the traditional mean-variance problem, as presented in Section 1.2. This means, we consider the problem of finding a portfolio process that generates an attainable terminal wealth with minimum variance for a given mean. There will be no consumption on [0, T]. For ease of exposition, we do this in a market with a riskless bond, a single stock, and assume constant market coefficients. In this situation, the corresponding optimisation problem reads

$$\begin{aligned} &\min_{\pi \in A_T'(y),\, y \le x} \quad \operatorname{Var}\left(X^{y,\pi}(T)\right) \\ &\text{s.t.} \quad E\left(X^{y,\pi}(T)\right) \ge K \end{aligned}, \tag{MV}$$

where K > 0 is a given constant and x > 0 is the initial wealth of the investor.

The first difficulty in attacking this problem is that

$$\operatorname{Var}\left(X^{y,\pi}(T)\right) = E\left(\left(X^{y,\pi}(T) - E\left(X^{y,\pi}(T)\right)\right)^2\right)$$

cannot be converted into the form of a utility function of Definition 3.1*, i.e. it cannot be written as

$$\operatorname{Var}\left(X^{y,\pi}(T)\right) = E\left(U\left(X^{y,\pi}(T)\right)\right).$$

To see this, simply replace $X^{y,\pi}(T)$ by x in the variance and observe that this will result in Var(x) = 0 for all $x \in \mathbf{R}$. To overcome this problem, we have to consider certain different settings of the market coefficients r and b separately.

Case 1 : $r \ge \frac{1}{T}\ln(\frac{K}{x})$

In this case, there exist pure bond strategies leading to a constant terminal wealth greater or equal to K. All these strategies have a variance of zero and satisfy the expectation constraint in (MV). Hence, they are all optimal strategies. One could perhaps pick out the one with the highest terminal wealth (i.e. X(0) = x, $\pi \equiv 0$) or the one with the lowest initial capital needed to generate a terminal wealth satisfying the expectation constraint (i.e. $X(0) = xe^{-rT}$, $\pi \equiv 0$).

Case 2 : $r < \frac{1}{T}\ln(\frac{K}{x})$ **and** $b \ne r$

Under the current parameter setting, every strategy that satisfies the expectation constraint must include stock investment, because all pure bond strategies do not satisfy the expectation constraint. Thus, the unconstrained minimum variance of zero cannot be attained. Hence, we know that the expectation constraint must be satisfied as an equality in an optimal solution. This observation will be crucial when we transform the mean-variance problem (MV) into a simpler one. To do so, consider the following problem

$$\begin{aligned} \min_{\pi \in A_T'(y),\, y \le x} \quad & \tfrac{1}{2} E\left(X^{y,\pi}(T) - K\right)^2 \\ \text{s.t.} \quad & E(X^{y,\pi}(T)) \ge K \end{aligned} \quad . \tag{36}$$

Due to our assumptions on r and K, the unconstrained minimum value of zero of the objective function cannot be reached by a solution of problem (36). Thus, the expectation constraint must be satisfied as an equality in every optimal solution, too. Consequently, solving problem (36) is equivalent to solving the mean-variance problem (MV). Note further that, due to the assumption "$xe^{rT} < K$", we can restrict ourselves to strategies $\pi \in A_T'(x)$ in both problems (MV) and (36). Using this fact and rewriting problem (36) in the standard form for the constrained portfolio problems considered in Theorem 2 as

$$\begin{aligned} \max_{\pi \in A_T'(x)} \quad & -\tfrac{1}{2} E\left(X^{x,\pi}(T) - K\right)^2 \\ \text{s.t.} \quad & K - E(X^{x,\pi}(T)) \le 0 \end{aligned} ,$$

we are able to solve it with the help of this theorem in the following way:

Step 1:
Fix $d \ge 0$ and solve the unconstrained problem

$$\begin{aligned} & \max_{\pi \in A_T'(x)} - E\left(\tfrac{1}{2}\left(X^{x,\pi}(T) - K\right)^2 + dK - dX^{x,\pi}(T) \right) \\ & = \max_{\pi \in A_T'(x)} - E\left(\tfrac{1}{2} X^{x,\pi}(T)^2 - (K+d) X^{x,\pi}(T) + \tfrac{1}{2} K^2 + dK \right) \\ & = \tfrac{1}{2} d^2 + \max_{\pi \in A_T'(x)} - E\left(\tfrac{1}{2}\left(X^{x,\pi}(T) - (K+d)\right)^2 \right) \\ & =: \max_{\pi \in A_T'(x)} L(\pi, d) , \end{aligned} \qquad 37)$$

Step 2:
Find the value $d^* \in [0, \infty)$ which yields the solution $\pi(d^*)$ of problem (37) with the smallest maximal value $L(\pi(d^*), d^*)$ among all $d \in [0, \infty)$.

Note that the optimisation problem in the second but last line of relation (37) is an unconstrained portfolio problem of the type considered in Example 3.20. The remark, made about its (possibly numerical) solution in Section 3.4, is here valid, too. Furthermore, the Lagrangian multiplier d now has a very apparent interpretation. It determines the auxiliary problem of the form (36) with the correct target value K + d. The solution of this auxiliary problem then yields the solution of the mean-variance problem.

As we do not obtain solutions of problem (37) as explicit functions of d, the optimisation in Step 2 must be done by an iteration procedure. However, we can greatly benefit from the fact that the set for the candidates d for d* in Step 2 can be further restricted.

Proposition 4

Let $xe^{rT} < K$, and let z* be defined by

$$z^* = \sup_{\pi \in A_T'(x)} -\tfrac{1}{2} E\left(X^{x,\pi}(T) - K\right)^2.$$

Then d*, yielding the minimum in Step 2 above, satisfies $d^* \in [0, \delta]$ with

$$\delta := \frac{Kxe^{rT} - (z^* + \tfrac{1}{2}(x^2e^{rT} + K^2))}{K - xe^{rT}}.$$

Proof:

Observe that we have

$$\max_{\pi \in A_T'(x)} L(\pi, d) \geq L(0, d) = \tfrac{1}{2}\left(d^2 - \left(xe^{rT} - (K+d)\right)^2\right)$$
$$= d\left(xe^{rT} - K\right) - \tfrac{1}{2}\left(x^2e^{2rT} + K^2\right) + Kxe^{rT},$$

$$\min_{d \geq 0} \max_{\pi \in A_T'(x)} L(\pi, d) \leq \max_{\pi \in A_T'(x)} L(\pi, 0) = z^*.$$

As a consequence of these two inequalities, a value of d exceeding δ cannot yield the infimum in Step 2 above. Note in particular that the inequality "$L(0, d) \leq z^*$" need not be satisfied for general $d \geq 0$, but for all $d \geq 0$ yielding the infimum in Step 2 above.

□

Before we present some numerical examples, we make a short comment on the trading strategy that generates the optimal terminal wealth for the continuous-time mean-variance problem (MV). As already said in Example 3.20 and in Section 3.6, the explicit form of the strategy has a tremendous length. However, by rewriting the optimal terminal wealth $X^{x,\pi^*}(T)$ for problem (MV) under the assumption "$xe^{rT} < K$" as

$$[(K + d^*) - Y(x)H(T)]^+ = [(K + d^*) - Y(x)H(T)] + [Y(x)H(T) - (K + d^*)]^+ ,(38)$$

we realise that it can be decomposed into the sum of the (unconstrained) terminal wealth

$$B^* = (K + d^*) - Y(x)H(T)$$

and the terminal payoff of a European put with strike zero on this terminal wealth B^*. This "put"-part (which "exists" only for reasons of interpretation) of the optimal terminal wealth ensures the non-negativity of the total value of the holdings at time t = T. It is non-zero if and only if B^* is negative. One could interpret this decomposition of the optimal terminal wealth as if the investor would split his initial holdings into two parts, one for ensuring non-negativity and one for unconstrained speculation. With the first part, the above described put is bought while the latter part is used to follow a portfolio strategy leading to a terminal wealth of B^*. If, for the moment, we assume that we are able to hold such a put then we can easily compute the additional numbers of shares held in stock and bond to replicate the optimal terminal wealth $X^{x,\pi^*}(T)$ for problem (MV).

Proposition 5
Let d^* be the Lagrangian multiplier that yields the minimum in Step 2 above. Let further $Y(x)$ be defined as in Example 3.20 (but with K replaced by $K + d^*$). Holding the above put with terminal payoff

$$[-((K + d^*) - Y(x)H(T))]^+$$

and additionally following the trading strategy (φ_0, φ_1) given by

$$\varphi_0(t) = (K + d^*)e^{-rT} - \frac{b+\sigma^2-r}{\sigma^2} Y(x)H(t)\exp\left(-rt + \left(\theta^2 - 2r\right)(T-t)\right) ,$$

$$\varphi_1(t) = \frac{b-r}{\sigma^2 P_1(t)} Y(x)H(t)\exp\left(\left(\theta^2 - 2r\right)(T-t)\right)$$

leads to a terminal wealth of $[(K + d^*) - Y(x)H(T)]^+$, i.e. this strategy generates the optimal terminal wealth of the mean-variance problem (MV) in the case "$xe^{rT} < K$".

Proof:
We use the representation (38) of the optimal terminal wealth for problem (MV). Obviously, $[Y(x)H(T) - (K + d^*)]^+$ will be replicated by holding the corresponding put. The deterministic terminal payment of $K + d^*$ will uniquely be obtained by holding $(K + d^*)e^{-rT}$ shares of the bond on $[0, T]$. Note further, that we have

$$X(t) := \frac{1}{H(t)} E\left(H(T)^2 \middle| F_t\right) = \exp\left(-\left(r + \tfrac{1}{2}\theta^2\right)t - \theta W(t) + \left(\theta^2 - 2r\right)(T-t)\right) =: f(t, W(t)),$$

$$f(0, W(0)) = \exp\left(\left(\theta^2 - 2r\right)T\right).$$

Application of Theorem 21 then yields that H(T) is generated by the strategies

$$\varphi_0^H(t) = \frac{b + \sigma^2 - r}{\sigma^2} H(t)\exp\left(-rt + \left(\theta^2 - 2r\right)(T-t)\right),$$

$$\varphi_1^H(t) = -\frac{b - r}{\sigma^2 P_1(t)} H(t)\exp\left(\left(\theta^2 - 2r\right)(T-t)\right).$$

Putting all these considerations together yields the assertions of the proposition. Note in particular that the trading strategy (φ_0, φ_1) is self-financing by construction.

□

One could also compute the generating strategy for the above put option, but we will not do so here. However, this is again recommended as an exercise in applying Theorem 21.

In addition to the above theoretical considerations, we will give some numerical results for Case 2 above. Therefore, let the initial wealth of the investor be $x = 1000$ and assume that we have $r = 0$ (i.e. the bond price is constant), $b = 0.1$, $\sigma = 0.05$ or $\sigma = 0.25$, and finally $T = 1$ or $T = 10$. For all four possible combinations of these values we consider four different values for K, the lower bound for the expected terminal wealth. The lowest of these values is always a little bit above the terminal wealth that can be attained by following the pure bond strategy; the third value is the same as the expected value of a pure stock strategy; the second values lies half way between the first and the third whereas the highest value lies a good amount above the expected value of a pure stock strategy. In all of the tables below we have computed the minimal variance (and its square root) for the continuous-time problem, $Var(X_c(T))$, and also minimal variance (and its square root) for the static, Markowitz setting, $Var(X_s(T))$ (where it is only allowed to trade at time $t = 0$ but also required to have a non-negative terminal wealth). Note especially, that for the highest required value for K, it is not possible to guarantee a non-negative terminal wealth in the static setting as short selling of the bond is required. Further, we have given the fraction of initial wealth invested in the stock, π, for the static case while for the continuous-time case, we also give the value d^* of the Lagrangian multiplier. The tables are completed by the quotient of the two minimal variances V_c / V_s in the continuous-time and the static setting, respectively.

In Table 4, we have chosen $T = 1$ and $\sigma = 0.25$. The value of the minimal variance in the continuous-time case is approximately 80 % of that of the static problem which can be read off from the last column of Table 4. The increase of the quotient to 85 % for $K = 1200$ can be explained by the fact that then we have dropped the non-negativity constraint in the static case.

K	$Var(X_s(T))$	std.dev.	π	$Var(X_c(T))$	std.dev.	d	V_c / V_s
1005	178	13.34	0.05	144	12.00	57	0.81
1050	17804	133.43	0.48	14409	120.04	576	0.81
1105	78517	280.21	1.00	63838	252.66	1228	0.81
1200	284872	482.94	1.90	241300	491.22	2570	0.85

Table 4: Comparison of solutions to a static and continuous-time mean variance problem, $T=1$, $\sigma = 0{,}25$.

In Table 5, we have chosen $T = 10$ and $\sigma = 0.25$. Compared to the results of Table 1, it is remarkable that the quotient between the minimal variance in the continuous-time and that in the static setting has decreased to values between 0.12 and 0.42. Hence, the advantage of the continuous-time method increases with increasing time horizon.

K	$Var(X_s(T))$	std.dev.	π	$Var(X_c(T))$	std.dev.	d	V_c / V_s
1005	54	7.37	0.003	6.32	2.51	2.52	0.12
1800	1390666	1179.27	0.46	274888	2524.30	851.56	0.20
2718	6413414	2532.47	1.00	1998871	1413.81	3159.69	0.31
3500	13580725	3685.20	1.45	5660876	2379.26	5570.96	0.42

Table 5: Comparison of solutions to a static and continuous-time mean variance problem, $T=10$, $\sigma = 0.25$.

These findings will be even more significant if we reduce σ to 0.05. Then, in the case of T = 1, the value of the minimum variance in the continuous-time model is only around 10% of the one of the static setting. In the case of T = 10 the gain of the continuous-time model is extremely dramatic: $Var(X_C(T))$ can be neglected compared to $Var(X_S(T))$. Nearly every desired expectation constraint seem to be attainable **without** variance. These values are not due to rounding errors or instabilities. The main reason for this is that, due to our choice of the market coefficients, it is very advantageous to invest in the stock. Holding negative bond positions to finance stock investment contains nearly no risk due to our extreme choice of market coefficients. This stock investment results in gains that exceed those obtained from bond investment with a very high probability. Immediate transfer of these gains into bond positions leads to a reduction of the variance of the terminal wealth. It is therefore also easy to explain why the variances are lower in the case of T = 10 than in the case of T = 1; there, the process of obtaining high gains from stock investment and putting them immediately into bond positions runs for a longer time. This results in a greater reduction of variance. The increase of the mean rate of stock return will lead to similar effects.

K	$Var(X_S(T))$	std.dev.	π	$Var(X_C(T))$	std.dev.	d	V_C / V_S
1005	7	2.63	0.05	0.5	0.69	0.19	0.07
1050	691	26.29	0.48	55	7.43	2.37	0.08
1105	3047	55.20	1.00	281	16.76	6.01	0.09
1200	11056	105.14	1.90	1237	35.18	14.62	0.11

Table 6: Comparison of solutions to a static and continuous-time mean variance problem, T=1, σ= 0.05.

K	$Var(X_S(T))$	std.dev.	π	$Var(X_C(T))$	std.dev	d	V_C / V_S
1005	1.6	1.26	0.003	≈ 0.00	≈ 0.00	≈ 0.00	≈ 0.00
1800	40547	201.36	0.46	0.0001	0.01	≈ 0.00	≈ 0.00
2718	186993	432.42	1.00	0.006	0.08	≈ 0.00	≈ 0.00
3500	395968	629.26	1.45	0.03	0.17	≈ 0.00	≈ 0.00

Table 7: Comparison of solutions to a static and continuous-time mean variance problem, T=10, σ= 0.05.

Remark 6

In recent times, some papers concerned with mean-variance problems in a continuous-time setting have been published. However, most of them leave aside the solvency constraint of a non-negative terminal wealth. This enables the authors to use Hilbert space methods (such as the Hilbert space projection theorem (see (Luenberger 1969)) and obtain closed form solutions (see e.g. (Duffie and Richardson 1991), (Richardson 1989), (Schweizer 1992) or (Kiesel 1996)) for both the optimal terminal wealth and the generating trading strategies. However, in their setting the probability of ending up with a negative terminal wealth is strictly positive. According to the size of the market coefficients, the size of this probability can be quite high. Further, the optimal terminal wealth in these unconstrained mean-variance problems is **not** bounded from below (see also (Korn 1997b)). A paper which is closely related to our presentation of the mean-variance problem under non-negativity constraints above is (Lioui 1997). Here, the author considers portfolio optimisation with a quadratic utility function under the additional constraint of a non-negative terminal wealth.

4.4 Portfolio Problems with Constrained Strategies

A further generalisation of the unconstrained portfolio problem of Chapter 3 lies in the introduction of constraints on the portfolio strategies (which is in fact a very realistic feature as traders are often restricted in their way of trading). Typical examples will be short-selling constraints (especially "no borrowing" constraints), leverage constraints or upper/lower bounds for the fractions of wealth invested in certain stocks. In this section, we will present results and methods given in (Cvitanic and Karatzas 1992) and (Xu and Shreve 1992 a, b).

The principal method of solving the constrained problems is the introduction of auxiliary markets which are complete ones. These markets typically differ from the original (constrained) market only in their market coefficients of the security prices. The unconstrained portfolio problem in each of these auxiliary markets can be solved with one of the methods presented in Chapter 3. It will be shown that there exists an auxiliary market such that the solution of our original constrained portfolio problem can be deduced from the corresponding optimal (unconstrained) solution in this auxiliary market. The main existence result for solutions of constrained problems will however be proved via the introduction of a dual (unconstrained) problem. This method again resembles methods from deterministic optimisation and is similar to the dual problem introduced in Section 3.5 ("Pliska's method").

For this section, we will assume that the utility functions $U_1(t, c)$ and $U_2(x)$ satisfy the following assumption **(A1)**:

$$\lim_{c\to\infty} U_2'(c) = \lim_{c\to\infty} U_1'(t, c) = 0 \quad \forall\, t\in[0,T],$$

$$\lim_{c\to 0} U_2'(c) = \lim_{c\to 0} U_1'(t, c) = \infty \quad \forall\, t\in[0,T].$$

We will denote by M the market model of Chapter 2 where we will not necessarily require that the number of stocks n and the dimension m of the driving Brownian motion W(t) coincide.

Definition 7
The **constrained portfolio problem** (in M) with respect to K is the optimisation problem

$$\max_{(\pi,c)\in A_K'(x)} J(x\,;\pi, c)$$

where $x > 0$ is the initial wealth of the investor, and the set of feasible strategies is given by

$$A_K'(x) := \{(\pi, c)\in A'(x) \mid \pi(t, \omega)\in K \text{ for } \mathcal{L}\otimes P\text{ - a. e. } (t, \omega)\in[0,T]\times\Omega\}$$

where K is a fixed, closed and convex subset of $\mathbf{R}^n$. Further, denote by V(x) the **value function** of this problem which is defined as

$$V(x) := \sup_{(\pi,c)\in A_K'(x)} J(x\,;\pi, c) \quad \forall\, x > 0$$

The set K represents the constraints put on the portfolio processes. To illustrate this, we take a look at the following examples.

Example 8 "No short selling of stocks"
Short selling constraints on the stocks, i.e.

$$\pi_i(t) \geq 0 \;\; \forall\, t\in[0,T] \text{ a.s., } i = 1,\dots, n,$$

correspond to the set $K = [0, \infty)^n$.

Example 9 "Bounded fraction for the wealth invested in risky securities"

Here, the investor is constrained to invest at most a given fraction $z \in [0,1]$ of his wealth into risky assets. Hence, the set corresponding set K is given by

$$K = \left\{ \pi \in \mathbf{R}^n : \sum_{i=1}^{n} \pi_i \le z \right\}.$$

If we also require short selling constraints as in Example 8 then K has the form

$$K = \left\{ \pi \in [0,\infty)^n : \sum_{i=1}^{n} \pi_i \le z \right\}.$$

We will start to develop the methods presented in (Karatzas and Cvitanic 1992) and (Xu and Shreve 1992 a, b) by introducing unconstrained auxiliary markets. With their help, we will be able to transform the solution of the constrained problem to suitable unconstrained ones.

i) Unconstrained Auxiliary Markets

The auxiliary markets M_μ (to avoid misunderstandings: these markets are just a theoretical construct to develop the solution method, they do not exist in the real economy !) consist of the same securities as our original market M but with different market coefficients. More precisely, the bond and stock prices $P_i^{(\mu)}(t)$ in M_μ are given as the unique solutions of the following (stochastic) differential equations

$$dP_0^{(\mu)}(t) = P_0^{(\mu)}(t)\,[\,r(t) + \mu_0(t)\,]\,dt\,, \qquad P_0^{(\mu)}(t)(0) = 1, \tag{39}$$

$$dP_i^{(\mu)}(t) = P_i^{(\mu)}(t)\left((b_i(t)+\mu_i(t)+\mu_0(t))dt + \sum_{k=1}^{n} \sigma_{ik}(t)\,dW_k(t) \right), \quad P_i^{(\mu)}(0) = p_i, \tag{40}$$

for $i=1,\dots,n$. Our goal will be to show that with the help of a suitable choice of the stochastic processes $\mu_0(t)$ and $\mu(t):=(\mu_1(t),\dots,\mu_n(t))'$, $t\in[0,T]$, we obtain:

1. The solutions of the **constrained problems in M_μ** all have a maximum expected utility that is at least as big as V(x) (see the proof of Proposition 12 below).

2. The constrained solution in a market M_λ with the smallest maximum expected utility among all auxiliary market can be computed as the solution of the unconstrained problem in M_λ, and it will also yield the optimal solution of the constrained problem in the original market M (see Theorem 2 below).

Provided that we choose $\mu_0(t)$, $\mu(t)$, $t\in[0,T]$, in such a way that the equations (39) and (40) have unique solutions then the wealth process X(t) **in $\mathbf{M}_\mu$** of an investor with an initial capital of $x > 0$, following the strategy (π, c), is given as the unique solution of the following stochastic differential equation:

$$\begin{aligned} dX(t) &= X(t)\,\pi(t)'[\,(b(t)+\mu(t)+\mu_0(t)\,\underline{1})\,dt+\sigma(t)\,dW(t)\,] \\ &\quad + (1-\pi(t)'\underline{1})\,X(t)\,[\,r(t)\ +\mu_0(t)\,]\,dt - c(t)\,dt \\ &= [\,r(t)\,X(t) - c(t)\,]\,dt + X(t)\,\pi(t)'\,[\,(b(t)-r(t)'\underline{1})\,dt+\sigma(t)\,dW(t)] \\ &\quad + X(t)\,[\,\mu_0(t)+\pi(t)'\mu(t)\,]\,dt \end{aligned} \tag{41}$$

By noting that equation (41) differs from the equation for a wealth process of an investor following the strategy (π, c) **in the original market M** only by the additional term

$$X(t)[\mu_0(t)+\pi(t)'\mu(t)]\,dt\,,$$

we can derive a sufficient condition for the choice of $\mu_0(t)$ to ensure the first of the above requirements on the auxiliary markets: Comparison of both the stochastic differential equations for the wealth process in M and in M_μ corresponding to a strategy $(\pi, c)\in A_K'$ implies that the wealth process in M_μ is bigger (in a pathwise sense) than the one in M if we would have

$$\mu_0(t,\omega) \geq \sup_{\pi\in K}(-\pi\,'\mu(t,\omega)) \quad (\text{for } \mathcal{L}\otimes P\text{ - a.e. } (t,\omega)\in[0,T]\times\Omega)\,.$$

In the following, we will therefore choose $\mu_0(t)$ to be equal to the right side of the above inequality.

Definition 10
Let K be a non-empty, closed, convex subset of $\mathbf{R}^n$. The function δ: $\mathbf{R}^n \rightarrow \mathbf{R}\cup\{\infty\}$ defined by

$$\delta(x) := \delta(x\,;K) := \sup_{\pi\in K}(-\pi\,'x)\ ,\quad x\in\mathbf{R}^n$$

is called the **support function** of $-K$ with **effective domain** $\hat{K}$ given by

$$\hat{K} := \left\{x\in R^n \,\middle|\, \delta(x;K)<\infty\right\}.$$

Remark 11

a) δ is a closed, convex function satisfying $\delta(x + y) \le \delta(x) + \delta(y)$ for all x, $y \in \mathbf{R}^n$, and $\hat{K}$ is a convex cone (see e.g. (Rockafellar 1970), p.114).

b) In the "No short selling"-case of Example 8, $K = [0, \infty)^n$, we have

$$\delta(x) = \begin{cases} 0, & x \in K \\ \infty, & x \notin K \end{cases},$$

while in the setting of Example 9 ("No short selling, bound on the fraction of wealth invested in risky assets"),

$$K = \left\{ \pi \in [0,\infty)^n : \sum_{i=1}^{n} \pi_i(t) \le z \right\},$$

the support function of – K is given by

$$\delta(x) = z \max\left(x_1^-, \ldots, x_n^-\right).$$

For technical reasons, we need the following two assumptions on the support function (which are in particular satisfied in the above examples)

$$\delta(\,.\,, K) \text{ is continuous on } \hat{K}, \qquad \textbf{(A2)}$$

$$\delta(\,.\,, K) \text{ is bounded from below on } \mathbf{R}^n. \qquad \textbf{(A3)}$$

Note that the first assumption is always satisfied if the pure bond strategy is feasible, i.e. if we have $0 \in K$ (then $\delta_0 = 0$ is a lower bound). With the aid of the support function and its effective domain, we are able to describe the space of possible candidates for the additional stock price drift processes $\mu(t)$ such that our two requirements made above are satisfied. Therefore, we introduce the two sets of processes,

$$H^n = \left\{ \mu \,\middle|\, \mu(t), t \in [0,T], \mathbf{R}^n\text{-valued}, F_t\text{-adapted}, \ E\left(\int_0^T \|\mu(t)\|^2 \, dt \right) < \infty \right\},$$

$$D = \left\{ \mu \in H^n \,\middle|\, E\left(\int_0^T \delta(\mu(t)) \, dt \right) < \infty \right\}.$$

In particular, note that $\mu \in D$ implies $\mu(t, \omega) \in \hat{K}$ for $L \otimes P$-a.e. $(t, \omega) \in [0,T] \times \Omega$. For notational convenience, we will also introduce the analogues to $\theta(t)$ and $H(t)$ in the auxiliary markets M_μ by

$$\theta_\mu(t) := \sigma^{-1}(t)\,[\,b(t) + \mu(t) - r(t)\underline{1}\,] = \theta(t) + \sigma^{-1}(t)\mu(t)\,,$$

$$H_\mu(t) := \exp\!\left(-\int_0^t (r(s) + \delta(\mu(s)) + \tfrac{1}{2}\|\theta_\mu(s)\|^2)ds - \int_0^t \theta_\mu(s)'dW(s)\right)$$

for stochastic processes $\mu \in D$. Note that, with the above definitions, $\theta_\mu(t)$, $t\in[0,T]$, is no longer uniformly bounded in $(t, \omega)\in[0,T] \times \Omega$. This is also the case for the bond interest rate and the mean rates of stock return in M_μ,

$$r_\mu(t) := r(t) + \delta(\mu(t)), \quad b_\mu(t) := b(t) + \mu(t) + \delta(\mu(t))\,\underline{1}\ .$$

However, by Theorem B15, the (stochastic) differential equations for the bond and stock prices still possess unique solutions for processes $\mu\in D$. Further, the unconstrained portfolio problems in the markets M_μ can still be solved with the help of the martingale method of Section 3.4. Note that there, we needed the uniform boundedness of r(t) and b(t) only in the proof of Proposition 3.15, but only in the case where the utility functions had a form that is prohibited here by Assumption **(A1)**. The stochastic differential equation for the wealth process $X_\mu(t)$, $t\in[0,T]$, of an investor using a strategy (π, c) in M_μ is now given as

$$\begin{aligned} dX_\mu(t) = {} & [r(t)X_\mu(t) - c(t)]\,dt + X_\mu(t)\,\pi(t)'[b(t)\,dt + \sigma(t)\,dW(t)] \\ & + X_\mu(t)[\delta(\mu(t)) + \pi(t)'\mu(t)]dt \qquad (42)\\ & X_\mu(t)(0) = x. \end{aligned}$$

Hence, the wealth process corresponding to (π, c) in M_μ coincides with the one in M if and only if the additional drift process vanishes, i.e. if we have

$$\delta(\mu(t)) + \pi(t)'\mu(t) = 0 \quad \text{for } \mathcal{L}\otimes P\text{-a.e. } (t, \omega)\in[0,T]\times\Omega. \qquad (43)$$

If we further denote by $A_\mu(x)$ and $A_\mu'(x)$ the analogues in M_μ of the definitions of A(x) and A'(x) and introduce the analogue of $\mathcal{X}(y)$ as

$$\mathcal{X}_\mu(y) = E\left[\int_0^T H_\mu(t)I_1(t, yH_\mu(t))\,dt + H_\mu(T)I_2(yH_\mu(T))\right],$$

then we are able to solve the unconstrained portfolio problem in M_μ with the help of Theorem 3.16. Under the assumption of finite values $\mathcal{X}_\mu(y)$ for all positive y, we obtain the following representations for the optimal consumption $c_\mu^*(t)$, $t\in[0,T]$, and optimal terminal wealth B_μ^* in the auxiliary market M_μ (compare also with Theorem 3.16)

$$c_\mu^*(t) = I_1(t, Y_\mu(x)H_\mu(t)) \quad \forall\, t\in[0,T], \tag{44}$$

$$B_\mu^* = I_2(Y_\mu(x)H_\mu(T)). \tag{45}$$

Once again, the theorem on complete markets yields the existence of a corresponding portfolio strategy $\pi_\mu^*(t)$ in M_μ with $(\pi_\mu^*, c_\mu^*) \in A_\mu'(x)$.

The following proposition (see also (Cvitanic and Karatzas 1992)) presents a sufficient condition for such an optimal strategy in M_μ being also optimal for the original constrained problem in M.

Proposition 12

Assume that for some $\lambda\in D$ (with $X_\lambda(y) < \infty$ for $y>0$) we have

$$\pi_\lambda^*(t)\in K \quad \text{for } L\otimes P\text{ - a.e. } (t,\omega)\in[0,T]\times\Omega\ , \tag{46}$$

$$\delta(\lambda(t)) + \pi_\lambda^*(t)'\,\lambda(t) = 0 \quad \text{for } L\otimes P\text{-a.e. } (t,\omega)\in[0,T]\times\Omega \tag{47}$$

Then $(\pi_\lambda^*, c_\lambda^*)\in A_\lambda'(x)$ implies $(\pi_\lambda^*, c_\lambda^*)\in A_K'(x)$ and

$$V(x) = J(x\,;\pi_\lambda^*, c_\lambda^*)\ . \tag{48}$$

Proof:

Note that equation (47) is just equation (43) with $\mu = \lambda$ which was the necessary and sufficient condition for the wealth processes corresponding to $(\pi_\lambda^*, c_\lambda^*)$ in M and in M_μ to coincide. Further, requirement (46) and $(\pi_\lambda^*, c_\lambda^*)\in A_\lambda'(x)$ together yield

$$(\pi_\lambda^*, c_\lambda^*)\in A_K'(x)\ . \tag{49}$$

Let $(\pi, c)\in A_K'(x)$. We then have

$$(\pi, c)\in A_\mu'(x) \tag{50}$$

for all $\mu\in D$ (i.e. every admissible strategy in the original market M is also admissible in M_μ). This can easily be seen by comparing the different wealth processes in M and in M_μ corresponding to (π, c) starting with an initial capital of x: By the definition of the support function, we have

$$\delta(\mu(t)) + \pi(t)'\mu(t) \ge 0 \quad \text{for } L\otimes P\text{-a.e. } (t,\omega)\in[0,T]\times\Omega$$

and thus

$$0 \le X(t) \le X_\mu(t) \quad \forall\, t\in[0,T] \text{ a.s.}.$$

Further, due to the monotonicity of the utility function $U_2(.)$, $(\pi, c)\in\hat{A}'(x)$ also implies that we have $(\pi, c)\in A_\mu'(x)$. From relation (50), we directly obtain

$$A_K'(x) \subseteq A_\mu'(x) \quad \forall\, x\in(0, \infty),\ \mu\in D$$

which leads to

$$V(x) \le V_\mu(x) := \sup_{(\pi,c)\in A_\mu'(x)} J(x\,;\pi, c) \quad \forall\, x\in(0, \infty), \mu\in D \tag{51}$$

Finally, relation (49) implies equality in relation (51), and everything is proved. □

ii) Optimality Conditions for the Constrained Problem

We will characterise the optimal solution by giving some equivalent optimality conditions for the constrained problem that relate the auxiliary markets with the original market. Before stating the main result, we will present an analogue to the theorem on complete markets in our constrained case. As the proof of its second assertion is highly technical, we will formulate parts of it as a separate lemma which is given in sub-section iv) below. In particular, we recommend to skip the proof at first reading.

Theorem 13 "Attainability in the constrained market"
a) Let $(\pi, c)\in A(x)$, $\mu\in D$. Then we have

$$E\left[\int_0^T H_\mu(s)c(s)ds + H_\mu(T)X(T)\right] \le x \quad . \tag{52}$$

b) Let $c(t)$, $t\in[0,T]$, be a consumption process, B a non-negative, F_T-measurable random variable. Assume further that there exists a process $\lambda\in D$ with

$$E\left[\int_0^T H_\mu(t)c(t)dt + H_\mu(T)B\right] \le E\left[\int_0^T H_\lambda(t)c(t)dt + H_\lambda(T)B\right] =: x < \infty \tag{53}$$

for all $\mu\in D$. Then there exists a portfolio process $\pi(t)$, $t\in[0,T]$, such that we have $(\pi, c)\in A_K'(x)$ and

$$X^{x,\pi,c}(T) = B \quad \text{a.s.}$$

(where $X^{x,\pi,c}(t)$ is the wealth process corresponding to (π, c) in M).

Proof :
a) Note first that $H_\mu(t)$ satisfies the sde

$$dH_\mu(t) = -H_\mu(t)[(r(t) + \delta(\mu(t)))dt + \theta_\mu(t)'dW(t)] .$$

Then, for $(\pi, c)\in A(x)$ and $\mu\in D$ we obtain

$$\begin{aligned}
&H_\mu(t)\,X(t) + \int_0^t H_\mu(s)c(s)ds \\
&= x + \int_0^t H_\mu(s)dX(s) + \int_0^t X(s)dH_\mu(s) + \langle H_\mu, X\rangle_t + \int_0^t H_\mu(s)c(s)ds \\
&= x + \int_0^t H_\mu(s)X(s)\{(1-\pi(s)'\underline{1})r(s) + \pi(s)'b(s)\}ds \\
&\quad + \int_0^t H_\mu(s)X(s)\pi(s)'\sigma(s)dW(s) - \int_0^t H_\mu(s)X(s)\{r(s)+\delta(\mu(s))\}ds \\
&\quad - \int_0^t H_\mu(s)X(s)\theta_\mu(s)'dW(s) - \int_0^t H_\mu(s)X(s)\pi(s)'\sigma(s)\theta_\mu(s)'ds \\
&= x + \int_0^t H_\mu(s)X(s)(\pi(s)'\sigma(s)-\theta_\mu(s)')dW(s) \\
&\quad - \int_0^t H_\mu(s)X(s)\{\delta(\mu(s))+\pi(s)'\mu(s)\}ds \\
&\le x + \int_0^t H_\mu(s)X(s)(\pi(s)'\sigma(s)-\theta_\mu(s)')dW(s) \quad \forall\, t\in[0, T] \text{ a.s.}, \qquad (54)
\end{aligned}$$

where we have used the definition of the support function to obtain the inequality. The left hand side of inequality (54) is non-negative. Hence, the local martingale on the right side must be a supermartingale (see Proposition A17). By taking expectations, this implies that for every $(\pi, c)\in A(x)$ we obtain relation (52) for $\mu\in D$.

b) By Theorem 2.7, there exists a portfolio process $\pi(t)$, $t\in[0,T]$, corresponding to B and c(t) such that we have $(\pi, c)\in A_\lambda'(x)$ and the corresponding wealth process $X(t)\equiv X_\lambda^{x,\pi,c}(t)$ in M_λ satisfies the initial and final conditions $X(0) = x$ and $X(T) = B$ a.s. in M_λ, respectively. Further, X(t) has the representation

$$\begin{aligned}
H_\lambda(t)\,X(t) + \int_0^t H_\lambda(s)c(s)ds &= E\left(\int_0^T H_\lambda(s)c(s)ds + H_\lambda(T)B \,\middle|\, F_t\right) \\
&= x + \int_0^t H_\lambda(s)X(s)[\sigma(s)'\pi(s) - \theta_\lambda(s)]dW(s) \qquad (55)
\end{aligned}$$

and satisfies the stochastic differential equation (42) in M_λ. It therefore remains to show that $\pi(t)$ only attains values in K $L\otimes P$-a.e. ("feasibility of π"), and that it satisfies the complementarity condition

$$\delta(\lambda(t)) + \pi(t)'\lambda(t) = 0$$

(which will guarantee that B will also be attained in M by following (π, c)). We will prove this via a variational argument which is split into four steps.

Step 1: "Some notations"
For $\nu \in D$ define

$$x(\nu) := E\left(\int_0^T H_\nu(s)c(s)ds + H_\nu(T)B\right)$$

$$L(t) := L^{(\nu)}(t) := \begin{cases} \int_0^t \delta(\nu(s) - \lambda(s))ds, & \text{if } \nu \neq 0 \\ -\int_0^t \delta(\lambda(s))ds & , \text{if } \nu = 0 \end{cases}$$

$$N(t) := N^{(\nu)}(t) := \int_0^t (\sigma^{-1}(s)'(\nu(s) - \lambda(s)))'(dW(s) + \theta_\lambda(s)ds)$$

$$\tau_n := T \wedge \inf\Big\{ t \in [0,T] \,\Big|\, |L(t)| \geq n \textbf{ or } |N(t)| \geq n \textbf{ or } \int_0^t \|\theta_\lambda(s)\|^2 ds \geq n \textbf{ or }$$
$$\int_0^t \left\|\sigma^{-1}(s)'(\nu(s) - \lambda(s))\right\|^2 ds \geq n \textbf{ or }$$
$$\int_0^t (X(s)H_\lambda(s))^2 \left\|\sigma^{-1}(s)'(\nu(s) - \lambda(s)) + (L(s) + N(s))\sigma(s)'\pi(s)\right\|^2 ds \geq n \Big\}$$

Step 2:

$$\lambda_{\varepsilon,n}^{(\nu)}(t) := \lambda(t) + \varepsilon\,[\nu(t) - \lambda(t)]\, 1_{\{t \leq \tau_n\}}$$

For both choices, $\nu \equiv \lambda + \rho$, $\rho \in D$, and $\nu \equiv 0$, we have $\tau_n \to T$ a.s. for $n \to \infty$. By Lemma 22 in sub-section iv) we have

$$\limsup_{\varepsilon \downarrow 0} \frac{1}{\varepsilon}\left(x(\lambda) - x\left(\lambda_{\varepsilon,n}^{(\nu)}\right)\right)$$
$$\leq E\left(H_\lambda(T)B(L^{(\nu)}(\tau_n) + N^{(\nu)}(\tau_n)) + \int_0^T H_\lambda(t)c(t)(L^{(\nu)}(t \wedge \tau_n) + N^{(\nu)}(t \wedge \tau_n))dt \right)$$
$$= E\left(\int_0^{\tau_n} H_\lambda(t)X(t)\left[\pi(t)'(\nu(t) - \lambda(t)dt + dL^{(\nu)}(t)\right]\right). \tag{56}$$

Step 3: "The complementarity condition"
The left hand side of relation (56) is non-negative by assumption (53). With the choice $\nu = \lambda + \rho$ this leads to

$$E\left(\int_0^{\tau_n} H_\lambda(t)X(t)[\pi(t)'\rho(t)+\delta(\rho(t))]dt\right) \geq 0 \quad \forall n\in\mathbf{N} \tag{57}$$

which implies

$$\pi(t)'\rho(t) + \delta(\rho(t)) \geq 0 \quad L\otimes P\text{-a.e.}. \tag{58}$$

Choosing $\nu = 0$ in relation (56), we obtain

$$E\left(\int_0^{\tau_n} H_\lambda(t)X(t)[\pi(t)'\lambda(t)+\delta(\lambda(t))]dt\right) \leq 0 \quad \forall n\in\mathbf{N}. \tag{59}$$

Choosing $\rho = \lambda$ in relation (57) together with relation (59) lead to the complementarity relation

$$\pi(t)'\lambda(t) + \delta(\lambda(t)) \geq 0 \quad L\otimes P\text{-a.e.}.$$

Step 4: "The constraints"
For every $v\in\tilde{K}$, relation (58) yields

$$-\pi(t,\omega)'v \leq \delta(v) \ \ \forall\ (t,\omega)\in A_r \tag{60}$$

for a set $A_r\subset[0, T]\times\Omega$ of full product measure. Define the set A by

$$A := \bigcap_{r\in\tilde{K}\cap Q^n} A_r .$$

The continuity assumption for the support function δ and relation (60) lead to

$$-\pi(t,\omega)'v \leq \delta(v) \ \ \forall\ (t,\omega)\in A,\ v\in\tilde{K} .$$

But under our assumptions on K, the above relation and Theorem 13.1 in (Rockafellar 1970), p. 112, all together imply that π only attains values in K $L\otimes P$-a.s. .

□

Remark 14
Theorem 13 is the analogue to the theorem on completeness of the market model in Chapter 2. The interpretation of H_μ, as the appropriate deflator process in M_μ to judge the payments resulting from investment according to an admissible strategy (π, c) in M_μ, is the same as in Chapter 2. The extra requirement (53) in part b) (compared to the theorem on complete markets) can be interpreted as a sufficient condition for the existence of a replicating portfolio process for B that only attains values in K

(i.e. that is feasible for the constrained problem). However, one has to admit that the explicit determination of the process $\lambda \in D$ satisfying requirement (53) is not easy at all. Also, its existence is not yet guaranteed. We will instead present an existence proof via a dual approach, as is done in (Cvitanic and Karatzas 1992). The use of this dual approach is again motivated by deterministic optimisation (and in some ways similar to Pliska's method given in Section 3.5). The reader not familiar with duality methods in optimisation of functionals is recommended to read Part E of the Appendix where we give a brief description of the deterministic method.

Remark 15

Let U be a utility function satisfying assumption **(A1)**. Then, in analogy to the concave conjugate functional (as defined in Part E of the Appendix), Cvitanic and Karatzas introduce the function $\tilde{U}: (0,\infty) \to \mathbf{R}$ given by

$$\tilde{U}(y) := \sup_{x>0} \left(U(x) - x\,y \right) = U(I(y)) - y\,I(y) \quad \forall\, y > 0$$

(which they refer to as the **Legendre - Fenchel transform** of $-U(-x)$). To see the analogy to the concave conjugate functional $U^*(y)$, note the equality

$$U^*(y) = \inf_{x>0} \left(x\,y - U(x) \right) = -\sup_{x>0} \left(U(x) - x\,y \right) = -\tilde{U}(y) \quad \forall\, y > 0 .$$

For later use, we list the following properties of $\tilde{U}(y)$ (see also (Karatzas, Lehoczky, Shreve, Xu 1991)) :

$$U(x) = \min_{y>0} \left(\tilde{U}(y) + x\,y\right) = \tilde{U}(U'(x)) + x\,U'(x)\,,$$

$\tilde{U}$ is strictly decreasing, strictly convex and differentiable with $\tilde{U}'(y) = -I(y)$,

$$\tilde{U}(U'(x)) + x\,(U'(x) - y) \le \tilde{U}(y) \quad \forall\, x, y > 0,$$

$$\tilde{U}(\infty) := \lim_{y\to\infty} \tilde{U}(y) = U(0+)\,, \qquad \tilde{U}(0+) := \lim_{y\downarrow 0} \tilde{U}(y) = U(\infty).$$

To state Theorem 16, we will need the following additional assumptions on U_i, $\tilde{U}_i$, $i = 1,2$:

Assumption (A4):

The functions $c \to c\,U_1'(t, c)$ and $x \to x\,U_2'(x)$ are non-decreasing on $(0, \infty)$.

Assumption (A5):

There exist $\alpha \in (0, 1)$, $\gamma \in (1, \infty)$ such that we have

$$\alpha\, U_1'(t, x) \geq U_1'(t, \gamma x) \quad \forall\, t \in [0,T]\,,\, x > 0, \qquad \alpha\, U_2'(x) \geq U_2'(\gamma x) \quad \forall\, x > 0.$$

Assumption (A6):

For all $y > 0$ there existents a process $\mu \in D$ with

$$E\left(\int_0^T \tilde{U}_1(t, yH_\mu(t))dt + \tilde{U}_2(yH_\mu(T))\right) < \infty\,.$$

To have an example, where all these assumptions are satisfied, we can look at the choices $U_1(t, x) = U_2(x) = \ln(x)$ or for $U_1(t, x) = U_2(x) = x^\alpha / \alpha$, $\alpha \in (0,1)$. We are now ready to formulate the main theorem of (Cvitanic and Karatzas 1992):

Theorem 16 : "Equivalent Optimality Conditions"

Let $x > 0$ be the initial capital of an investor who follows a strategy $(\hat{\pi}, \hat{c}) \in A'(x)$ with corresponding wealth process $\hat{X}(t)$, $t \in [0,T]$. Let further be $\lambda \in D$ with

$$\mathcal{X}_\lambda(y) < \infty \quad \forall\, y > 0. \tag{61}$$

We also assume that the assumptions **(A 1), (A 4) – (A 6)** are satisfied.
Then, the following conditions (B) – (E) are equivalent and imply condition (A) (i.e. the optimality of the strategy $(\hat{\pi}, \hat{c})$ for the constrained problem).

(A) <u>Optimality of</u> $(\hat{\pi}, \hat{c})$

There exists a pair $(\hat{\pi}, \hat{c}) \in A'(x)$ satisfying

$$J(x\,;\,\pi, c) \leq J(x\,;\,\hat{\pi}, \hat{c}) \quad \forall\, (\pi, c) \in A'(x)\,.$$

(B) <u>Financiability of</u> $(c_\lambda{}^*, B_\lambda{}^*)$

There exists a portfolio process $\pi_\lambda{}^*$ corresponding to the pair $(c_\lambda{}^*, B_\lambda{}^*)$ of equations (44), (45) such that we have $(\pi_\lambda{}^*, c_\lambda{}^*) \in A'(x)$ and

$$\pi_\lambda{}^*(t) \in K \quad \text{for } \mathcal{L}\otimes P\text{-a.e. } (t, \omega) \in [0,T]\times\Omega,$$

$$X(T) = B_\lambda{}^* \quad \text{a.s.}\,,$$

$$\delta(\lambda(t)) + \pi_\lambda{}^*(t)'\,\lambda(t) = 0 \quad \text{for } \mathcal{L}\otimes P\text{-a.e. } (t, \omega) \in [0,T]\times\Omega\,.$$

(C) <u>Minimality of</u> λ

$$E\left(\int_0^T U_1(t, c_\lambda{}^*(t))dt + U_2(B_\lambda{}^*)\right) = V_\lambda(x) \leq V_\mu(x) \;\; \forall \mu \in D \tag{62}$$

(D) <u>Dual optimality of</u> λ

$$E\left(\int_0^T \tilde{U}_1(t, \mathcal{Y}_\lambda(x)H_\lambda(t))dt + \tilde{U}_2(\mathcal{Y}_\lambda(x)H_\lambda(T))\right)$$

$$\leq E\left(\int_0^T \tilde{U}_1(t, \mathcal{Y}_\lambda(x)H_\nu(t))dt + \tilde{U}_2(\mathcal{Y}_\lambda(x)H_\nu(T))\right) \quad \forall \nu \in D \qquad (63)$$

(E) <u>Parsimony of</u> λ

$$E\left(\int_0^T H_\mu(s)c_\lambda{}^*(s)ds + H_\mu(T)B_\lambda{}^*\right) \leq x \quad \forall \mu \in D$$

Proof :
We will prove the theorem by showing the following chain of implications :

$$\textbf{(A)} \Leftarrow \textbf{(B)} \Rightarrow \textbf{(C)} \Rightarrow \textbf{(D)} \Rightarrow \textbf{(B)} \Rightarrow \textbf{(E)} \Rightarrow \textbf{(B)}.$$

"**(A)**⇐ **(B)**" : follows from Proposition 12 with the choice $(\hat{\pi}, \hat{c}) = (\pi_\lambda{}^*, c_\lambda{}^*)$.

"**(B)** ⇒**(C)**" : follows from (the proof and the claims of) Proposition 12, in particular with the help of equations (48) and (51).

"**(B)** ⇒**(E)**" : Due to $(\pi_\lambda{}^*, c_\lambda{}^*) \in A'(x)$ condition **(E)** will be implied by condition **(B)** with the choice $B_\lambda{}^* = X(T)$ a.s. (see equation (52)).

"**(E)** ⇒**(B)**" : follows from Theorem 13 with the choices $c(t) = c_\lambda{}^*(t)\ \forall\, t \in [0,T]$ and $B = \xi_\lambda$.

"**(C)** ⇒**(D)**" : For $\nu \in D$ and $(\pi, c) \in A_\nu'(x)$, $x, y \in (0,\infty)$, $X(t) = X^{x,\pi,c}(t)$ we have

$$U_1(t, c(t)) \leq \tilde{U}_1(t, yH_\nu(t)) + yH_\nu(t)c(t), \qquad (64)$$

$$U_2(X(T)) \leq \tilde{U}_2(yH_\nu(T)) + yH_\nu(T)X(T) \qquad (65)$$

by the definition of $\tilde{U}_i$, $i = 1, 2$. These inequalities imply:

$$E\left(\int_0^T U_1(t, c(t))dt + U_2(X(T))\right)$$

$$\leq E\left(\int_0^T \tilde{U}_1(t, yH_\nu(t))dt + \tilde{U}_2(yH_\nu(T))\right) + yE\left(\int_0^T H_\nu(t)c(t)dt + H_\nu(T)X(T)\right)$$

$$\leq f_\nu(y) := E\left(\int_0^T \tilde{U}_1(t, yH_\nu(t))dt + \tilde{U}_2(yH_\nu(T))\right) + yx.$$

This relation yields $V_\nu(x) \leq f_\nu(y)$ for all positive x, y and by relation (62), we obtain

$$f_\nu(Y_\lambda(x)) \geq V_\nu(x) \geq E\left(\int_0^T U_1(t, c_\lambda{}^*(t))dt + U_2(B_\lambda{}^*)\right)$$

$$= E\left(\int_0^T \tilde{U}_1(t, Y_\lambda(x)H_\lambda(t))dt + \tilde{U}_2(Y_\lambda(x)H_\lambda(T))\right)$$

$$+ Y_\lambda(x)\, E\left(\int_0^T H_\lambda(t) I_1(t, Y_\lambda(x)H_\lambda(t))ds + H_\lambda(T) I_2(Y_\lambda(x)H_\lambda(T))\right)$$

$$= E\left(\int_0^T \tilde{U}_1(t, Y_\lambda(x)H_\lambda(t))dt + \tilde{U}_2(Y_\lambda(x)H_\lambda(T))\right) + Y_\lambda(x)\, x\,.$$

But this equation and the definition of $f_\nu(Y_\lambda(x))$ allow us to conclude

$$E\left(\int_0^T \tilde{U}_1(t, Y_\lambda(x)H_\lambda(t))dt + \tilde{U}_2(Y_\lambda(x)H_\lambda(T))\right) \leq f_\nu(y) - Y_\lambda(x)\, x$$
$$= E\left(\int_0^T \tilde{U}_1(t, Y_\lambda(x)H_\nu(t))dt + \tilde{U}_2(Y_\lambda(x)H_\nu(T))\right).$$

"**(D) ⇒(B)**" : This assertion can be proved in a similar way as part b) of Theorem 13. In deed, setting

$$c(t) = c_\lambda(t) = I_1(t,\ Y_\lambda(x)H_\lambda(t)), \quad B = \xi_\lambda = I_2(Y_\lambda(x)H_\lambda(T)),$$

we can parallel this proof in the following way: If instead of assumption (53) we use the dual optimality relation (63) then as an analogue to Lemma 22 (which essentially represents Step 2 of the proof), we can show

$$\limsup_{\varepsilon \downarrow 0} \frac{1}{\varepsilon} \left\{ E\left(\int_0^T \tilde{U}_1(t, yH_{\lambda_{\varepsilon,n}}(t))dt + \tilde{U}_2\left(yH_{\lambda_{\varepsilon,n}}(T)\right)\right)\right.$$
$$\left. - \ E\left(\int_0^T \tilde{U}_1(t, yH_\lambda(t))dt + \tilde{U}_2(yH_\lambda(T))\right)\right\}$$
$$\leq y\, E\left(BH_\lambda(T)(L(\tau_n) + N(\tau_n)) + \int_0^T H_\lambda(t)c(t)(L(t \wedge \tau_n) + N(t \wedge \tau_n))dt \right)$$

(see Lemma 23 in sub-section v)) for $y > 0$. If we use this relation with the choice $y = Y_\lambda(x)$ then the remaining parts of the proof are identical to Steps 3 and 4 of the proof of part b) of Theorem 13.

□

Remark 17

a) In (Cvitanic and Karatzas 1992) it is also proved that under an additional requirement on the utility functions the implication "**(A)⇒(B)**" is valid, too. As this proof is highly technical and as this implication is the non-interesting one (from the point of the application), we will omit it.

b) As a consequence of relations (64), (65), the expectations in relation (63) are finite.

c) There will be no problem in treating the case of a pure terminal wealth maximisation in the constrained market. One could simply set $U_1 \equiv 0$, $c \equiv 0$, and the whole analysis will go through (meaning both the results in sub-sections i), ii) above and sub-section iii) below are still valid). With the change of the norm in the Hilbert space H to

$$\|v\| = \sqrt{E\left(\int_0^T (T-t)\|v(t)\|^2 dt\right)}$$

and that of the definition of D to

$$D = \left\{ v \in H \middle| E\left(\int_0^T (T-t)\delta(v(t))dt\right) < \infty \right\},$$

the results of sub-sections i) and ii) above also remain valid for the case of $U_2 \equiv 0$, i.e. that of a pure consumtion problem. For the results of sub-section iii) to go through, we need the additional assumptiom that $\tilde{U}_1$ has the form

$$\tilde{U}_1(t, y) = g(t)\, \tilde{U}(y)$$

where $\tilde{U}(y)$ satisfies the assumptions for a utility function made in this section, g(.) is positive and continuous on [0, T] (see Chapter 16 of (Cvitanic and Karatzsas 1992)).

The main difficulty to solve the constrained portfolio problem with the help of one of the optimality conditions **(B)**, **(C)**, or **(E)** of Theorem 16 is the determination of the process $\lambda_.$, i.e. the determination of the appropriate auxiliary market in which the solution of the unconstrained portfolio problem is also a solution to the constrained problem in the original market M. The most promising approach for determining λ is given by condition **(D)** via the introduction of the dual problem to the constrained problem. We will deal with it in the following sub-section.

iii) Solving the Constrained Optimisation Problem with the Help of its Dual Problem

The dual optimisation problem to our original, constrained problem of Definition 7 is already mentioned in part (D) of Theorem 16. However, as $Y_\lambda(x)$ already depends on the optimal λ, it must have be formulated for a fixed, but arbitrary, positive real number y. It is defined in the following way :

Definition 18
The optimisation problem

$$\min_{\mu \in D} \tilde{J}(y;\mu) := \min_{\mu \in D} E\left(\int_0^T \tilde{U}_1(t, yH_\mu(t))dt + \tilde{U}_2(yH_\mu(T)) \right), \quad y > 0, \quad (66)$$

will be called the **dual problem** to the constrained portfolio problem of Definition 7. Its value function $\tilde{V}(y)$ is given by

$$\tilde{V}(y) := \inf_{\mu \in D} \tilde{J}(y;\mu), \quad y > 0 .$$

Remark 19 "The form of the dual problem"
a) In contrast to the dual problem in Section 2, the dual problem (66) is a dynamic problem with respect to time. This can be explained by the fact that in Section 2, we considered constraints "in the mean", i.e. static constraints, while the constraints in the problem of Definition 7 are present at every time instant $t \in [0,T]$ and for (almost) all $\omega \in \Omega$. 15. The dual problem (66) was introduced in (Xu 1990). Note that this problem is again an unconstrained one (apart from the integrability requirements on μ). It is motivated by deterministic analogues in the usual methods for optimising functionals with the help of the Fenchel-Duality-Theorem (see e.g. (Luenberger 1969), p.201 or Theorem E1).
b) To discuss the explicit form of the dual problem (66), we will restrict ourselves to the case of a pure terminal wealth maximisation problem. In this case, by using Remark 15, the objective function of problem (66) is given as

$$E(\tilde{U}_2(yH_\mu(T))) = E(U_2(I(yH_\mu(T)) - yH_\mu(T)I(yH_\mu(T))).$$

The right hand side of this equation has a familiar form. It is equal to the maximum of the function $h(B, y) := L(B, y) + xy$ for all non-negative, F_T-measurable B with $E(H_\mu(T)B) \leq x$ where L(B, y) is the (analogue of the) Lagrangian function defined in Section 3.4. Then, a minimisation over all positive numbers y of h(B, y) would yield the optimal utility for the unconstrained portfolio problem in M_μ. In view of the preceeding sub-section, we could now minimise $E(\tilde{U}_2(yH_\mu(T)))$ in y, obtain the solution of the unconstrained problem in M_μ, and then perform a minimisation over all $\mu \in M_\mu$. However, the main idea behind the formulation of the dual problem (66) is instead to do the minimisation over all $\mu \in M_\mu$ (for a fixed positive positive y) first, hoping that this minimisation and the one in y can be interchanged. Thus, by Theorem 16, we would also obtain the constrained solution. The definite advantage of this method becomes clear when we treat some explicit examples in sub-section 5 ii): under some restrictions on the market coefficients, problem (66) can be solved by conventional stochastic control methods where $H_\mu(t)$ is the controlled process and $\mu(t)$ will be identified as a control strategy. Having solved this problem (for fixed y), we are left with the remaining task of a deterministic minimisation in $y > 0$ (of course, for every y, we have to solve a dual problem (66) to get the value of the objective function of this deterministic problem). Note the similarity of this algorithm to the one presented in Section 2 to solve the problem with terminal wealth constraints. All these considerations will be made rigorous below.

The following Theorem indicates how to solve the constrained problem ("the primal problem") with the help of its dual problem. We will also write this in form of an algorithm in Remark 21 below.

Theorem 20
Let the assumptions **(A 1) – (A 3)** and **(A 5)** be satisfied. Assume further :

$$\inf_{0\le t\le T} U_1(t, 0+) > -\infty, \quad U_2(0+) > -\infty, \quad U_2(\infty) = \infty , \tag{67}$$

$$\forall\, y\in(0, \infty)\ \exists\, \lambda(y)\in D \text{ with } \tilde{V}(y) = \tilde{J}\,(y; \lambda(y)) . \tag{68}$$

Then for all $x\in(0, \infty)$ there exists a number $y(x)\in(0, \infty)$ with

$$V(x) = \inf_{y>0} \left(\tilde{V}(y) + x\, y \right) = \tilde{V}(y(x)) + x\, y(x) , \tag{69}$$

$$x = \mathcal{X}_{\lambda(y(x))}(y(x)), \tag{70}$$

and the strategy $(c_{\lambda(y(x))}, \pi_{\lambda(y(x))})$ (i.e. the optimal unconstrained strategy in the market $M_{\lambda(y(x))}$) is an optimal one for the constrained problem in M.

Proof :
a) For $(\pi,c)\in A'(x)$, x, $y > 0$, the relations (54), (63), (64), and (65) imply

$$J(x; \pi,c) \le \tilde{J}\,(y; \nu) + yE\left(\int_0^T H_\nu(t)c(t)dt + H_\nu(T)X(t)\right) \le \tilde{J}\,(y; \nu) + yx . \tag{71}$$

By relations (54), (64), and (65), we have equality in (71) if and only if we have

$$c(t) = I_1(t, yH_\nu(t)) \;\; \forall\, t \in [0, T], \qquad X(T) = I_2(yH_\nu(T)), \tag{72}$$

$$\delta(\nu(t)) + \pi(t)'\nu(t) = 0 \quad \text{for } L\otimes P\text{-a.e. } (t, \omega) \in [0,T]\times\Omega\,. \tag{73}$$

Specifically, for all positive x, y, relation (71) implies

$$V(x) \le \tilde{V}(y) + xy \;\; . \tag{74}$$

b) For a given positive number y, we set $x = X_{\lambda(y)}(y)$ (with $\lambda(y)$ as in assumption (68)). Then assumption (68) and assertion **(D)** of Theorem 16 (with $\lambda = \lambda(y)$) imply the existence of an optimal pair $(\hat{\pi}, \hat{c})$ for the constrained problem in M with

$$\hat{c}(t) = c_{\lambda(y)}(t), \;\; \hat{\pi}(t) = c_{\lambda(y)}(t) \;\; \forall\, t \in [0, T],$$

such that $(\hat{\pi}, \hat{c})$ and the corresponding wealth process X(t) satisfy relations (72), (73). Hence, we have equality in relation (71) for this strategy (with the choice $\nu = \lambda(y)$) which together with inequality (74) yield

$$\tilde{V}(y) = \tilde{J}(y; \lambda(y)) = J(x; \hat{\pi}, \hat{c}) - xy = \sup_{\xi>0}(\, V(\xi) - y\xi) \;\; \forall\, y > 0. \tag{75}$$

In particular, $\tilde{V}(y)$ is convex as supremum of linear functions (in y).

c) Let x be an arbitrary positive number, $\nu \in D$. By our assumptions on the market coefficients in M_ν and the definition of $H_\nu(t)$, we have that $E(H_\nu(t))$ is uniformly bounded in t by a constant C (of course, we only need an upper bound as $H_\nu(t)$ is strictly positive). By using the fact that the functions $\tilde{U}_1(t,.), \tilde{U}_2(.)$ are decreasing and convex, an application of Jensen's inequality yields

$$\begin{aligned} \tilde{J}(y; \nu) &\ge \int_0^T \tilde{U}_1(t, yE(H_\nu(t)))dt + \tilde{U}_2(yE(H_\nu(T))) \\ &\ge \int_0^T \tilde{U}_1(t, yC)dt + \tilde{U}_2(yC)\,. \end{aligned}$$

But due to assumption (67) and relations (64), (65), for $y \downarrow 0$ the right hand side of the above inequality tends to infinity. In particular, we have thus shown $\tilde{V}(0+) = \infty$. Further, assumption (67) also implies $V(\infty) = \infty$. Using these considerations in connection with relation (75) demonstrates that the convex function $f_x(y) := \tilde{V}(y) + xy$, $y > 0$, satisfies $f_x(0+) = f_x(\infty) = \infty$. Consequently, $f_x(y)$ attains its infimum at some positive number y(x). Assumption (68) with y = y(x) yields

$$\inf_{\xi>0} \{\xi y(x)x + \tilde{J}(\xi y(x); \lambda(y(x)))\} = \inf_{y>0} \{yx + \tilde{J}(y; \lambda(y(x)))\} \geq \inf_{y>0} \{yx + \tilde{V}(y)\}$$
$$= f_x(y(x)) = xy(x) + \tilde{V}(y). \tag{76}$$

With the help of the dominated convergence theorem, the identities

$$\tilde{U}_1'(t,y) = -I_1(t,y), \quad \tilde{U}_2'(y) = -I_2(y),$$

and a similar argument as given in the proof of Theorem 3.23, we see that the function

$$G_y(\xi) := \tilde{J}(\xi y; \lambda(y)), \quad 0 < \xi < \infty,$$

is well defined, finite and continuously differentiable at $\xi = 1$ with

$$G_y'(1) = -yX_{\lambda(y)}(y).$$

On the other hand, relation (76) implies that the function

$$D_x(\xi) := \xi y(x)x + G_{y(x)}(\xi), \quad 0 < \xi < \infty,$$

attains its infimum at $\xi = 1$. Hence, the derivative of $D_x(\xi)$ with respect to ξ must vanish at $\xi = 1$ (of course, due to the preceding considerations it also exists !) which yields assertion (70).

d) Finally, choosing $y = y(x)$ in equation (75) yields

$$\tilde{V}(y(x)) + xy(x) = J(x; \hat{\pi}, \hat{c})$$

which together with inequality (74) implies equality (69).

□

Remark 21

a) Assumption (68) is a very strong one. It states that for every positive number y there exists a process $\lambda(y)$ solving the dual optimisation problem (66). By Theorem 13.1 in (Cvitanic and Karatzas 1992), assumption **(A 4)** implies requirement (68). Thus, existence of the solution of the "primal problem" is proved via existence of the solution of its dual. We will not give this result here as in our examples given in Section 5 below, we can directly verify that assumption (68) is satisfied.

b) The main implication of Theorem 20 is that it indicates the following algorithm to compute the optimal strategy $(\hat{\pi}, \hat{c}) \in \hat{A}'(x)$ of the primal optimisation problem with the help of the dual one (note that by the proof of Theorem 20, the infimum in step 2 will always be attained):

Algorithm:"Solving the constrained poblem by the dual approach"

1. Solve the dual optimisation problem (66) in dependence of $y > 0$.
2. Minimise the function
$$f(y) := \tilde{V}(y) + x\,y$$
in $y > 0$, where x represents the fixed initial capital of the investor .
3. Let $\bar{y}$ yield the minimum of f(y). Then find a process $\lambda(\bar{y}) \in D$ with
$$\tilde{V}(\bar{y}) = \tilde{J}(\bar{y};\lambda(\bar{y})) .$$
4. Solve the unconstrained portfolio problem in $\mathbf{M}_{\lambda(\bar{y})}$ (by the method of Section 3.4). The optimal strategy for this problem then coincides with the optimal strategy $(\hat{\pi},\hat{c})$ of the constrained problem in $\mathbf{M}$.

iv) Some Extensions and Comments

The above methodology can be extended to cover situations in which the constraints are given by random processes. This would correspond to a random set K(t) describing the constraints. The interested reader is recommended to read the appropriate sections of (Cvitanic and Karatzas 1992). This paper also contains the interesting case of (unconstrained) portfolio optimisation with different interest rates for borrowing and lending. The then obvious extension to the case of constrained portfolio optimisation in the presence of different interest rates is given in (Korn 1992). Further examples of the treatment of problems with constraints on the portfolio process (but using different methods) are given in (Cox and Huang 1991), (Fleming and Zariphopoulou 1991), (He and Pearson 1991), (Karatzas, Lehoczky, Shreve and Xu 1991).

v) Two Technical Lemmas

Lemma 22
With the notations introduced in the proof of Theorem 13, we have

$$\limsup_{\varepsilon\downarrow 0}\frac{1}{\varepsilon}\left(x(\lambda)-x\left(\lambda_{\varepsilon,n}^{(\nu)}\right)\right)$$

$$\leq E\left(H_\lambda(T)B(L^{(\nu)}(\tau_n)+N^{(\nu)}(\tau_n))+\int_0^T H_\lambda(t)c(t)(L^{(\nu)}(t\wedge\tau_n)+N^{(\nu)}(t\wedge\tau_n))dt\right)$$

$$=E\left(\int_0^{\tau_n} H_\lambda(t)X(t)\left[\pi(t)'(\nu(t)-\lambda(t)dt+dL^{(\nu)}(t)\right]\right). \tag{56}$$

for both the choices $\nu\equiv\lambda+\rho$, $\rho\in D$, and $\nu\equiv 0$.

Proof:
a) By the definition of the support function, for $\nu\equiv 0$ we have

$$\delta((1-\varepsilon)\lambda(t))-\delta(\lambda(t))=-\varepsilon\delta(\lambda(t)),$$

whereas for $\nu\equiv\lambda+\rho$ we obtain

$$\delta(\lambda(t)+\varepsilon(\nu(t)-\lambda(t)))-\delta(\lambda(t))\leq\varepsilon\delta(\nu(t)-\lambda(t)).$$

In both these cases, we have

$$\frac{H_{\lambda_{\varepsilon,n}}(t)}{H_\lambda(t)}=\exp\left(-\int_0^{t\wedge\tau_n}\left(\delta\left(\lambda(s)+\varepsilon(\nu(s)-\lambda(s))\right)-\delta(\lambda(s))\right)ds\right.$$
$$\left.-\varepsilon N_{t\wedge\tau_n}-\frac{\varepsilon^2}{2}\int_0^{t\wedge\tau_n}\left\|\sigma^{-1}(s)(\nu(s)-\lambda(s))\right\|^2ds\right)$$
$$\geq\exp\left(-\varepsilon\left(L_{t\wedge\tau_n}+N_{t\wedge\tau_n}\right)-\frac{\varepsilon^2}{2}\int_0^{t\wedge\tau_n}\left\|\sigma^{-1}(s)(\nu(s)-\lambda(s))\right\|^2ds\right)$$
$$\geq e^{-3\varepsilon n}. \tag{76}$$

by the construction of τ_n and the definition of L_t. This specifically implies

$$\limsup_{\varepsilon\downarrow 0}\frac{1}{\varepsilon}\left\{1-\frac{H_{\lambda_{\varepsilon,n}}(t)}{H_\lambda(t)}\right\}\leq L_{t\wedge\tau_n}+N_{t\wedge\tau_n}. \tag{77}$$

We further have

$$\frac{1}{\varepsilon}\left(x(\lambda)-x\left(\lambda_{\varepsilon,n}^{(\nu)}\right)\right)=E\left(Q_n^\varepsilon\right)$$

where Q_n^{ε} is given by

$$Q_n^{\varepsilon} := \frac{1}{\varepsilon}\left[H_\lambda(T)B\left(1-\frac{H_{\lambda_{\varepsilon,n}}(T)}{H_\lambda(T)}\right)+\int_0^T H_\lambda(t)c(t)\left(1-\frac{H_{\lambda_{\varepsilon,n}}(t)}{H_\lambda(t)}\right)dt\right].$$

But the set $\{Q_n^{\varepsilon},\ 0<\varepsilon<1\}$ is dominated by the random variable Q_n with

$$Q_n := K_n\left[H_\lambda(T)B+\int_0^T H_\lambda(t)c(t)dt\right],$$

$$K_n := \sup_{0<\varepsilon<1}\frac{1-e^{-3\varepsilon n}}{\varepsilon}.$$

By noting that we have $E(Q_n) = K_n\, x(\lambda) < \infty$, application of Fatou's lemma in the following relation together with inequality (77) yields

$$\limsup_{\varepsilon\downarrow 0}\frac{1}{\varepsilon}\left(x(\lambda)-x\left(\lambda_{\varepsilon,n}^{(\nu)}\right)\right) \le E\left(\limsup_{\varepsilon\downarrow 0} Q_n^{\varepsilon}\right)$$

$$\le E\left(H_\lambda(T)B(L(\tau_n)+N(\tau_n)) + \int_0^T H_\lambda(t)c(t)(L(t\wedge\tau_n)+N(t\wedge\tau_n))dt\right).$$

b) It remains to prove the equality in relation (56). Using Itô's formula we get (recall equation (55) and the definition of L(t), N(t) !) :

$$\begin{aligned}
&d(\, H_\lambda(t)X(t)[L(t)+N(t)]\,) \\
&\quad = H_\lambda(t)X(t)d(L(t)+N(t)) + (L(t)+N(t))d(H_\lambda(t)X(t)) + d\langle H_\lambda X, N\rangle_t \\
&\quad = H_\lambda(t)X(t)d(L(t)+N(t)) + (L(t)+N(t))\{H_\lambda(t)X(t)[\sigma(t)'\pi(t)-\theta_\lambda(t)]dW(t) \\
&\qquad - H_\lambda(t)c(t)dt\} + H_\lambda(t)X(t)[(\sigma(t)'\pi(t)-\theta_\lambda(t))'\sigma^{-1}(t)(\nu(t)-\lambda(t))]dt \\
&\quad = H_\lambda(t)X(t)[dL(t) + \pi(t)'(\nu(t)-\lambda(t))]dt + (L(t)+N(t))[\sigma(t)'\pi(t)-\theta_\lambda(t)]dW(t) \\
&\qquad\qquad - (L(t)+N(t))H_\lambda(t)c(t)dt
\end{aligned} \tag{78}$$

By definition of τ_n the appropriate condition is satisfied such that the expectation of the stochastic integral over $[0, \tau_n]$ in (78) vanishes. This implies

$$E\left(H_\lambda(\tau_n)X(\tau_n)(L(\tau_n)+N(\tau_n)) + \int_0^{\tau_n} H_\lambda(t)c(t)(L(t)+N(t))dt \right)$$

$$= E\left(\int_0^{\tau_n} H_\lambda(t)X(t)\left[\pi(t)'(v(t)-\lambda(t)dt + dL^{(v)}(t)\right]\right). \tag{79}$$

By construction of $X(\tau_n)$, we have got the representation

$$H_\lambda(\tau_n)X(\tau_n) = E\left(H_\lambda(T)B + \int_{\tau_n}^T H_\lambda(t)c(t)dt \middle| f_{\tau_n} \right).$$

Substituting this into the left hand side of equation (79) completes the proof.

□

Lemma 23
Under the assumptions of Theorem 16 we have

$$\limsup_{\varepsilon \downarrow 0} \frac{1}{\varepsilon}\left\{ E\left(\int_0^T \tilde{U}_1(t, yH_{\lambda_{\varepsilon,n}}(t))dt + \tilde{U}_2(yH_{\lambda_{\varepsilon,n}}(T)) \right.\right.$$

$$\left.\left. - \int_0^T \tilde{U}_1(t, yH_\lambda(t))dt + \tilde{U}_2,(yH_\lambda(T)) \right)\right\}$$

$$\leq y\, E\left(BH_\lambda(T)(L(\tau_n)+N(\tau_n)) + \int_0^T H_\lambda(t)c(t)(L(t\wedge\tau_n)+N(t\wedge\tau_n))dt \right)$$

for $y > 0$ (For the notations see the proof of Theorem 16).

Proof :
We first claim that

$$Y_\varepsilon^{(n)} := \frac{1}{\varepsilon}\left(\int_0^T \tilde{U}_1(t, yH_{\lambda_{\varepsilon,n}}(t))dt + \tilde{U}_2\left(H_{\lambda_{\varepsilon,n}}(T)\right) \right.$$

$$\left. - \int_0^T \tilde{U}_1(t, yH_\lambda(t))dt + \tilde{U}_2(yH_\lambda(T)) \right)$$

is bounded from above by the random variable

$$Y^{(n)} := yK_n\left(\int_0^T H_\lambda(t)I_1(t, ye^{-3n}H_\lambda(t))dt + H_\lambda(T)I_2(ye^{-3n}H_\lambda(T)) \right)$$

with

$$K_n := \sup_{0<\varepsilon<1} \frac{1-e^{-3\varepsilon n}}{\varepsilon}.$$

To see this, we have to use inequality (76), the convexity of the functions $\tilde{U}_1(t,.)$, $\tilde{U}_2(.)$, the relation $\tilde{U}'(y) = -I(y)$ (see Remark 12), and the fact that $I_1(t,.)$, $I_2(.)$ are decreasing functions. Due to assumption (61), $E(Y^{(n)})$ is finite, and therefore, Fatou's lemma implies

$$\limsup_{\varepsilon \downarrow 0} E(Y_\varepsilon^{(n)}) \le E(\limsup_{\varepsilon \downarrow 0} Y_\varepsilon^{(n)}).$$

One can also check (in a similar way as above, but using a different part of inequality (76)) that $Y_\varepsilon^{(n)}$ satisfy the following (a.s.) inequality

$$Y_\varepsilon^{(n)} \le V_\varepsilon^{(n)}$$

$$=: y\left(\int_0^T H_\lambda(t) I_1(t, ye^{-3\varepsilon n} H_\lambda(t)) \Lambda_\varepsilon^{(n)}(t)dt + H_\lambda(T) I_2(ye^{-3\varepsilon n} H_\lambda(T)) \Lambda_\varepsilon^{(n)}(T)\right)$$

with the notation

$$\Lambda_\varepsilon^{(n)}(t) := \frac{1}{\varepsilon}\left(1 - \exp\left(-\varepsilon\left(L_{t\wedge\tau_n} + N_{t\wedge\tau_n}\right) - \frac{\varepsilon^2}{2}\int_0^{t\wedge\tau_n} \left\|\sigma^{-1}(s)(\nu(s) - \lambda(s))\right\|^2 ds\right)\right).$$

Now for $\varepsilon \downarrow 0$, we will have $\Lambda_\varepsilon^{(n)}(t) \to L_{t\wedge\tau_n} + N_{t\wedge\tau_n}$ a.s. and thus

$$V_\varepsilon^{(n)} \xrightarrow{\varepsilon \downarrow 0} y\left(\int_0^T H_\lambda(t) I_1(t, yH_\lambda(t))(L_{t\wedge\tau_n} + N_{t\wedge\tau_n})dt \right.$$
$$\left. + H_\lambda(T) I_2(yH_\lambda(T)) L_{\tau_n} + N_{\tau_n}\right), \text{ a.s.}\,.$$

Putting all these inequalities together and using the definition of B, we obtain

$$\limsup_{\varepsilon \downarrow 0} E\left(Y_\varepsilon^{(n)}\right) \le E\left(\lim_{\varepsilon \downarrow 0} V_\varepsilon^{(n)}\right) =$$

$$= y\, E\left(BH_\lambda(T)(L(\tau_n) + N(\tau_n)) + \int_0^T H_\lambda(t)c(t)(L(t\wedge\tau_n) + N(t\wedge\tau_n))dt\right)$$

which is the desired result.

□

4.5 Some Examples of Constrained Problems

In this section, we present some explicit examples of portfolio problems with constraints on the portfolio process $\pi(t)$ that can be solved by the method developed in Section 4. We will concentrate on the popular examples of the logarithm and the power functions.

i) General Constraints and Logarithmic Utility

In the case of the logarithmic utility function,

$$U_1(t, x) = U_2(x) = \ln(x), \quad \forall\, t\in[0,T]\,,\, x>0\,,$$

we can explicitly calculate all the necessary ingredients to solve a constrained problem:

$$I_1(t, x) = I_2(x) = 1/x \quad \forall\, t\in[0,T]\,,\, x>0\,,$$

$$\tilde{U}_1(t, y) = \tilde{U}_2(y) - (1+\ln(y)) \quad \forall\, t\in[0,T]\,,\, y>0\,,$$

$$\mathcal{X}_\mu(y) = E\left(\int_0^t (1/t)dt + 1/y\right) = \frac{T+1}{y} = \mathcal{Y}_\mu(y) \quad \forall\, y>0\ .$$

Note that $\mathcal{X}_\mu(y)$ and $\mathcal{Y}_\mu(y)$ are independent of $\mu\in D$. The optimal consumption and the optimal terminal wealth of an investor in the market M_μ for $\mu\in D$ are thus given by (recall the representations (44) and (45))

$$c_\mu(t) = I_1(t,\mathcal{Y}_\mu(x)H_\mu(t)) = \frac{x}{T+1}\frac{1}{H_\mu(t)} \qquad \forall\, t\in[0,T],$$

$$B_\mu = I_2(\mathcal{Y}_\mu(x)H_\mu(T)) = \frac{x}{T+1}\frac{1}{H_\mu(T)}\ .$$

Due to the independence of $\mathcal{Y}_\mu(y)$ of $\mu\in D$, we obtain for $\lambda,\ \mu\in D$:

$$\tilde{J}\,(\mathcal{Y}_\lambda(x);\,\mu) = E\left(\int_0^T \tilde{U}_1(t,\mathcal{Y}_\lambda(x)H_\mu(t))dt + \tilde{U}_2(\mathcal{Y}_\lambda(x)H_\mu(T))\right)$$

$$= -E\left(\int_0^T \left(1+\ln\left(\tfrac{T+1}{x}H_\mu(t)\right)\right)dt + 1 + \ln\left(\tfrac{T+1}{x}H_\mu(T)\right)\right).$$

But due to

$$E\left(\ln(1/H_\mu(t))\right) = E\left(\int_0^t (r(s)+\delta(\mu(s))+\tfrac{1}{2}\|\theta_\mu(s)\|^2)ds\right)$$

(and the finiteness of this expectation for $\mu \in D$), $\tilde{J}(Y_\lambda(x); \mu)$ will be minimal for $\lambda(t)$ which is determined by **pointwise minimisation** of the function

$$\delta(u) + \tfrac{1}{2}\| \theta(t) + \sigma^{-1}(t)u \|^2, \qquad u \in \tilde{K} ,$$

i.e.

$$\lambda(t) = \arg\min_{u \in \tilde{K}} \left(\delta(u) + \tfrac{1}{2}\| \theta(t) + \sigma^{-1}(t)\, u \|^2\right) \tag{80}$$

Due to a measurable selection theorem such as in (Schäl 1974), $\lambda(t)$, $t \in [0, T]$, can be chosen F_t-adapted. By construction of $\lambda(t)$, $\delta(0) = 0$ and $0 \in \tilde{K}$, we have

$$E\left(\int_0^T\left(\delta(\lambda(t)) + \tfrac{1}{2}\left\|\theta(t) + \sigma^{-1}(t)\lambda(t)\right\|^2\right)dt\right) \leq \tfrac{1}{2}E\left(\int_0^T \|\theta(t)\|^2\, dt\right) < \infty .$$

Because $\delta(.)$ is bounded from below, this inequality also implies

$$E\left(\int_0^T \delta(\lambda(t))dt\right) < \infty .$$

If we further have $\lambda(t) \in D$ (which must be verified for every particular form of K) then due to Theorem 16 and the above considerations, $\lambda(t)$, as in equation (80), yields the optimal value of the dual problem. In the case of the logarithmic utility function, we can even compute the corresponding optimal portfolio process $\pi(t)$ explicitly: As in Example 3.19 we have

$$\begin{aligned} H_\lambda(t)\, X_\lambda(t) &= E\left(\int_t^T H_\lambda(s)c_\lambda(s)ds + H_\lambda(T)B_\lambda \middle| F_t\right) \\ &= \frac{x}{T+1}\,(T - t + 1) = x\left(1 - \frac{t}{T+1}\right) \end{aligned}$$

and

$$\begin{aligned} x = x\left(1 - \frac{t}{T+1}\right) + x\frac{t}{T+1} &= H_\lambda(t)\, X_\lambda(t) + \int_t^T (H_\lambda(t)c_\lambda(s))ds \\ &= x + \int_0^t H_\lambda(s)X_\lambda(s)[\pi_\lambda{}^*(s)'\sigma(s) - \theta_\lambda(s)']ds . \end{aligned}$$

As $H_\lambda(s)X_\lambda(s)$ is strictly positive for all $s > 0$, we must have

$$\sigma(t)'\pi_\lambda(t) = \theta_\lambda(t) \quad \forall\, t \in [0,T] \text{ a.s.} ,$$

hence,

$$\pi_\lambda(t) = (\sigma(t)\sigma(t)')^{-1}(\lambda(t) + b(t) - r(t))) \quad \forall\, t \in [0,T] \text{ a.s.} \,. \tag{81}$$

Of course, the optimal consumption is given as

$$c_\lambda(t) = \frac{x}{T+1}\frac{1}{H_\lambda(t)} \qquad \forall\, t\in[0,T] \text{ a.s.} \,.$$

Having completed these preparations, we are ready to consider some specific examples of constrained problems:

Example 24 "No short selling of stocks"
In this case, we have $K = \tilde{K} = [0, \infty)^n$ (see Example 8 in Section 4) and the minimum in the optimisation problem (80) will be found by simply minimising

$$\| \theta(t) + \sigma^{-1}(t)x \|^2$$

in x over $[0, \infty)^n$ which can easily be achieved by solving a quadratic programming problem at every time instant $t\in[0,T]$. Particularly, in the case of $n = 1$, we obtain

$$\lambda(t) = \left(r(t) - b(t)\right)^- \quad \forall\, t\in[0,T] \text{ a.s.},$$

$$\pi(t) = \frac{\left(b(t) - r(t)\right)^+}{\sigma(t)^2} \quad \forall\, t\in[0,T] \text{ a.s.},$$

i.e. if the mean rate of stock return b(t) exceeds the riskless rate r(t) then the investor follows the unconstrained optimal portfolio process (see Example 3.19). Otherwise, he will invest his total money in the bond.

Example 25 "Bounded fraction of wealth in risky security, no short selling of stocks"
We now look at the situation of Example 9 of Section 4 (including short selling constraints) and restrict ourselves to the case of $n = 1$ (the general case is notationally more involved but in principle the same). Then, we have

$$K = [0, z], \quad \tilde{K} = \mathbf{R}, \quad \delta(x) = z\,x^- \quad \forall\, x\in R\,.$$

Hence, the relations (80) and (81) yield (we assume $\sigma(t) > 0$) :

$$\lambda(t) = \begin{cases} \sigma(t)(\sigma(t)z - \theta(t)), & \text{if } \sigma(t)z < \theta(t) \\ r(t) - b(t) & \text{, if } \theta(t) < 0 \\ 0 & \text{, else} \end{cases}$$

$$\pi(t) = \begin{cases} z & \text{, if } \sigma(t)z < \theta(t) \\ 0 & \text{, if } \theta(t) < 0 \\ (b(t) - r(t)) / \sigma(t)^2 & \text{, else} \end{cases}$$

i.e. the upper bound on the fraction of wealth invested in the risky asset is only binding if the mean rate of stock return clearly exceeds the riskless rate, more precisely, if we have $r(t) + z\sigma(t)^2 < b(t)$. In the other two cases in the above representations of the optimal strategy, the comments of Example 24 apply, too. It is a recommended exercise to carry out the calculations for the case $n = 2$ or for the even more general case of the presence of an additional lower bound $v \leq 0$ for the fraction of wealth invested in the risky asset (see Section 14 of (Cvitanic and Karatzas 1992)).

Example 26 "Incomplete markets"
In this example (see (Karatzas, Lehoczky, Shreve and Xu 1991)), the investor is constrained to invest in the "first" m of n stocks, i.e. we have

$$K = \left\{\pi \in R^n \middle| \pi_i = 0, i = m+1,..,n\right\}$$

and therefore

$$\tilde{K} = \left\{x \in R^n \middle| \pi_i = 0, i = 1,..,m\right\},$$

$$\delta(x) = \begin{cases} 0, & x \in \tilde{K} \\ \infty, & x \notin \tilde{K} \end{cases}.$$

For convenience, we will assume that $\sigma(t)$ is a diagonal matrix and denote by $\sigma_1(t)$, $\sigma_2(t)$ the first m and the last n–m rows of $\sigma(t)$, respectively. We further set $B_1(t) := (b_1(t), .., b_m(t))'$, $B_2(t) := (b_{m+1}(t), .., b_n(t))'$. Then relations (80), (81) lead to

$$\lambda(t) = \begin{pmatrix} 0 \\ r(t)\underline{1} - B_2(t) \end{pmatrix},$$

$$\pi(t) = \begin{pmatrix} (\sigma_1(t)\sigma_1(t)')^{-1}(B_1(t) - r(t)\underline{1}) \\ 0 \end{pmatrix}$$

(where the 0- and $\underline{1}$-vectors are always assumed to be of appropriate dimension). We can obtain a similar result for a general matrix $\sigma(t)$ with the help of a QR-factorisation of $\sigma(t)$.

ii) Constant Coefficients and Power Utility

The foregoing example of logarithmic utility is of course an exceptional one. One cannot assume to acquire such explicit results for general utility functions. In contrast to this situation, we will now consider an example where we actually have to solve the constrained problem in M via solving the corresponding dual problem introduced in Sub-section iii) of Section 4. For simplicity, we will assume that the market coefficients r(t), b(t), $\sigma(t)$ are constant (see Section 15 of (Cvitanic and Karatzas 1992) for the more general case of deterministic market coefficients). Under this assumption we can regard the corresponding dual problem

$$\min_{\mu \in D} \tilde{J}(y;\mu) := \min_{\mu \in D} E\left(\int_0^T \tilde{U}_1(t, yH_\mu(t))dt + \tilde{U}_2(yH_\mu(T))\right), \quad y>0 \qquad \textbf{(DP)}$$

as a stochastic control problem (see part C of the **Appendix** and Section 3.3 for more information and results on stochastic control problems). The crucial point for solving (DP) is to choose the process $yH_\mu(t)$, $t\in[0,T]$, as the controlled process. By noting that we have the following representation for $yH_\mu(t)$

$$d(yH_\mu(t)) = -yH_\mu(t)\,[\,(r+\delta(\mu(t)))\,dt + (\theta+\mu(t))'dW(t)\,],$$

we can set up the corresponding HJB-Equation for **(DP)**:

$$\inf_{x\in\tilde{K}} \left\{ \tfrac{1}{2} y^2 Q_{yy}(t,y) \left\| \theta + \sigma^{-1}x \right\|^2 - y\delta(x)Q_y(t,y) \right\}$$

$$+\, Q_t(t,y) - r\,y\,Q_y(t,y) + \tilde{U}_1(t,y) = 0 \quad \forall\,(t,y)\in[0,T)\times(0,\infty), \qquad (82)$$

$$Q(T,y) = \tilde{U}_2(y) \quad \forall\, y\in(0,\infty). \qquad (83)$$

If there exists a solution $Q\in C^{1,2}([0,T)\times(0,\infty))$ to the equations (82) and (83) that satisfies a polynomial growth condition in y and a (process) λ that attains the infimum in (83) then we obtain the value function of the dual problem via

$$\tilde{V}(y) = Q(0,y),$$

and we get $\lambda(t)$ (the process that determines the "correct" auxiliary market) as

$$\lambda(t) = \arg \inf_{x \in \tilde{K}} \left\{ \tfrac{1}{2}(yH_\lambda(t))^2 Q_{yy}(t, yH_\lambda(t)) \left\| \theta + \sigma^{-1}x \right\|^2 - yH_\lambda(t)\delta(x) Q_y(t, yH_\lambda(t)) \right\} \quad (84)$$

for $y > 0$, $t \in [0,T]$ (if the infimum will be attained). It is important to note that the process $\lambda(s)$ for $0 \le s < t$ is already known at time t, and the actual value of $H_\lambda(t)$ is independent of $\lambda(t)$. Thus, the above representation (84) for $\lambda(t)$ is really an explicit one. We will now further restrict ourselves by assuming that we use the power utility functions of the form

$$U_1(t, x) = U_2(x) = x^\alpha / \alpha, \qquad (\alpha \in (0,1) \text{ fixed}).$$

Noting that we have $\tilde{U}_1(t, y) = \tilde{U}_2(y) = y^{-\rho} / \rho$, $0 < y < \infty$, with $\rho := \alpha / (1-\alpha)$, we can explicitly calculate the solution of the HJB-equation (82), (83) as

$$Q(t, y) = \frac{1}{\rho} y^{-\rho} v(t), \quad (85)$$

$$v(t) = \frac{1}{h}\left((1+h)e^{h(T-t)} - h\right), \quad (86)$$

$$h = \rho \inf_{x \in \tilde{K}} \left\{ \frac{1+\rho}{2} \left\| \theta + \sigma^{-1}x \right\|^2 + \delta(x) \right\} + r\rho . \quad (87)$$

To obtain this solution, we have first inserted the conjectured form (85) for $Q(t, y)$ into equation (82), solved the minimisation problem in (82) (which was then independent of y !), and finally, we transformed the resulting partial differential equation into the ordinary differential equation $v'(t) + hv(t) + 1 = 0$, $v(T) = 1$ for $v(t)$. This last equation can easily be solved, and we can then verify that $Q(t, y)$ given by (85) is indeed a solution to the HJB-Equation satisfying all required smoothness and growth conditions. Furthermore, from equations (84)–(87) we obtain $\lambda(t)$ as

$$\lambda(t) \equiv \lambda = \arg \inf_{x \in \tilde{K}} \left\{ \left\| \theta + \sigma^{-1}x \right\|^2 + 2(1-\alpha)\delta(x) \right\} \quad (88)$$

which — as a consequence of the constant coefficient assumption — is constant over time and independent of y. Hence, for different forms of the constraints, i.e. for different forms of $\tilde{K}$, we only have to compute λ via equation (88), and then we can solve the unconstrained problem in M_λ which yields the optimal pair (π_λ, c_λ) by Theorem 16. We now look at the examples already studied in the logarithmic case:

Example 27 "No short selling of stocks"
In the case of n = 1 we obtain (as in the logarithmic case)

$$\lambda(t) = (r-b)^- \quad \forall\, t\in[0,T] \text{ a.s.}.$$

Solving the unconstrained problem in M_λ as in Example 3.19 yields

$$\pi(t) = \frac{1}{\sigma^2(1-\alpha)}(b-r)^+ \quad \forall\, t\in[0,T] \text{ a.s. }.$$

As in the logarithmic case, if the mean rate of stock return b exceeds the riskless rate r then the investor follows the unconstrained optimal portfolio process. Otherwise a pure bond strategy is optimal.

Example 28 "Bounded fraction of wealth in risky security, no short selling of stocks"
In this setting, equation (88) yields

$$\lambda(t) = \begin{cases} \sigma(\sigma z - \theta) & \text{, if } \sigma z < \theta \\ r-b & \text{, if } \theta < 0 \\ 0 & \text{, else} \end{cases},$$

$$\pi(t) = \begin{cases} z & \text{, if } \sigma z < \theta \\ 0 & \text{, if } \theta < 0 \\ (b-r)/(\sigma^2(1-\alpha)) & \text{, else} \end{cases},$$

i.e. the upper bound on the fraction of wealth invested in the risky asset is only binding if we have $r + z\sigma^2 < b$. The comments of the logarithmic case applies here, too. The only difference is caused by the different optimal unconstrained portfolio.

Example 29 "Incomplete markets"
With the same notation and assumptions as in the logarithmic case, we obtain

$$\lambda(t) = \begin{pmatrix} 0 \\ r\underline{1} - B_2 \end{pmatrix},$$

$$\pi(t) = \begin{pmatrix} (\sigma_1\sigma_1')^{-1}(B_1 - r\underline{1}) \\ 0 \end{pmatrix}$$

(where the 0- and $\underline{1}$-vectors are always assumed to be of appropriate dimension).

iii) Further Examples

Further examples treated with the same methodology, such as problems with different riskless rates for borrowing and lending, problems including general utility functions and constant coefficients, random upper and lower bounds on the fraction of wealth hold in the risky assets can be found in (Cvitanic and Karatzas 1992), (Karatzas, Lehoczky, Shreve and Xu 1991), (Xu and Shreve 1992 a, b).

CHAPTER 5

PORTFOLIO OPTIMISATION IN THE PRESENCE OF TRANSACTION COSTS

So far, we have seen a lot of nice and sophisticated mathematics in action, thereby yielding portfolio and consumption strategies that often have a surprisingly simple structure. Even more, to solve the continuous-time portfolio problem, we do essentially not need more information on (parameters of) the future securities prices as for the simple one-period Markowitz mean-variance approach, namely, (functions of) the first and second moments of the price processes. However, we have often remarked that even a constant optimal portfolio process does not mean that there is no trading necessary. In deed, it means that we have to do the opposite extreme, trading at every time instant. And even more, the optimal trading strategies have infinite variation in every time interval! Hence, following such a strategy in the presence of (any reasonable form of) transaction costs would certainly lead to the ruin of the trader. But as trading is the central action in the portfolio problem, the impact of transaction costs cannot be neglected. We will therefore look at three different approaches to the portfolio problem under transaction costs. They will not only differ by the ways the problems are solved, but also by the form of the optimisation problems and by the form of the transaction cost. However, they all have one common feature, the trade-off between gaining a better position by rebalancing the position on one hand and the occurrence of additional transaction costs as a consequence of such an action on the other hand. The principle to decide on trading or not can be more or less formulated as "trade only if the gain from trading pays the transaction costs".

5.1 Optimal Life-Time Consumption with Proportional Transaction Costs

As a first example of a portfolio selection problem in the presence of transaction costs, we will take up the optimum life-time consumption problem of Section 3.3. We will present its treatment in the presence of proportional transaction costs as it is was first given in (Davis and Norman 1990). The interested reader will also be referred to a generalisation of the Davis and Norman problem by Soner and Shreve who use viscosity solution techniques instead of the more classic stochastic control approach of Davis and Norman (see (Soner and Shreve 1994)).

Before we start with the presentation of the solution in the presence of transaction costs let us recall its solution in the case without transaction costs. We consider a simple financial market consisting of a bond and a single stock with price dynamics given by

$$dP_0(t) = P_1(t)\, r\, dt, \quad P_0(0) = 1,$$

$$dP_1(t) = P_1(t)(bdt + \sigma dW(t)), \quad P_1(0) = p,$$

where r, b and σ are real constants with $\sigma > 0$ and where $\{W(t), F_t, t \geq 0\}$ is a one-dimensional, standard Brownian motion with its natural filtration. The optimal strategy for the maximum life-time consumption problem

$$\max_{(\pi,c)} E\left(\int_0^\infty e^{-\delta t}(c(t))^\gamma\, dt\right)$$

(with $\delta > 0$, $0 < \gamma < 1$, and (π, c) a pair of portfolio and consumption rate processes) is given by

$$c^*(t) = \frac{1}{1-\gamma}\left(\delta - r\gamma - \frac{\gamma}{2(1-\gamma)}\left(\frac{b-r}{\sigma}\right)^2\right) X(t),$$

$$\pi^*(t) = \frac{b-r}{(1-\gamma)\sigma^2}$$

if we have

$$\delta > r\gamma + \frac{\gamma}{2(1-\gamma)}\left(\frac{b-r}{\sigma}\right)^2$$

(see Section 3.3 for an explicit derivation of this result). Note that it is always optimal to invest the same constant fraction of wealth π^* in the stock at every time instant. Furthermore, one always consumes at a rate $c^*(t)$ which is proportional to the current wealth of the investor. The first fact can also be expressed graphically. Therefore, let

$$B(t) = \varphi_0(t)\, P_0(t), \quad S(t) = \varphi_1(t)\, P_1(t)$$

be the amount of money of the investor invested in the stock and in the bond, respectively (where the pair $(\varphi_0(t), \varphi_1(t))$ is the trading strategy corresponding to the pair $(\pi^*(t), c^*(t))$). Then the form of $\pi^*(t)$ implies that the process $(B(t), S(t))$ lies on the so-called Merton line for all non-negative times t. This is a straight line through the origin with slope $\pi^*/(1-\pi^*)$ (see Figure 6). Note that the investor has to choose trading strategies $(\varphi_0(t),\varphi_1(t))$ with infinite variation on every finite time interval to keep the process $(B(t), S(t))$ on the Merton line. In particular, he has to rebalance his stock and bond holdings at every time instant. So if transaction costs occur, the investor will be ruined after a short amount of time (at least if the transaction costs behave in a "realistic" way). Hence, in the presence of transaction costs, another class of strategies must be considered.

As it is not possible to keep the process (B(t), S(t)) on the Merton line in the presence of transaction costs, one could expect a strategy to be optimal in which the investor only rebalances his holdings if (B(t), S(t)) is "far away" from the Merton line. However, this raises at least two non-trivial questions: What do we mean by "far away" ? How does the optimal rebalancing action look like ?

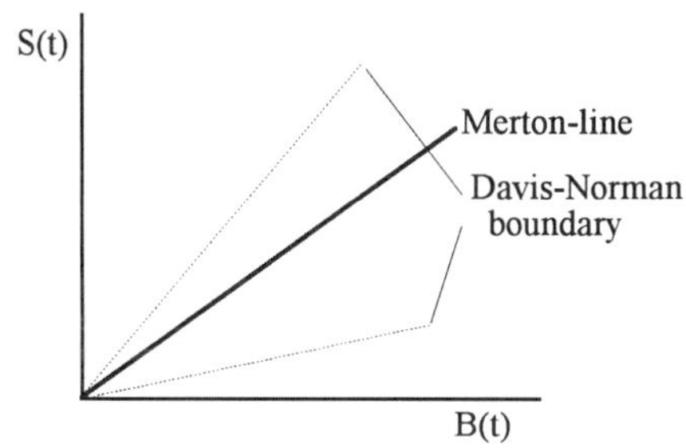

Figure 6: Merton-line and Davis-Norman boundary

Answers to this questions are given in (Davis and Norman 1990). Their main result states that it is optimal to consume at a constant rate as long as the process (S(t), B(t)) is inside the wedge given by the dotted lines in Figure 6. Everytime the boundaries of the wedge are reached by (B(t), S(t)), the minimal action (selling or purchase of stock) to keep the process inside the wedge is done. In particular, it is not optimal to transact back to the Merton line. This will be derived in detail below.

Let us first give a precise definition of the transaction cost structure. Purchasing and selling the risky security result in transaction costs that are proportional to the actual amount traded. More precisely, the investor has to pay the fraction $\lambda > 0$ of the value of the amount of stock purchased and the fraction $\mu > 0$ of the value of the amount of stock sold. The transaction costs and the consumption are always assumed to be deducted from the bond holdings (however, there is still only consumption at a rate c(t) and no consumption of lump sums). If (L_t, U_t) denote the cumulative amount of purchase and sale of stock up to time t, respectively, and c(t) denotes the consumption rate at time t then the dynamics of the bond and stock holdings are given by

$$dB(t) = [B(t)r - c(t)]\,dt - (1+\lambda)dL_t + (1-\mu)dU_t\,,\quad B(0) = x, \tag{1}$$

$$dS(t) = S(t)[bdt + \sigma dW(t)] + dL_t - dU_t\,,\quad S(0) = y\,. \tag{2}$$

By introducing the "solvency region"

$$\mathcal{S}_{\lambda,\mu} := \{(x, y)\in\mathbf{R}^2 \mid x + (1-\mu)y \geq 0,\ \ x + (1+\lambda)y \geq 0\}$$

(i.e. the region where selling of all shares of the risky asset or closing of the short position in the risky asset leads to non-negative bond holdings after transaction costs) and using results from (Doléans-Dade 1976), we know that equations (1) and (2) have a unique solution (at least up to $\tau = \inf\{t > 0 \mid (B(t),S(t))\notin\mathcal{S}_{\lambda,\mu}\}$, the first occurrence of a negative wealth after selling the remaining holdings). Here, L_t and U_t have to be increasing processes which they are by definition.

An **admissible strategy** (c, L, U) will be a strategy such that we have $\tau = \infty$ a.s. (after starting inside $\mathcal{S}_{\lambda,\mu}$). It can now be shown that for all $(x, y)\in\mathcal{S}_{\lambda,\mu}$, there exists an admissible strategy: simply take an arbitrary strategy (c, L, U) that does not jump out of $\mathcal{S}_{\lambda,\mu}$ (by making non-financiable transactions), and "stop" it by selling all stock holdings and stopping consumption at the first time the boundary of $\mathcal{S}_{\lambda,\mu}$ is reached.

As in the problem without transaction costs, the investor's goal will be to maximise his utility from life-time consumption, i.e. to maximise

$$J_{x,y}(c, L, U) := E^{x,y}\left(\int_0^{\infty} e^{-\delta t}\,\frac{1}{\gamma}(c(t))^{\gamma}\,dt\right)$$

for some $\gamma\in(0,1)$, $\delta > 0$ and $(x, y)\in\mathcal{S}_{\lambda,\mu}$. Note that in the presence of transaction costs, the initial wealth x + y is not sufficient to describe the investor's situation. It is necessary to know both bond and stock holdings. The value function v(x, y) is defined as

$$v(x, y) := \sup_{(c,L,U)\in\mathcal{U}(x,y)} J_{x,y}(c, L, U) \tag{3}$$

where $\mathcal{U}(x, y)$ denotes the set of admissible strategies starting in $(x, y)\in\mathcal{S}_{\lambda,\mu}$. In (Davis and Norman 1990), the choice of the logarithm as utility function in the definition of $J_{x,y}$ is also considered. We will not present it here as it does not produce new results and additional insight, but needs some additional notations .

The following theorem which states two properties of the value function, will prove to be very useful in the sequel. In particular, it will allow a reduction of the dimension of our problem later on.

Theorem 1

a) v is concave.

b) v has the homothetic property, i.e. $v(\rho x,\rho y) = \rho^\gamma v(x, y)$ for $\rho > 0$, $(x, y)\in \mathcal{S}_{\lambda,\mu}$.

Proof :

a) To prove the concavity, let

$$F^{(i)} := \left(B^{(i)},\ S^{(i)},\ c^{(i)},\ L^{(i)},\ U^{(i)}\right),\ i = 1,2,$$

be feasible solutions of equations (1), (2) with initial conditions $B^{(i)} = x_i$, $S^{(i)} = y_i$. With $\theta\in[0, 1]$ the convex combination $\theta F^{(1)}+(1-\theta)F^{(2)}$ is also a solution to (1), (2) with initial holdings $x_\theta = \theta x_1+(1-\theta)x_2$, $y_\theta = \theta y_1+(1-\theta)y_2$. This leads to the following relation :

$$v(\theta x, \theta y) \geq \sup_{\theta F^{(1)}+(1-\theta)F^{(2)}} E^{x_\theta, y_\theta}\left(\int_0^\infty e^{-\delta t}\frac{1}{\gamma}(c(t))^\gamma\, dt\right)$$

$$= \sup_{F^{(1)}, F^{(2)}} E^{x_\theta, y_\theta}\left(\int_0^\infty e^{-\delta t}\frac{1}{\gamma}\left(\theta c^{(1)}(t) + (1-\theta)c^{(2)}\right)^\gamma dt\right)$$

$$\geq \theta \sup_{F^{(1)}} E^{x_1,y_1}\left(\int_0^\infty e^{-\delta t}\frac{1}{\gamma}\left(c^{(1)}(t)\right)^\gamma dt\right) + (1-\theta)\sup_{F^{(2)}} E^{x_2,y_2}\left(\int_0^\infty e^{-\delta t}\frac{1}{\gamma}\left(c^{(2)}(t)\right)^\gamma dt\right)$$

$$= \theta\, v(x_1, y_1) + (1-\theta)\, v(x_2, y_2)\,.$$

b) For $\rho > 0$, $(x, y)\in\mathcal{S}_{\lambda,\mu}$, it can easily be checked that we have

$$\mathcal{U}(\rho x, \rho y) = \{(\rho c, \rho L, \rho U) \mid (c, L, U)\in \mathcal{U}(x, y)\}.$$

Hence,

$$v(\rho x,\rho y) = \sup_{(c,L,U)\in\mathcal{U}(x,y)} E^{x,y}\left(\int_0^\infty e^{-\delta t}\frac{1}{\gamma}(\rho c(t))^\gamma\, dt\right) = \rho^\gamma v(x, y)\,.$$

□

The method used by Davis and Norman is similar to the stochastic control approach of Section 3.3. We will first derive a formal Bellman equation, show that a sufficiently smooth solution of it coincides with the value function and then prove existence of such a solution. The main idea consists of a partition of $\mathcal{S}_{\lambda,\mu}$ into three different regions where it will be optimal to sell shares, purchase shares or do no transaction at all, respectively. First, we present a heuristic derivation of a free boundary problem which will completely determine the optimal strategy and the value function.

i) Heuristic Derivation of the Equation of Optimality

In (Davis and Norman 1990), the form of the optimality equation is heuristically derived with the help of a sub-problem of our life-time consumption problem (3) where only absolutely continuous trading strategies of the form

$$L_t = \int_0^t l(s)ds \quad U_t = \int_0^t u(s)ds$$

with bounded derivatives $0 \leq l(s), u(s) \leq \kappa$ are allowed. Under this assumption, the controlled stochastic differential equations (1), (2) for the holdings can be viewed as one vector sde. Then, the optimisation problem (3), restricted to the absolutely continuous controls above, corresponds to the HJB-Equation

$$\max_{c\geq 0, 0\leq l, u\leq \kappa} \left\{ \tfrac{1}{2}\sigma^2 y^2 v_{yy} + byv_y + rxv_x + \frac{1}{\gamma}c^\gamma - cv_x + (v_y - (1+\lambda)v_x)l \right.$$
$$\left. +((1-\mu)v_x - v_y)u - \delta v \right\} = 0 .$$

To solve the maximisation problem inside this HJB-Equation, note that v_x and v_y must be positive as the potential for future consumption clearly increases in both the initial stock and bond holdings (always assuming that the (appropriate) solution of the HJB-Equation and the value function coincide!). Then, the maximising triple (c, l, u) is given by

$$c = (v_x)^{\frac{1}{\gamma-1}}, \quad l = \begin{cases} \kappa \text{ if } v_y \geq (1+\lambda)v_x \\ 0 \text{ if } v_y < (1+\lambda)v_x \end{cases}, \quad u = \begin{cases} 0 \text{ if } v_y > (1-\mu)v_x \\ \kappa \text{ if } v_y \leq (1-\mu)v_x \end{cases},$$

i.e. we sell at the maximum rate in the region given by "$v_y \leq (1-\mu)v_x$", purchase at the maximum rate in the region "$v_y \geq (1+\lambda)v_x$", and do no transaction if (B(t), S(t)) lies in the remaining part of $S_{\lambda,\mu}$, the "no transaction region". Therefore, before we could solve the above HJB-Equation, we have to determine the boundaries between these regions. We will not do this here, but take the above considerations of the restricted problem as the basis for a guess on the form of the optimal solution of the original problem (3).

To continue with our heuristic derivation, we will always assume that the value function v(x, y) satisfies every desired regularity condition needed to justify our considerations. Further, we assume that $S_{\lambda,\mu}$ can be decomposed into the "buy region" B (where the optimal action is to buy additional shares of stock), the "sell region" S (where the optimal action is to sell some shares of stock), and the "no transaction region" NT as indicated in the subproblem above (see also Figure 7). In the buy region B, given by "$v_y \geq (1+\lambda)v_x$", the investor buys at maximum rate, having the interpretation that he makes an instantaneous purchase of stock that takes him (or more precisely, his pair of holdings (B(t),S(t))) to the boundary ∂B between B and NT, the no transaction region. In the sell region S, given by "$v_y \leq (1-\mu)v_x$", he sells as much shares of stock instantaneously such that he will reach the boundary ∂S between S and NT. And finally, he makes no stock transaction at all inside NT.

We will make multiple use of the homothetic property of the value function to obtain the shape of the boundaries ∂B and ∂S and to derive a simpler form of the resulting HJB- Equation. By assuming that v is continuously differentiable, we find

$$v_x(\rho x, \rho y) = \rho^{\gamma-1} v_x(x, y), \quad v_y(\rho x, \rho y) = \rho^{\gamma-1} v_y(x, y).$$

If thus the equations "$v_y(x, y) = (1+\lambda)v_x$" or "$v_y(x, y) = (1-\mu)v_x$" hold for some (x, y) then they must hold along the ray through (x, y). Hence, we have good reasons to conjecture that the boundaries between NT, B, and S are straight lines through the origin (see also Figure 7 where we have drawn the solvency region $S_{\lambda,\mu}$, its decomposition into S, NT and B and some further lines to be explained later).

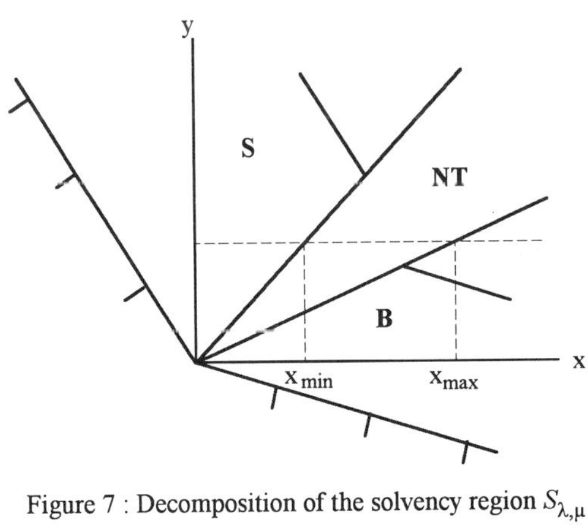

Figure 7 : Decomposition of the solvency region $S_{\lambda,\mu}$

Due to the assumed transaction cost structure, any finite transaction in the regions S and B will move the pair (B(t),S(t)) down a line with slope $-(1/(1-\mu))$, respectively up along a line with slope $-(1/(1+\lambda))$ (indicated in Figure 7 by the lines starting in S

and B, respectively, and leading to the corresponding border of NT). To see this, simply remind yourself of the fact that transaction costs are paid from the bond holdings. After a possible initial transaction from outside NT to the border of NT, all further transactions must take part at the boundaries, as inside NT the process (B(t),S(t)) is continuous (note that we only allow for consumption at a rate c(t), no consumption of lump sums). As a consequence, inside NT, we have $u = l = 0$, and this suggests that v(x, y) should satisfy the HJB-type equation

$$\begin{aligned} 0 &= \max_{c\geq 0}\left\{ \tfrac{1}{2}(\sigma y)^2 v_{yy} + (rx - c)v_x + byv_y + \tfrac{1}{\gamma}c^\gamma - \delta v \right\} \\ &= \tfrac{1}{2}(\sigma y)^2 v_{yy} + rxv_x + byv_y + \tfrac{1-\gamma}{\gamma}(v_x)^{-\frac{\gamma}{1-\gamma}} - \delta v \end{aligned} \tag{4}$$

inside NT. By introducing the function

$$\psi(x) := v(x, 1)$$

and making use of the homothetic property of v(x, y), we can simplify the pde (4). Note therefore that due to the relations

$$v(x, y) = y^\gamma \psi\left(\tfrac{x}{y}\right), \quad v_x(x, 1) = \psi'(x), \quad v_y(x, 1) = \gamma\psi(x) - x\psi'(x),$$
$$v_{yy}(x, 1) = -\gamma(1-\gamma)\psi(x) + 2(1-\gamma)\psi'(x) + x^2\psi''(x),$$

v(x, y) is totally determined by $\psi(x)$, and the pde (4) is equivalent to the following ordinary differential equation for $\psi(x)$,

$$\beta_3 x^2\psi''(x) + \beta_2 x\psi'(x) + \beta_1\psi(x) + \tfrac{1-\gamma}{\gamma}(\psi')^{-\frac{\gamma}{1-\gamma}} = 0, \quad x\in[x_{min}, x_{max}], \tag{5}$$

with

$$\beta_1 = \tfrac{1}{2}\sigma^2\gamma(1-\gamma) + b\gamma - \delta, \quad \beta_2 = \sigma^2(1-\gamma) + r - b, \quad \beta_3 = \tfrac{1}{2}\sigma^2 \tag{6}$$

where x_{min}, x_{max} coincide with the inverse of the slope of the boundaries ∂S and ∂B, respectively (this fact can also be read off from Figure 7). Of course, these values still have to be determined. Further, if our conjectured optimal strategy is indeed optimal then inside S, the value function v(x, y) must be constant along lines of slope $-(1/(1-\mu))$. Also, inside B, v(x, y) must be constant along lines of slope $-(1/(1+\lambda))$. Using the homothetic property of v(x, y) once more results in the two representations of $\psi(x)$,

$$\psi(x) = \tfrac{1}{\gamma}A(x + 1 - \mu)^\gamma \quad , \qquad x \leq x_{min}, \tag{7}$$

$$\psi(x) = \frac{1}{\gamma} B(x + 1 + \lambda)^{\gamma} \quad , \qquad x \geq x_{max} \tag{8}$$

for suitable constants A, *B*. Note that the sets "$x \leq x_{min}$", "$x \geq x_{max}$" (with $y = 1$) correspond to the sell and the buy region (in the transformed variables), respectively.

If all these considerations can be made rigorous and all conjectures prove to be correct then the remaining task to solve the optimal life-time consumption problem under transaction costs has the following form :

> "Find positive constants A, B, x_{min}, x_{max} (with $x_{min} < x_{max}$) such that there exists a C^2 - function $\psi(x)$ satisfying the set of equations (5), (7), and (8) "

The optimal controlled process (B(t), S(t)) will then be a reflected diffusion (see Theorem 2 below for an exact definition of (B(t), S(t))), defined in the wedge *NT* as given in Figure 7. The above problem is a so called "free boundary problem" because the boundary points x_{min}, x_{max} have to be determined as part of the solution of the whole problem

ii) Theoretical Justification of the Optimality Equation

In the foregoing section, we have often described the form of the (conjectured) optimal controlled process. However, it is not immediately clear if such a process really exists. Therefore, we give a theorem on the existence of reflected diffusions which ensures this existence. It also gives a precise description of the strategies (L, U) that consist of "the minimal control action preventing the process (B(t), S(t)) from leaving *NT*".

Theorem 2
Let $0 < x_{min} < x_{max}$ and *NT* be the closed wedge shown in Figure 7 with upper and lower boundaries ∂S and ∂B, respectively. Further, let c: $NT \to [0, \infty)$ be a Lipschitz continuous function and let $(x, y) \in NT$. Then there exist unique increasing processes L, U such that for $t < \tau = \inf\{ t > 0 \mid (B(t),S(t)) = (0, 0)\}$ we have

$$dB(t) = [r\,B(t) - c(B(t),S(t))]\,dt - (1+\lambda)\,dL_t + (1-\mu)\,dU_t\ , \quad B(0) = x,$$

$$dS(t) = S(t)\,[b\,dt + \sigma\,dW(t)] + dL_t - dU_t\ , \quad S(0) = y, \tag{9}$$

$$L_t = \int_0^t 1_{\{(B(s),S(s)) \in \partial B\}}(s)\,dL_s\,, \quad U_t = \int_0^t 1_{\{(B(s),S(s)) \in \partial S\}}(s)\,dU_s\ .$$

Remark 3

a) Theorem 2 can be proved by using results of (Varadhan and Williams 1985) on Brownian motion in a wedge, as is indicated in (Davis and Norman 1990). As mentioned in Section VIII.7 of (Fleming and Soner 1993), one could also use a result of (Lions and Sznitman 1984) on stochastic differential equations with reflecting boundaries. In particular, we could conclude from the last source that L, U are the local times of the process (B(t), S(t)) spent at ∂B and ∂S, respectively. However, it is not possible to give a detailed description of the techniques and results of (Varadhan and Williams 1985) or of (Lions and Sznitman 1984) without filling some pages. Therefore, we have to refer the interested readers to these two sources.

b) From the form of the equations for B(t) and S(t), we can read off the direction of the reflection of the diffusion inside *NT*. If (B(t),S(t)) reaches ∂B then it will be reflected in the direction of $(1, -(1+\lambda))$. If it reaches ∂S then the direction of the reflection is given by $(-1, (1-\mu))$. Both these directions coincide with the directions of the optimal action inside *B* and *S*, respectively, a fact from which continuity of the derivative of the value function at the boundary of *NT* can be explained (if the conjectured strategy is indeed optimal).

Having established the existence of a controlled process of the form conjectured for the optimal one, our next task is to show that it is indeed optimal. The first step to do so is the following verification theorem. It will show that a solution of the (precisely formulated) task that we heuristically derived in the last sub-section really yields the optimal utility and the optimal strategy of our consumption problem (3). However, as we will follow the method of (Davis and Norman 1990), we have to restrict the class of admissible strategies further by not allowing for short selling of the stock. Therefore, we introduce the following two sets:

$$\mathcal{S}_\mu := \{(x, y)\in \mathcal{S}_{\lambda,\mu} \mid y \geq 0 \},$$

$$\mathcal{U}'(x, y) := \{(c, L, U)\in \mathcal{U}(x, y) \mid (B(t), S(t))\in \mathcal{S}_\mu \ \forall\, t \geq 0 \},$$

Theorem 4 "Verification Theorem"

Let $\gamma\in(0,1)$ and suppose that the market coefficients satisfy

$$\delta > \gamma\left[r + \frac{1}{2(1-\gamma)}\left(\frac{b-r}{\sigma}\right)^2\right].$$

Let β_i, i = 1, 2, 3 be defined as in relation (6), and suppose that there are constants A, B, x_{min}, x_{max}, and a function ψ: $[-(1-\mu), \infty) \to \mathbf{R}$ such that we have

$$0 < x_{min} < x_{max} < \infty,$$

$$\psi \in C^2 \text{ with } \psi'(x) > 0 \text{ for } x \in (-(1-\mu), \infty),$$

$$\beta_3 x^2 \psi''(x) + \beta_2 x \psi'(x) + \beta_1 \psi(x) + \frac{1-\gamma}{\gamma}(\psi')^{-\frac{\gamma}{1-\gamma}} = 0 \quad \text{for } x \in [x_{min}, x_{max}],$$

$$\psi(x) = \frac{1}{\gamma} A (x + 1 - \mu)^{\gamma}, \quad x \le x_{min}, \tag{10}$$

$$\psi(x) = \frac{1}{\gamma} B (x + 1 + \lambda)^{\gamma}, \quad x \ge x_{max}. \tag{11}$$

Define

$$c^*(x, y) = y \left[\psi'\left(\frac{x}{y}\right) \right]^{-\frac{1}{1-\gamma}} \quad \forall\, (x, y) \in NT \setminus \{(0,0)\}.$$

Then the strategy $(c^*(t), L^*(t), U^*(t))$ with $c^*(t) = c^*(B(t),S(t))$, where $(B(t), S(t), L^*, U^*)$ is the solution of the set of equations (9) with $c = c^*$, is optimal among all strategies in $\mathcal{U}(x, y)$. The corresponding maximum expected utility is given by

$$v(x, y) = y^{\gamma} \psi\left(\frac{x}{y}\right). \tag{12}$$

Proof :
We will give the proof of this verification theorem in the next sub-section as it is rather lengthy and technical.

□

Remark 5
a) One can rephrase the form of the optimal holdings in Theorem 4 as: The optimal strategy is to hold the fraction of wealth invested in the stock,

$$\pi(t) = \frac{S(t)}{B(t)+S(t)},$$

between the constants $\pi_1^* := (1 + x_{max})^{-1}$ and $\pi_2^* := (1 + x_{min})^{-1}$.
b) It only remains to show the existence of such a function $\psi(x)$ as required in Theorem 4. We sketch the proof below for the case "$r + (1-\gamma)\sigma^2 > b > r$" (i.e. the mean rate of return of the stock exceeds the riskless interest rate, but not too much). In the case "$b < r$", we would expect short-selling of stocks to be optimal which is prohibited in our setting. Therefore, not investing into the stock would be the candidate for the optimal stock position. This fact directly follows from a result of (Davis and Norman 1990) where it is proved that, even in the case "$b = r$", it is optimal not to invest into the stock at all. We do not present this result here.

Theorem 6 "Existence of a solution of the free boundary problem"
Define β_i, i = 1, 2, 3, as in the relations (6). Let r, b, σ, γ, δ, λ, μ be real constants with

$$0 \le \mu < 1, \quad 0 \le \lambda < \infty, \quad \mu + \lambda > 0, \quad \delta > 0, \quad 0 < \gamma < 1,$$

$$0 < r < b < r + (1-\gamma)\sigma^2,$$

$$\delta - \gamma r - \frac{\gamma}{2(1-\gamma)}\left(\frac{b-r}{\sigma}\right)^2 > 0 .$$

Then there exists a C^2-function ψ: $[-(1-\mu), \infty) \to \mathbf{R}$ and positive constants x_{min}, x_{max}, A, B satisfying all the assumptions of Theorem 4 and further

$$\gamma\psi(x)\psi''(x) + (1-\gamma)\left(\psi'(x)\right)^2 \le 0$$

for $x \in [-(1-\mu), \infty)$ where we have equality for $x \le x_{min}$ and for $x \ge x_{max}$.

Idea of proof:
As the detailed proof is far too long to present it here, we will only present its underlying idea which consists of "smooth pasting", i.e. of determining the unknown constants A, B and the free boundary points x_{min}, x_{max} with the help of the smoothness requirements for the function $\psi(x)$. Further, we will solve a similar but easier existence problem in Section 6.3 in great detail and therefore omit a detailed description of the proof of Theorem 6, here. However, the basic idea to prove existence is as follows :

• The general form of $\psi(x)$ for $x \le x_{min}$ and for $x \ge x_{max}$ is given by equations (7) and (8). By construction of the corresponding controls (c, L, U) we have continuity of $\psi(x)$ in the points $x = x_{min}$ and $x = x_{max}$.

• We then show that for every 0 < x* there exists a constant A(x*) such that the differential equation (5) has a unique solution with a derivative in x* that coincides with the derivative of $\psi(x)$ given by equation (7) (where, for the moment, we have chosen A = A(x*) and x_{min} = x*). Then we determine the lower boundary x_{min} by the requirement that the second derivatives in x * of the solution to (5) and of $\psi(x)$ given by equation (7) should also coincide. Finally, we set x_{min} = x* and A = A(x*) (where x* is the unique positive constant satisfying the two smoothness conditions).

• Having determined x_{min}, we determine x_{max} in a similar way: For every x* > x_{min} there exists a constant B(x*) such that the derivative of the solution of the differential equation (5) (as computed in the previous step) and the derivative of $\psi(x)$ given by equation (8) (where we have chosen $B = B(x^*)$ and $x_{max} = x^*$) coincide in x^*. Again, there is a unique x^* such that also the second derivatives in x^* of the solution to (5) and of $\psi(x)$ given by equation (8) coincide. We then set $x_{max} = x^*$ and $B = B(x^*)$.

□

iii) Proof of the Verification Theorem

In this section, we will prove Theorem 4. To do so, we first state three lemmas.

Lemma 7
Let $v(x, y)$ be given by equation (12). Then for all $(x, y)\in NT$ we have

$$y\, v_y(x, y) = \gamma\left(1 - f\left(\frac{x}{y}\right)\right) v(x, y)$$

with $f(x) := \frac{x\psi'(x)}{\gamma\, \psi(x)}$ for $x\in[x_{min}, x_{max}]$ with $\psi(x)$ as in Theorem 4.

Proof:
Note first that, due to the assumptions on $\psi(x)$ (and by $\mu < 1$), f is well defined, continuous and bounded. Using the homothetic property of $v(x, y)$ (which follows from the explicit form of $v(x, y)$ given in equation (12)), we obtain

$$v_y(x, y) = \left(y^\gamma v\left(\frac{x}{y}, 1\right)\right)_y = \gamma y^{\gamma-1} v\left(\frac{x}{y}, 1\right) - y^{\gamma-2} x v_x\left(\frac{x}{y}, 1\right).$$

Substituting $\psi(x)$ into this relation results in

$$y\, v_y(x, y) = \gamma y^\gamma \psi\left(\frac{x}{y}\right) - y^{\gamma-1} x \psi'\left(\frac{x}{y}\right)$$

$$= \gamma v(x, y) - \gamma\psi\left(\frac{x}{y}\right) y^\gamma \frac{\frac{x}{y}\psi'\left(\frac{x}{y}\right)}{\gamma\psi\left(\frac{x}{y}\right)} = \gamma\left(1 - f\left(\frac{x}{y}\right)\right) v(x, y).$$

□

Lemma 8 "Representation of the value process"
Let the assumptions of Theorem 4 be satisfied. Let (c^*, L^*, U^*) be the strategy described in Theorem 4 and let $(B^*(t), S^*(t))$ be the corresponding solution of the set of equations (9). Define $\xi^*(t) = B^*(t) / S^*(t)$. Then, we have the following representation of the "value process" $v(B^*(t), S^*(t))$:

$$\gamma\, v(B^*(T),S^*(T)) = \gamma\, v(x,y)\exp\left(\int_0^T \left(\delta - \frac{\gamma}{1-\gamma} h(\xi^*(t))\right) dt\right)$$
$$\cdot \exp\left(\int_0^T \gamma\sigma(1-f(\xi^*(t)))dW(t) - \tfrac{1}{2}\int_0^T (\gamma\sigma(1-f(\xi^*(t))))^2\, dt\right)$$

with f(x) given as in Lemma 5 for $T > 0$ and h(x) defined on $[x_{min}, x_{max}]$ by

$$h(x) = \frac{(1-\gamma)(\psi'(x))^{-\frac{\gamma}{1-\gamma}}}{\gamma^2 \psi(x)}.$$

Proof :

Note first that for $(x, y)\in NT$ we have $v(x, y) > 0$ if and only if $(x, y) \neq (0,0)$. Set

$$\tau := \tau(x, y) := \inf\{\, t > 0 \mid (B^*(t),S^*(t)) = 0,\ (B^*(0),S^*(0)) = (x, y)\,\}.$$

By continuity of the wealth process, we get $\tau > 0$ a.s. for all $(x, y)\in NT\backslash\{(0,0)\}$. We then apply Itô's formula to $\ln(\exp(-\delta(T\wedge\tau))\gamma v(B^*(T\wedge\tau),S^*(T\wedge\tau)))$ which yields

$$\ln\left(e^{-\delta(T\wedge\tau)}\gamma v(B^*(T\wedge\tau),S^*(T\wedge\tau))\right) - \ln(\gamma v(B^*(0),S^*(0)))$$
$$= \int_0^{T\wedge\tau} \frac{1}{v}\left[-\delta v + rB^*(t)v_x + bS^*(t)v_y + \tfrac{1}{2}\sigma^2 S^*(t)^2 v_{yy} - c^* v_x\right]dt$$
$$- \int_0^{T\wedge\tau} \frac{1}{2v^2}\sigma^2 S^*(t)^2 {v_y}^2 dt + \int_0^{T\wedge\tau} \frac{1}{v}\left[-(1+\lambda)v_x + v_y\right]dL^*(t)$$
$$+ \int_0^{T\wedge\tau} \frac{1}{v}\left[(1-\mu)v_x - v_y\right]dU^*(t) +$$
$$= \int_0^{T\wedge\tau} \frac{1}{v}\left[Gv - \frac{1}{\gamma}(v_x)^{-\frac{\gamma}{1-\gamma}}\right]dt + \int_0^{T\wedge\tau}\frac{1}{v}\sigma S^*(t) v_y dW(t) - \int_0^{T\wedge\tau}\frac{1}{2v^2}\sigma^2 S^*(t)^2 {v_y}^2 dt \tag{13}$$

Note that by construction of the optimal strategy, the dL*- and dU*-integrals vanish, and also that the operator G is given by

$$Gv := \tfrac{1}{2}(\sigma y)^2 v_{yy} + rxv_x + byv_y - \delta v + \frac{1-\gamma}{\gamma}(v_x)^{-\frac{\gamma}{1-\gamma}}.$$

By noting further that we have

$$Gv(B^*(t),S^*(t)) = 0 \qquad \forall\, t \geq 0 \text{ a.s.},$$

$$\gamma\sigma(1-f(\xi^*(t))) = \sigma\,\frac{\gamma\psi(\xi^*(t)) - \xi^*(t)\psi'(\xi^*(t))}{\psi(\xi^*(t))}$$

$$= \sigma\,\frac{\gamma(S^*(t))^{\gamma}\,\psi(\xi^*(t)) - (S^*(t))^{\gamma-1}B^*(t)\psi'(\xi^*(t))}{(S^*(t))^{\gamma}\,\psi(\xi^*(t))}$$

$$= \frac{\sigma S^*(t)v_y(S^*(t),B^*(t))}{v(B^*(t),S^*(t))},$$

$$\frac{1}{1-\gamma}\,h(\xi^*(t)) = \frac{(v_x(\xi^*(t),1))^{-\frac{\gamma}{1-\gamma}}}{\gamma^2 v(\xi^*(t),1)} = \frac{(v(B^*(t),S^*(t)))^{-\frac{\gamma}{1-\gamma}}}{\gamma v(B^*(t),S^*(t))},$$

we arrive at the following simplified form of equation (13) :

$$\ln\left(\frac{e^{-\delta(T\wedge\tau)}v(B^*(T\wedge\tau),S^*(T\wedge\tau))}{v(x,y)}\right)$$

$$= -\int_0^{T\wedge\tau}\frac{\gamma}{1-\gamma}h(\xi^*(t))dt + \int_0^{T\wedge\tau}\gamma\sigma(1-f(\xi^*(t))dW(t) - \int_0^{T\wedge\tau}(\sigma\gamma(1-f(\xi^*(t)))^2dt$$

Application of the exponential function to both sides of this representation will prove Lemma 8 if we can show that we have $\tau = \infty$ a.s.. Let us therefore rewrite this simplified form of equation (13) by applying the exponential function:

$$\gamma\, v(B^*(T\wedge\tau),S^*(T\wedge\tau)) = \gamma\, v(x, y)\exp\left(\int_0^{T\wedge\tau}\left(\delta - \frac{\gamma}{1-\gamma}h(\xi^*(t))\right)dt\right)$$

$$\cdot\exp\left(\int_0^{T\wedge\tau}\gamma\sigma(1-f(\xi^*(t)))dW(t) - \tfrac{1}{2}\int_0^{T\wedge\tau}(\gamma\sigma(1-f(\xi^*(t))))^2dt\right). \quad (14)$$

Note that f(x) and h(x) are positive and bounded on $[x_{min}, x_{max}]$. Hence, all the integrals on the right hand side of equation (14) are bounded. In particular, v(x, y) is finite for all $(x, y)\in NT$, due to the fact that it is bounded from above by the value function of the life-time consumption problem without transaction costs. But this value function is finite (see also Section 3.3). Consequently, the right hand side of equation (14) is positive a.s.. If we assume that τ would be finite on a set of positive probability, the positivity of the right hand side of equation (14) would lead to a contradiction, as for all ω with a finite value $\tau(\omega)$, we would have

$$v(B^*(\tau(\omega)),S^*(\tau(\omega))) = 0.$$

Hence, Lemma 8 is proved.

□

Lemma 9
Let v(x, y) be defined as in equation (12) with $\gamma \in (0,1)$. Then there exist constants K and K_ε (where the latter depends on $\varepsilon > 0$) with

$$0 \le v(x, y) \le K\,(x+y)^\gamma \quad \forall\,(x, y) \in S_\mu, \tag{15}$$

$$0 \le y v_y(x, y) \le K_\varepsilon\,(1 + x + y) \quad \forall\,(x, y) \in S_\mu \setminus \left\{(x,y) \in S_\mu \,\middle|\, x + (1-\mu)y > \varepsilon\right\}.$$

Proof:
By definition of S_μ, we have

$$y \le \frac{1}{\mu}(x+y) \quad \forall\,(x, y) \in S_\mu. \tag{16}$$

Using representations (12) and (10) for v(x, y) in S (or more precisely in $S \cap S_\mu$) we obtain (note also that we have $\psi'(x) > 0$ for $x > -(1-\mu)$) the inequality

$$v(x, y) = y^\gamma \psi\left(\tfrac{x}{y}\right) \le \mu^{-\gamma}(x+y)^\gamma\,\psi\left(\mu\frac{x}{x+y}\right)$$

$$\le \mu^{-\gamma}(x+y)^\gamma\,\frac{A}{\gamma}\left(\frac{x+(1-\mu)y}{x+y}\right)^\gamma \le \mu^{-\gamma}\frac{A}{\gamma}(x+y)^\gamma$$

which implies relation (15) in $S \cap S_\mu$. In a similar way, relation (15) can be proved in $B \cap S_\mu$ (of course, we have to use representation (11) instead of (10) while in *NT* relation (15) follows from the inequality

$$v(x, y) = y^\gamma \psi\left(\tfrac{x}{y}\right) \le M_0\,(x+y)^\gamma$$

with $M_0 := \mu^{-\gamma} \max\limits_{\xi \in [x_{min}, x_{max}]} \psi(\xi)$. The positivity of $\psi'(x)$ implies

$$y v_y(x, y) = \gamma\, y^\gamma \psi\left(\tfrac{x}{y}\right) - x y^{\gamma-1} \psi'\left(\tfrac{x}{y}\right)\gamma \le \gamma\, M_0\,(x+y)^\gamma$$

in *NT*. By using the decomposition of $S \cap S_\mu$ into $S_+ := S \cap S_\mu \cap [0, \infty)^2$ and $S_- := (S \cap S_\mu) \setminus S_+$ we have the following relations

$$y v_y(x, y) = B\,y\,(x + (1+\lambda)y)^{\gamma-1} \le B\,(x+y)^\gamma, \quad (x, y) \in B \cap S_\mu,$$

$$y v_y(x, y) = A\,y\,(x + (1-\mu)y)^{\gamma-1} \le A(1-\mu)^{\gamma-1}(x+y)^\gamma, \quad (x, y) \in S_+,$$

$$yv_y(x, y) = A\, y\,(x+(1-\mu)y)^{\gamma-1} \le \frac{A}{\mu\varepsilon^{1-\gamma}}\,(x+y), \quad (x, y)\in S,$$

in the remaining parts of S_μ for all pairs (x, y) with (x + (1−μ) y) > ε. We can now choose an appropriate constant K_ε such that the right hand side of the last four inequalities are all dominated by $K_\varepsilon(1+x+y)$ which completes the proof of Lemma 9.

□

We are now in the position to prove the verification theorem, Theorem 4.

Proof of Theorem 4:
a) By construction of v(x, y) via equation (12) and by the assumptions on ψ(x), it can directly be verified that

1. v is concave,
2. the form of v in S and B is given by equations (7) and (8), respectively,
3. $(1-\mu)v_x - v_y \le 0$ in S_μ (with equality in S),
   $-(1+\lambda)v_x + v_y \le 0$ in S_μ (with equality in B),
4. $Gv(x, y) = \max_{c\ge 0}\left\{\tfrac{1}{2}(\sigma y)^2 v_{yy} + (rx-c)v_x + byv_y + \tfrac{1}{\gamma}c^\gamma - \delta v\right\}$ in S_μ,
   $Gv = 0$ in NT, $Gv \le 0$ in $S\cup B$,

(where G is the operator defined in the proof of Lemma 8). We will use these properties to prove the remaining claims of the theorem.
b) Let (c, L, U)∈U(x, y), (B(t), S(t)) be the corresponding solution of equations (1), (2) starting in (x, y)∈S_μ, and let χ(x, y) be a C^2-function. Define

$$M_t^\chi := \int_0^T e^{-\delta t}\frac{1}{\gamma}c(t)^\gamma\,dt - e^{-\delta T}\chi(B(T), S(T))$$

We will show that M is a martingale if we choose χ(x, y) = v(x, y) and a supermartingale for all other choices of χ(x, y). From this, we will deduce that v(x, y) is the value function of our problem (this is usually called the "martingale optimality principle", see (Rogers and Williams 1987)). Application of the generalised Itô's formula (see Theorem B25 in the Appendix) leads to

$$M_T^\chi - M_0^\chi$$
$$= \int_0^T e^{-\delta t}\left\{\tfrac{1}{2}(\sigma S(t))^2\chi_{yy} + (rB(t)-c(t))\chi_x + bS(t)\chi_y + \tfrac{1}{\gamma}c(t)^\gamma - \delta\chi\right\}dt +$$

$$+\int_0^T e^{-\delta t}\left\{-(1+\lambda)\chi_x+\chi_y\right\}dL(t)+\int_0^T e^{-\delta t}\left\{(1-\mu)\chi_x-\chi_y\right\}dU(t)$$
$$+\sum_{0\le t\le T} e^{-\delta t}\{\chi(B(t),S(t))-\chi(B(t-),S(t-))$$
$$-\chi_x(B(t-),S(t-))(B(t)-B(t))-\chi_y(B(t-),S(t-))(S(t)-S(t))\}$$
$$+\int_0^T e^{-\delta t}\sigma S(t)\chi_y dW(t)=:I_1+I_2+I_3+I_4+I_5 \qquad (17)$$

Let now (c, L, U) = (c*, L*, U*), (x, y)∈*NT*, and let v(x, y) be defined by equation (12). We then have

$$c^*=\left(v_x\right)^{-\frac{1}{1-\gamma}},\ c^*_y(\rho x,\rho y)=c^*_y(x,y),\ c^*_x(\rho x,\rho y)=c^*_x(x,y)$$

which yields the Lipschitz continuity of c*. Hence, by Theorem 2, the set of equations (9) has a unique solution. Now choose χ(x, y) = v(x, y), (c, L, U) = (c*, L*, U*) in equation (17). By part a) of the proof we can conclude

$$I_1=I_2=I_3=I_4=0$$

(to see that I_4 vanishes note that (B(t), S(t)) = (B*(t), S*(t)) is continuous). Hence,

$$v(x,y)=M_0^v=M_T^v-I_5$$
$$=\int_0^T e^{-\delta t}\frac{1}{\gamma}c(t)^\gamma\,dt+e^{-\delta T}v(B(T),S(T))-\int_0^T e^{-\delta t}\sigma S(t)v_y dW(t) \qquad (18)$$

and by Lemma 8, we have the following representation for some bounded, positive functions f, h:

$$v(B^*(T),S^*(T))=$$
$$v(x,y)e^{\int_0^T\left(\delta-\frac{\gamma}{1-\gamma}h(\xi^*(t))\right)dt}\,e^{\int_0^T\gamma\sigma(1-f(\xi^*(t)))dW(t)-\frac{1}{2}\int_0^T(\gamma\sigma(1-f(\xi^*(t))))^2dt}$$
$$=:v(x,y)e^{\int_0^T\left(\delta-\frac{\gamma}{1-\gamma}h(\xi^*(t))\right)dt}H(T). \qquad (19)$$

Lemma 5 yields

$$y\,v_y(x,y)=\gamma(1-f\left(\tfrac{x}{y}\right))\,v(x,y). \qquad (20)$$

By the boundedness of f and h, we have the finiteness of $E_{x,y}(H(T)^2)$. Then, the representations (19) and (20) imply that the stochastic integral in equation (18) is a martingale which leads to the equation

$$v(x, y) = E_{x,y}\left(\int_0^T e^{-\delta t}\frac{1}{\gamma}c(t)^{\gamma}\,dt\right) + e^{-\delta T}E_{x,y}\,(v(B(T), S(T)))$$

For $T\to\infty$ the last term converges to zero due to representation(19) (note that H(T) is a martingale with mean equal to one). Hence, we have

$$v(x, y) = E_{x,y}\left(\int_0^{\infty} e^{-\delta t}\frac{1}{\gamma}c(t)^{\gamma}\,dt\right)\ ,\ \ (x, y)\in \mathcal{NT}.$$

For $(x, y)\in\mathcal{S}_\mu \setminus \mathcal{NT}$ we have

$$J_{x,y}(c^*, L^*, U^*) = J_{x',y'}(c^*, L^*, U^*)$$

where (x', y') is the point on the boundary of $\mathcal{NT}$ to where the initial transaction is made. We thus have

$$v(x, y) = J_{x,y}(c^*, L^*, U^*)$$

for all $(x, y)\in\mathcal{S}_\mu$.

c) Next, we show that for every $(c, L, U)\in\mathcal{U}(x, y)$ we have

$$v(x, y) \geq J_{x,y}(c, L, U)\ \ \forall\ (x, y)\in\mathcal{S}_\mu .$$

Therefore, let $(c, L, U)\in\mathcal{U}(x, y)$ with corresponding holdings process (B(t), S(t)). As usual, define its corresponding wealth process X(t) = B(t) + S(t) and the portfolio process $\pi(t) = S(t)/X(t)$. Then, (π, c) is admissible for the problem without transaction costs, too. X(t) satisfies the equation

$$dX(t) = [\ [r + (b-r)\pi(t)]\ X(t) - c(t)\ ]\ dt + \sigma\pi(t)X(t)\ dW(t) - dA(t)$$

where we have set $A := \lambda L + \mu U$. By comparing this equation with the one for the wealth process $X_0(t)$ corresponding to (π, c) **without** transaction costs, we can conclude that we must have

$$X(t) \leq X_0(t)\quad \forall\ t\geq 0\ \text{ a.s.}$$

(note that A(t) is a non-decreasing process and that X(t) is always non-negative). For fixed $\varepsilon > 0$ define

$$v^{\varepsilon}(x,y) := v(x+\varepsilon, y)$$

and choose $\chi(x, y) = v^{\varepsilon}(x,y)$ in equation (17). By Lemma 9, we have

$$E_{x,y}\left(\int_0^T \left[v^{\varepsilon}{}_y(B(t),S(t))S(t)\right]^2 dt\right) \le K_\varepsilon E_{x,y}\left(\int_0^T [1+X(t)]^2 dt\right)$$

$$\le K_\varepsilon E_{x,y}\left(\int_0^T [1+X_0(t)]^2 dt\right) < \infty$$

(see also Section 3.3 for the finiteness of the last expression). Again, I_5 in equation (17) is a martingale. Property 3 in part a) of the proof and the fact that we have

$$v^{\varepsilon}{}_x(x,y) = v_x(x+\varepsilon, y), \quad v^{\varepsilon}{}_y(x,y) = v_y(x+\varepsilon, y), \quad v^{\varepsilon}{}_{yy}(x,y) = v_{yy}(x+\varepsilon, y)$$

yield

$$Gv^{\varepsilon}(x,y) = Gv(x+\varepsilon, y) - r\varepsilon\, v_x(x+\varepsilon, y)\,.$$

Using Property 4 given in part a) of the proof and the non-negativity of $\psi'(x)$, we obtain for any $\varepsilon \ge 0$, $(x, y) \in S_\mu$

$$\frac{1}{\gamma}c^\gamma + (rx - c)v^{\varepsilon}{}_x + byv^{\varepsilon}{}_y + \tfrac{1}{2}\sigma^2 y^2 v^{\varepsilon}{}_{yy} - \delta v^{\varepsilon} \le 0.$$

Thus, I_1 is a decreasing process in T. By part a) of the proof, I_2, I_3, I_4 are also decreasing processes which yields that $M_t^{v^\varepsilon}$ is a supermartingale. Hence:

$$v^{\varepsilon}(x,y) = M_0^{v^\varepsilon} \ge E_{x,y}\left(\int_0^T e^{-\delta t}\frac{1}{\gamma}c(t)^\gamma dt\right) + e^{-\delta T}E_{x,y}\left(v^{\varepsilon}(B(T),S(T))\right)$$

$$\ge E_{x,y}\left(\int_0^T e^{-\delta t}\frac{1}{\gamma}c(t)^\gamma dt\right). \qquad (21)$$

By taking the limit for $T \to \infty$ in inequality (21) (using monotone convergence), we obtain

$$v^{\varepsilon}(x,y) \ge J_{x,y}(c, L, U)\,.$$

Finally $v^{\varepsilon}(x,y) \downarrow v(x, y)$ for $\varepsilon \to 0$ yields the desired result.

□

iv) Some Remarks on Extensions, Numerical Solution and Practical Relevance

In Section 7 of (Davis and Norman 1990) the authors present an algorithm for a numerical solution of the free boundary problem of Theorem 6. As it needs the introduction of a huge amount of further notation, we do not give it here. It mainly consists of a one-dimensional search coupled with the solution of a sequence of differential equations (by finite difference methods).

In (Taksar, Klass, Assaf 1988), the problem of maximising the asymptotic growth rate of the wealth process in the presence of exactly the same transaction cost structure as in the Davis and Norman setting is treated by a different method as the one given in the previous sub-sections. However, the optimal strategy is of the same structure (keeping the holdings process in a wedge, using local time-type controls).

A finite horizon multi-dimensional portfolio problem with the same transaction cost structure as in the current setting is considered in (Akian, Séquier, Sulem 1995). By using viscosity solution techniques, the authors were able to justify the use of numerical methods to solve a multi-asset problem. As an application they solved a three asset terminal wealth maximisation problem using finite-difference methods.

In general, it seems hopeless to deal with problems of m risky assets in the Davis and Norman setting if m has a realistic size (i.e. at least around 20–30). Note that the presence of m risky assets leads to 3^m different transaction regions ! Hence, increasing the number of risky assets leads to an explosion of the computational effort.

As the application of this method to real world problems with a number m of risky assets clearly exceeding $m = 3$ is thus very limited (or better: impossible), there remains the question if the results of this section have any practical importance. From a non-formal point of view, the answer to this question should be positive, despite all the criticisms raised above. The Davis and Norman results imply that a rule of thumb of the form "Keep the risky fraction in an interval [a, b], trade only if its boundary is reached, and then do the minimal transaction to prevent it from leaving the interval" is a reasonable one (if the choice of a and b is not "too bad": at least the interval should contain the Merton solution). It is also an important finding that it is not optimal to do "big transactions", i.e. to transact back to the Merton line. However, we will also see in the two following sections that this last finding crucially depends on the transaction cost structure.

5.2 Impulse Control Methods and Portfolio Optimisation with Strictly Positive Transaction Costs

Although transaction costs are included into the model of Section 1, the model still lacks some closeness to reality. The main problem in the Davis-Norman approach is that the strategies still consist of doing infinitesimally small transactions (which is, of course, not the case in real world). In the presence of fixed costs, no matter how small they would be, such a strategy will still lead to enormous transaction costs if the process of the holdings reaches the boundary of the no transaction region.

Therefore, the situation completely changes when the transaction costs will contain such a fixed cost component K (i.e. a positive lower bound K for the transaction costs). Then, at every intervention time, the process of the holdings (S(t), B(t)) must have a point of discontinuity due to the fact that the wealth of the total holdings **after** the transaction is at least by an amount of K smaller than the wealth of the total holdings **before** the transaction. Thus, in contrast to the local time type strategies, the intervention times of an optimal strategy will not accumulate at any time instant. The appropriate mathematical framework for this situation is the concept of impulse control. By including a fixed cost part in the transaction costs, in (Eastham and Hastings 1988) an impulse control approach to portfolio optimisation which has exactly these features, was used. Also, it is possible to include constraints (on the holdings process) in this approach. Unfortunately, in their main theorem, Eastham and Hastings required a regularity assumption on the value function that could only be satisfied if the optimal strategy consists of no intervention. Further, their formulation of the optimisation problem was in some sense too restrictive. Their approach was taken up again and refined in various aspects in (Korn 1994) which forms the basis for the presentation of this section. In particular, we offer a new solution approach via optimal stopping which resembles methods already given in (Davis 1993) for the case of optimal control of piecewise deterministic processes. Of course, we will also present the usual approach of solving an impulse control problem via solving the corresponding quasi-variational inequalities (see (Bensoussan and Lions 1984)) which is the analogue to the stochastic control approach to the (unconstrained) portfolio problem.

As it is not hard to imagine, the possibility to obtain an analytic solution to the portfolio problem including fixed and proportional transaction costs is very limited. To overcome this problem, we will also present an approximation method for small but non-zero transaction costs, the method of asymptotic analysis. This method will also be applied in the next section.

i) Impulse Control Strategies for Portfolio Problems

As in Section 2, we consider a securities market consisting of a bond and a single stock with constant market coefficients (for the bond price we will even assume that it is constant over time). Later, we will make some remarks about a generalisation of our method to more complicated market models (including both more complicated price processes and more stocks). The fundamental price processes in the following considerations will again be the value of the stock and bond holdings (S(t), B(t)) as defined in Section 1. Under our assumptions, their dynamics between intervention times of the investor (i.e. time instants where the investor changes his holdings due to rebalancing the stock and bond position and/or consumption) are the same as that (of a multiple) of the bond and the stock price, i.e.

$$dB(t) = 0 \quad \text{"Bond holdings"}$$
$$dS(t) = S(t)(b\,dt + \sigma\,dW(t)) \quad \text{"Stock holdings"} \tag{22}$$

Further, we assume that at every intervention time θ_i the following transaction costs must be paid by the investor :

$$K + k\,|\,\Delta S_i\,|\ ,\ \ 0 < K,\ \ 0 < k < 1$$

where ΔS_i is the change in stock holdings at time θ_i. K is the **fixed cost** component that occurs at every intervention time (independent of the investor's action). The cost component $k|\Delta S_i|$ is proportional to the absolute volume of the stock transaction made within the i th action of the investor. We assume that transaction costs must be paid from the bond holdings. Thus, if B_i denotes the value of the bond holdings after the i th intervention $(\Delta S_i, c_i)$, consisting of the change of stock holdings ΔS_i and the amount c_i consumption, we have the following balance equation:

$$B_i = B_{i-1} - \Delta S_i - k\,|\Delta S_i| - K - c_i\ . \tag{23}$$

Hence, the i th action of an investor is completely described by his change in stock holdings and his consumption$(\Delta S_i, c_i)$.

We will now define an impulse control strategy for the portfolio problem to formalise the description indicated in the above motivation (Here, (B(t),S(t)) will always denote the value of these processes at time t **after all** actions of the investor at time t are made) :

Definition 10

An **impulse control strategy** $\{(\theta_i, \Delta S_i, c_i), i \in \mathbf{N}\}$ is a sequence of intervention times θ_i and control actions $(\Delta S_i, c_i)$ with

i) $0 \le \theta_i \le \theta_{i+1}$ a.s. $\forall\, i \in \mathbf{N}$.

ii) θ_i is a stopping time with respect to the filtration $\sigma((S(s-), B(s-)), s \le t, (\theta_n, \Delta S_n, c_n), n < i), t \ge 0$.

iii) $(\Delta S_i, c_i)$ are measurable with respect to $\sigma((S(\theta_i -), B(\theta_i -)), (\Delta S_n, c_n), n < i)$. (24)

iv) $c_i \geq 0$, $B(\theta_i) \geq 0$, $S(\theta_i) \geq 0$. (if $\theta_{i-1} < \theta_i < \theta_{i+1}$)

An impulse control strategy will be called **admissible** if we have

v) $$P(\lim_{i\to\infty} \theta_i \leq T) = 0 \quad \forall\, T \geq 0 .$$

Condition (24) iv) becomes notationally more complicated if more than one intervention at the same time instant occur, i.e. $\theta_i = \theta_{i-1} = \ldots = \theta_{i-k}$ for some $k \leq i$. We then have to replace $B(\theta_i)$, $S(\theta_i)$ by their values after the i th action of the investor. Of course, in a portfolio problem where the goal is to maximise the utility of terminal wealth (with a strictly increasing utility function), a strategy including coinciding intervention times will never be optimal. This is in general not the case in the subsequently considered problem of optimal life-time consumption. Further, condition (24) iv) implies that the maximum consumption should be the total wealth of the investor after selling of all stocks, that the wealth of the total holdings after transaction costs should be non-negative, and the investor is not allowed to sell bonds or stocks short. Moreover, condition (24) iv) can be written in the following form which we will use in the proof of Lemma 12:

$$\Delta S_i \in \left[-S(\theta_i -), \frac{B(\theta_{i-1})-K}{1+k} \right], \quad c_i \in [0, B(\theta_{i-1}) + (1-k)\, S(\theta_i -) - K]$$

$$\Delta S_i + c_i \leq B(\theta_{i-1}) - K - k\, |\Delta S_i| \qquad (\text{if} \quad \theta_{i-1} < \theta_i)$$

If we want to ensure a non-negative wealth of the holdings after transaction costs at **every** time instant (and not just at the intervention times as in (24) iv)), we have to require that the gains of a total sale,

$$\hat{X}(t) := B(t) - K + (1-k)S(t),$$

is non-negative for all t. This imposes a further constraint on an admissible impulse control strategy. One always has to intervene when the "net wealth process" $\hat{X}(t)$ reaches zero (for the first time). Then, the only admissible action is to sell all the holdings and to end up with a zero wealth. Of course, from now on we will always assume

$$\hat{X}(0) > 0 . \tag{N}$$

Under the additional requirement of a non-negative net wealth process we have to modify Definition 10:

Definition 10*

An admissible impulse control strategy $\{(\theta_i, \Delta S_i, c_i), i \in \mathbf{N}\}$ **with a non-negative net wealth process** is a sequence of **intervention times** θ_i and **actions** $(\Delta S_i, c_i)$ satisfying conditions (24) i), ii), iv), v) and

(24) ii*) θ_i is a stopping time w. r. t. $\sigma((S(t-),B(t-)), (\theta_n,\Delta S_n,c_n), n<i)$

$$\text{with } \theta_i \begin{cases} \leq \inf\left\{t \geq \theta_{i-1} \middle| \hat{X}(t)=0\right\}, & \text{if } \hat{X}(\theta_{i-1})>0 \\ =+\infty & \text{, else} \end{cases}$$

If, in contrast, one is not interested in a non-negative wealth process, one can also drop condition (24) iv) in Definition 10. Thus the constraints on $(\Delta S_i, c_i)$ disappear. On one hand this will make the whole situation more risky for the investor, but will on the other hand simplify the optimisation problem (as we will see in one of our following examples below).

We can now state impulse control versions of some typical problems of continuous trading in our setting. Here, U, U_1, U_2 will denote the utility functions which (unless otherwise stated) are required to be non-negative, continuous and monotonous or bounded. Moreover, we assume $U(0) = 0$ (unless otherwise stated). Let further Z be the set of admissible impulse controls (where "admissible" can have the meaning of Definitions 10 or 10* according to the investor's intentions) corresponding to the considered problems.

Problem 1 : "Optimal Terminal Wealth"

$$\max_{(\theta_i,\Delta S_i)\in Z} E\,(U(X(T))$$
$$\text{s.t. } X(T) = B(T) + (1-k)\,S(T) - K$$

i.e. we maximise expected utility of terminal wealth **after** selling all stocks. If we are just interested in maximising the utility from terminal paper wealth, we can drop the additional constraint on X(T).

Problem 2 : "Optimal Life-Time Consumption"

$$\max_{(\theta_i,\Delta S_i,c_i)\in Z} E\left(\sum_{i=1}^{\infty} e^{-\alpha\theta_i}\, U(c_i)\,1_{\{\theta_i<\infty\}}\right)$$
$$\text{s.t. } c_i = -(\Delta B_i + \Delta S_i) - (K + k\,|\Delta S_i|)$$

i.e. we consider the impulse control analogue to the problem presented in Sections 3.3 (without transaction costs) and 5.1 (with proportional transaction costs). Note that the consumption c_i can be interpreted as the excess of transaction gains over costs.

Problem 3 : "Optimal Consumption and Terminal Wealth"

$$\max_{(\theta_i,\Delta S_i,c_i)\in Z} E\left(\sum_{i=1}^{\infty} e^{-\alpha\theta_i} U_1(c_i) 1_{\{\theta_i<T\}} + U_2(X(T))\right)$$

$$\text{s.t.}\quad c_i = -(\Delta B_i + \Delta S_i) - (K + k\,|\Delta S_i|\,),\quad X(T) = B(T) + (1-k)\,S(T) - K$$

i.e. the combination of Problem 1 and a finite-time horizon version of Problem 2.

For convenience, we will concentrate on Problem 2 for the moment (where "admissible" should be understood in the sense of Definition 10), but extend the methods developed below in sub-section iv) to cover the other two problems. Before starting with the presentation of our theoretical results, we would like to point out one striking difference to the impulse control problems usually dealt with in the literature as in (Bensoussan and Lions 1984) or (Harrison e.a. 1983): There is **no additive separation** in the objective function between the utility one receives from steering the underlying process to a better position and the costs due to control actions in the portfolio problem. This separation was an essential feature in these sources allowing to solve some problems explicitly (see Section 6.3 for an example where the separation between transaction costs and transaction gains allows for a semi-explicit solution of an impulse control problem). Further, we now have to put a lot of additional constraints on the control actions and the underlying process.

ii) Characterisation of the Value Function via an Optimal Stopping Approach

Let

$$v(B, S) := \sup_{(\theta_i,\Delta S_i,c_i)\in Z} E_{B,S}\left(\sum_{i=1}^{\infty} e^{-\alpha\theta_i} U(c_i) 1_{\{\theta_i<\infty\}}\right)$$

be the value function of Problem 2. The main idea behind the solution method given in this section is the comparison of the value function with the following maximum operator:

$$Mv(B, S) := \max_{(c,\Delta S)\in A(B,S)} [\, v(B - K - c - \Delta S - k\,|\Delta S|,\ S + \Delta S) + U(c)\,] \tag{25}$$

where A(B, S) is the compact (perhaps empty) subset of $\mathbf{R}^2$ consisting of all pairs (c, ΔS) that satisfy the constraints (24) iv). For the maximum in (25) to exist, we assume the value function to be upper semi-continuous. Note that the maximum operator Mv(B, S) represents the value of the strategy that consists of doing the **best immediate action** (if our actual bond and stock holdings are (B, S)) and **optimal behaviour afterwards** (if A(B, S) is empty we define Mv(B, S):= 0). More precisely, the gains from following the strategy corresponding to Mv(B, S) consist of the immediate gain of utility U(c) from consumption and the gain from future optimal behaviour with the new holdings which is reflected in $v(B-K-c-\Delta S-k|\Delta S|, S+\Delta S)$. Note also that the transaction costs directly enter the value function v. Thus, their effect is only given implicitly and not explicitly as an additional additive term.

As it need not always be the best strategy to do an immediate action, we have

$$v(B, S) \geq Mv(B, S),$$

but at the first time (after the start in (B, S)) when it is optimal to intervene, v and Mv must coincide, which suggests the following form of the Bellman principle to hold :

$$v(B, S) = \sup_{\tau \in \Sigma} E_{B,S}\,(e^{-\alpha\tau}\, Mv(B, S(\tau))) =: G\,v(B, S) \qquad (26)$$

where Σ is the set of finite stopping times. Observe that equation (26) has reduced the impulse control Problem 2 to an optimal stopping problem. This is however only a formal gain, as we do not know the reward function $e^{-\alpha\tau}Mv(B, S(\tau))$ of our optimal stopping problem, and therefore the usual way of solving it by constructing the least superharmonic majorant fails. However, we will solve it by a value iteration procedure of the form: "Start with an initial estimate v_0 for v and define $v_n = G\,v_{n-1}$." The main characteristics of this procedure are given in the following theorem :

Theorem 11 "Optimal stopping and a Bellman principle"
Assume

$$v(B, S) < \infty \quad \forall\ (B, S) \subset \mathbf{R}_+^2$$

and set

$$v_0(B, S) \equiv 0 \ , \ v_n(B, S) \equiv G\,v_{n-1}(B, S) \quad \forall\ n \in \mathbf{N}\ .$$

Then, we have

$$v_n(B, S) \uparrow v(B, S) \ \text{ as } n \to \infty,$$
$$v(B, S) = G\,v(B, S)$$

i.e. the value function v is a fixed point of the operator G and therefore, Bellman's principle (26) holds.

Proof :

The proof parallels some of the ideas presented in Section 54 of (Davis 1993) who instead considered the optimal control of piecewise deterministic Markov processes. Before starting the proof we need two lemmas .

Lemma 12

Let $g:[0, \infty)^3 \to [0, \infty)$ be a continuous function. Then under either of the following two conditions

(b) g is bounded,

(d) g(t, B, S) is decreasing in t and increasing in B and in S and $\forall\, s \in [0, \infty)$ the expectation $E_{t,B,S}\,(g(t+s, B, S(t+s))$ exists,

we have that $G(t, B, S) = E_{t,B,S}\,(g(t+s, B, S(t+s))$ is a continuous function for every fixed $s \in [0, \infty)$ (where S(s) has dynamics given by equation (22)).

Proof (of Lemma 12):

Under condition (b), the lemma is a special case of Lemma 8.4 in (Oksendal 1992). In case of condition (d), let (t_n, B_n, S_n) be a sequence converging to (t, B, S). Then, by the requirements on g and by the explicit form of the process S(t+s), there exists a triplet (t', B', S') such that g(t', B', S') dominates $g(t_n, B_n, S_n)$ and g(t, B, S) in the sense needed to apply the dominated convergence theorem. With this, the proof of Lemma 8.4 in (Oksendal 1992) goes through.

□

Lemma 13

Let A(B, S) be the feasible region for the actions (c, ΔS) if (B, S) are the current bond and stock holdings of the investor. Then, A(B, S) is Lipschitz-continuous (as a set- valued function) in the topology which is induced by the Hausdorff metric.

Proof (of Lemma 13):

From the alternative description of condition (24) iv) given after the Definition 10, it can immediately be seen that the feasible region has two typical forms depending on the value of the bond holdings which are given in Figure 8 below. In the first case, it is possible to purchase additional shares of stock (up to a certain amount), because the transaction costs can be covered by changing the bond holdings. In the second case this is no longer possible; every admissible ΔS_i must be negative.

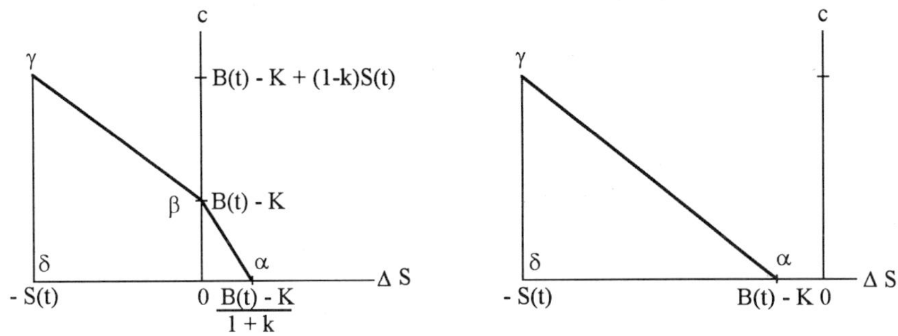

Figure 8: Typical forms of the feasible regions

Even more, we have to finance the costs of consumption by changing both the bond and stock holdings in a suitable way. In either of the two cases, the Hausdorff distance between two sets A(B, S) and A(B', S') is bounded from above by the maximum of the distances between corresponding pairs of points (α, α'), (β, β'), (γ, γ') and (δ, δ') (see Figure 8 for the meaning of these points). To see this, note that the boundaries are given by linear functions. Explicit calculation of these distances shows

$$d_H(A(B, S), A(B', S')) \leq q\, d_E((B, S), (B', S'))$$

where q is a real constant and d_H and d_E are the Hausdorff and Euclidean distances, respectively. Thus, Lemma 13 is proved.

□

Now, back to the proof of Theorem 11:

Step 1: Well-definedness of the value iteration
We show by induction that the optimal stopping problems can be solved at every step of the iteration $v_n(B, S) = G\, v_{n-1}(B, S)$.

n = 1 :
By defining Mv(B, S) = 0 if the feasible region for the actions is empty, we obtain

$$Mv_0(B, S(\tau)) = U((B + (1-k)S(\tau) - K)^+)$$

With the notations of

$$g_n(t, B, S) := \max_{s \in \zeta_n} E_{t,B,S}\big(g_{n-1}(t+s, B, S(t+s))\big) \;,\; n \in \mathbf{N}$$

$$g_0(t, B, S) := e^{-\alpha t}\ U((B + (1-k)\,S - K)^+),$$

(with $\zeta_n := \{k \cdot 2^{-n} \mid 0 \le k \le 4^n\}$), we can solve the problem

$$v_1(B, S) = \sup_{\tau \in \Sigma} E_{B,S}\left(e^{-\alpha\tau} M v_0(B, S(\tau))\right)$$

with the help of Theorem 10.7 of (Oksendal 1992) and obtain

$$v_1(B, S) = \hat{g}(0, B, S) = \lim_{n \to \infty} g_n(t, B, S).$$

Also, $v_1(B, S)$ is non-negative. It is increasing in B and in S if U is an increasing function. It is bounded if U is bounded. Further, $v_1(B, S)$ is continuous. To see this, observe that by definition of $\hat{g}$ we have

$$v_1(B, S) - v_1(B', S') = \hat{g}(0, B, S) - \hat{g}(0, B', S') \le g_n(0, B, S) + \varepsilon - g_n(0, B', S') \quad (27)$$

for a suitable $n \in \mathbf{N}$. Now, by Lemma 12 and an induction argument, one can show that the functions g_n are continuous as maxima of finitely many continuous functions. Therefore, the right hand side of (27) converges to ε if (B', S') converges to (B, S). By interchanging the roles of (B', S') and (B, S) in (27) and using the fact that ε is arbitrary, we obtain continuity of v_1.

n–1 → n :

Note that M conserves the non-negativity, boundedness or monotonicity properties of v_{n-1}. It remains to show that Mv_{n-1} is continuous, and then the optimal stopping problem at stage n can be solved as in the case n = 1. Let therefore $(B, S) \neq (B', S')$. This implies :

$$\begin{aligned}
&Mv_{n-1}(B, S) - Mv_{n-1}(B', S') \\
&\quad = \max_{(c,\Delta S) \in A(B,S)} [\, v_{n-1}(B - K - c - \Delta S - k\,|\Delta S|, S + \Delta S) + U(c)\,] - \\
&\quad - \max_{(c,\Delta S) \in A(B',S')} [\, v_{n-1}(B' - K - c - \Delta S - k\,|\Delta S|, S' + \Delta S) + U(c)\,] \\
&\qquad \le U(\bar{c}) - U(c) + v_{n-1}(B - K - \bar{c} - \overline{\Delta S} - k|\overline{\Delta S}|, S + \overline{\Delta S}) \\
&\qquad\quad - v_{n-1}(B' - K - c - \Delta S - k|\Delta S|, S' + \Delta S)
\end{aligned}$$

where $(\bar{c}, \overline{\Delta S})$ is a maximiser of the optimisation problem in $Mv_{n-1}(B, S)$ and where $(c, \Delta S) \in A(B', S')$. Next, choose δ_1, δ_2 such that we have

$$|U(\bar{c})-U(c)| < \frac{\varepsilon}{2} \quad \forall\, (c, \Delta S) \in A(B, S) \text{ with } |\bar{c}-c| < \delta_1,$$

$$v_{n-1}(B-K-\bar{c}-\overline{\Delta S}-k|\overline{\Delta S}|, S+\overline{\Delta S}) - v_{n-1}(B'-K-c-\Delta S-k|\Delta S|, S'+\Delta S)| < \frac{\varepsilon}{2}$$

$\forall\, (c, \Delta S) \in A(B, S)$ with $|(\bar{c}, \overline{\Delta S}) - (c, \Delta S)| < \delta_2$ for $\varepsilon > 0$. By Lemma 12, this is always achievable because (B', S') can be chosen in such a way that we have

$$d_H(A(B, S), A(B', S')) < \min(\delta_1, \delta_2)$$

for any (B', S') sufficiently close to (B, S). Consequently, we have

$$Mv_{n-1}(B, S) - Mv_{n-1}(B', S') < \varepsilon .$$

Interchanging of (B, S) and (B', S') yields

$$Mv_{n-1}(B', S') - Mv_{n-1}(B, S) < \varepsilon ,$$

hence continuity of Mv_{n-1}.

Step 2: Convergence of v_n

By induction, it can be shown that $v_n(B, S)$ is the value function corresponding to our impulse control problem if at most n interventions of the investor are allowed. Therefore, $v_n(B, S)$ is an increasing sequence which is bounded above by v(B, S). Hence, we have

$$v(B, S) \geq \lim_{n\to\infty} v_n(B, S) =: w(B, S) .$$

To show the opposite inequality, let $I = \{(\theta_i, \Delta S_i, c_i)\}_{i \in N}$ be an admissible impulse control strategy and define its corresponding n-step variant

$$I_n = \{(\theta_i, \Delta S_i, c_i)\}_{i=1,\dots,n} .$$

Thus, I_n is identical to I up to the n th intervention, and then no further action is allowed. Define

$$J_{B,S}(I) := E_{B,S}\Big(\sum_{i=1}^{\infty} e^{-\alpha\theta_i}\, U(c_i)\, 1_{\{\theta_i<\infty\}}\Big),$$

$$J_{B,S}(I_n) := E_{B,S}\Big(\sum_{i=1}^{n} e^{-\alpha\theta_i}\, U(c_i)\, 1_{\{\theta_i<\infty\}}\Big),$$

and observe

$$J_{B,S}(I) - J_{B,S}(I_n) = E_{B,S}\Big(\sum_{i=n+1}^{\infty} e^{-\alpha\theta_i} U(c_i) 1_{\{\theta_i < \infty\}} \Big) \to 0 \quad \text{as } n \to \infty . \quad (28)$$

Let A_n be the set of all admissible n-step strategies I_n. Then relation (28) implies

$$v(B, S) = \sup_{I \in \bigcup_{n \in N} A_n} J_{B,S}(I) \le w(B, S),$$

and this completes Step 2 of the proof.

Step 3 : Fixed point property of v
The combination of the two relations

$$v(B, S) \xleftarrow{n\to\infty} v_{n+1}(B, S) = G\, v_n(B, S) = \sup_{\tau \in \Sigma} E_{B,S}\big(e^{-\alpha\tau} Mv_n(B, S(\tau))\big) \quad (29)$$

$$\begin{aligned} Mv(B, S) &= \max_{(c,\Delta S) \in A(B,S)} [\, v(B - K - c - \Delta S - k\,|\Delta S|, S + \Delta S) + U(c) \,] \\ &= \max_{(c,\Delta S) \in A(B,S)} [\, \sup_{n \in N} v_n(B-K-c-\Delta S-k\,|\Delta S|, S + \Delta S) + U(c) \,] \\ &= \sup_{n \in N} \max_{(c,\Delta S) \in A(B,S)} [\, v_n(B-K-c-\Delta S-k\,|\Delta S|, S + \Delta S) + U(c) \,] \\ &= \sup_{n \in N} Mv_n(B, S) \end{aligned}$$

and application of the monotone convergence theorem to the right most term in relation (29) yields the fixed point property $v(B, S) = G\, v(B, S)$. To see that the interchange of "max" and "sup" in relation (29) is allowed, note that the equality

$$\begin{aligned} & \max_{(c,\Delta S) \in A(B,S)} [\, \sup_{n \in \{1,..,k\}} v_n(B-K-c-\Delta S-k\,|\Delta S|, S + \Delta S) + U(c) \,] \\ = \; & \sup_{n \in \{1,..,k\}} \max_{(c,\Delta S) \in A(B,S)} [\, v_n(B-K-c-\Delta S-k\,|\Delta S|, S + \Delta S) + U(c) \,] \end{aligned}$$

is valid for every $k \in \mathbf{N}$ (as the supremum over the finite set is always attained which is also the case for the maximum over the compact set A(B, S)).

□

Remark 14

a) Note that v(B, S) is lower semi-continuous as the limit of the increasing sequence of continuous functions $v_n(B, S)$. If we further have that v(B, S) is upper semi-continuous then Theorem 11 also yields the existence of an optimal impulse control strategy of the form:

"Intervene at every time instant when v(B, S) and Mv(B, S) coincide, and choose a pair (c, ΔS) that is a maximiser in the optimisation problem corresponding to Mv(B, S)."

Continuity of the value function v(B, S) can be proved under some additional Lipschitz assumptions on the utility functions U (for a similar result see (Korn 1997c)). In any case, one can at least obtain ε-optimal strategies by using a sequence of strategies that are optimal in the sub-problems characterised by v_n. These can be constructed with the help of the least superharmonic majorants (see Chapter 10 of (Oksendal 1992)).

b) Theorem 11 also yields a computational method to solve the impulse control problem:

"Solve a sequence of optimal stopping problems by standard methods (as presented in Chapter 10 of (Oksendal 1992)) until this sequence converges."

However, each step of this iteration involves a large amount of computations. So the practicability of this solution method relies crucially on the possibility to solve conventional optimal stopping problems in an efficient way.

iii) Quasi-Variational Inequalities and Optimal Impulse Control

Another approach to solving the optimal life-time consumption problem is that of solving the corresponding quasi-variational inequalities (see (Korn 1997d) for the statement and explicit solution of some (generalised) impulse control problems with the help of the characterisation of the value function via quasi-variational inequalities). A standard reference for the relations between quasi-variational inequalities and impulse control problems is the monograph (Bensoussan and Lions 1984). Unfortunately, our situation is not covered by their results (as already indicated in the introduction to this section).

Definition 15

The following relations are called the **quasi-variational inequalities** (for brevity: qvi) for the optimal life-time consumption problem of sub-section ii)

$$
\begin{aligned}
&\text{i)} && Lw(B,S) := \tfrac{1}{2}\sigma^2 S^2 w_{SS}(B,S) + bSw_S(B,S) - \alpha\, w(B,S) \le 0 \\
&\text{ii)} && w(B,S) \ge Mw(B,S) \qquad (30)\\
&\text{iii)} && (w(B,S) - Mw(B,S))\, Lw(B,S) = 0
\end{aligned}
$$

for $(B, S) \in [0,\infty)^2$ where the operator Mw(B, S) is defined as in (25).

Remark 16

The subscripts in (30) i) denote partial derivatives with respect to S (of the relevant order). However, we only require these partial derivatives to exist as left limits.

One can construct a special impulse control strategy from a solution of the quasi-variational inequalities .

Definition 17

A **qvi-control** with respect to a solution w of the qvi (if it exists !) is given by

i) $$(\theta_0; \Delta S_0, c_0) := (0 ; 0,0)$$

ii) $$\theta_i = \inf\{t \geq \theta_{i-1} \mid w(B(t), S(t)) = Mw(B(t), S(t))\}$$

iii) $$(\Delta S_i, c_i) = \arg \max_{(c,\Delta S)\in A(B,S)} \{w(B(\theta_i)-K-c-\Delta S-k|\Delta S|, S(\theta_i)+\Delta S)+U(c)\}$$

i.e. at every time instant when w and Mw coincide, the investor intervenes. Then, his action is the maximiser of the optimisation problem corresponding to Mw(B, S) (compare the argument leading to the formal Bellman principle in sub-section iii).

Remark 18

a) Both the definitions of the qvi and of a qvi-control have to be suitably modified in the case of the requirement of a non-negative net wealth process $\hat{X}(t)$. In particular, in part ii) of Definition 17, θ_i must be given by

$$\theta_i = \begin{cases} \inf\left\{t\geq\theta_{i-1}\middle|\hat{X}(t)=0 \textbf{ or } w(B(t),S(t))=Mw(B(t),S(t))\right\}, & \text{if } \hat{X}(\theta_{i-1})>0 \\ +\infty & \text{, else} \end{cases}.$$

b) As will be seen in Theorem 19 below, the role of the qvi in impulse control problems is the same as that of the HJB-Equation in stochastic control problems. Before stating this main result, we will introduce the following manner of speaking. We will say that "a function f satisfies Itô's formula (with respect to the process S(t))" if it is continuous and satisfies in "a certain sense"

$$f(t,S(t)) = f(0,S(0)) + \int_0^t \left(f_t(s,S(s)) + f_S(s,S(s))bS(s) + \tfrac{1}{2} f_{SS}(s,S(s))\sigma^2 S(s)^2\right)ds$$
$$+ \int_0^t f_S(s,S(s))\sigma S(s)dW(s) \qquad (*)$$

on time intervals where S(s) is continuous. This could be the usual meaning where f is a $C^{1,2}$- function. It will also include the cases where the partial derivatives are assumed to be left derivatives or generalised derivatives (compare for example (Krylov 1980) or (Haussmann 1994)).

Theorem 19 "Optimality of the qvi-control"
If there exists a solution v* of the qvi (30) that satisfies Ito's formula (in the sense of Remark 18 b) or Remark 20 a) below) and the growth conditions

$$E\left(\int_0^\infty (e^{-\alpha t}\sigma S(t) v^*_S (B(t),S(t)))^2\, dt\right) < \infty \tag{31}$$

$$E\left(e^{-\alpha T} v^*(B(T),S(T))\right) \xrightarrow{T\to\infty} 0 \tag{32}$$

for all processes (B(t),S(t)) corresponding to admissible impulse controls then we have

$$v(B, S) \le v^*(B, S) \quad \forall\, (B, S)\in[0,\infty)^2. \tag{33}$$

Further, if there exists an admissible qvi-control then it is an optimal impulse control, and v* is identical to the value function v.

Proof:
Let $\{(\theta_i, \Delta S_i, c_i)\}_{i\in N}$ be an admissible impulse control. Because we cannot guarantee that there are no intervention times that coincide, we have to introduce additional notation. Let

$$m_1 := 1, \quad m_i := \inf\{k\in\mathbf{N} : \theta_k > \theta_{m_{i-1}}\}.$$

Let further $T\in[0,\infty)$ be fixed. Define

$$\hat\theta_k := \theta_k \wedge T .$$

Let (B(t), S(t)) be the pair of bond and stockholdings at time t after all interventions at time t have been made while (B_i, S_i) denotes the pair after the i th action of the investor. Moreover, let S(t–) be the left limit of the S-process at time t. Then, we have

$$\begin{aligned}
&e^{-\alpha\hat\theta_{m_i}} v^*(B(\hat\theta_{m_i}), S(\hat\theta_{m_i})) - e^{-\alpha\hat\theta_{m_{i-1}}} v^*(B(\hat\theta_{m_{i-1}}), S(\hat\theta_{m_{i-1}}))\\
&= e^{-\alpha\hat\theta_{m_i}} v^*(B_{m_i-1}, S(\hat\theta_{m_i}-)) - e^{-\alpha\hat\theta_{m_{i-1}}} v^*(B_{m_i-1}, S(\hat\theta_{m_{i-1}}))\\
&+ 1_{\{\theta_{m_i}<T\}} \sum_{k=m_i+1}^{m_{i+1}-1} \left(e^{-\alpha\theta_{m_i}} \left(v^*(B_k,S_k) - v^*(B_{k-1},S_{k-1})\right)\right)\\
&+ 1_{\{\theta_{m_i}<T\}} e^{-\alpha\theta_{m_i}} \left(v^*\left(B_{m_i}, S_{m_i}\right) - v^*\left(B_{m_i-1}, S_{m_i}-\right)\right)
\end{aligned} \tag{34}$$

To see this, consider the following two cases

i) $\theta_{m_i} < T$:

The only terms not vanishing in square brackets at the right side of equation (34) form the difference

$$e^{-\alpha\theta_{m_i}} \left(v^*\left(B_{m_{i+1}-1}, S_{m_{i+1}-1}\right) - v^*\left(B_{m_i-1}, S_{m_i}-\right)\right).$$

But the identities

$$v^*(B(\hat{\theta}_{m_i}), S(\hat{\theta}_{m_i})) = v^*(B_{m_{i+1}-1}, S_{m_{i+1}-1}),$$

$$v^*(B(\hat{\theta}_{m_{i-1}}), S(\hat{\theta}_{m_{i-1}})) = v^*(B_{m_i-1}, S(\hat{\theta}_{m_{i-1}}))$$

yield equation (34) in this case.

ii) $\theta_{m_i} \geq T$:

If we also have $\theta_{m_{i-1}} \geq T$ then (34) reduces to the equation "0 = 0". Therefore, assume $\theta_{m_{i-1}} < T$. But then both sides of (34) are equal to

$$e^{-\alpha T}\, v^*(B_{m_i-1}, S(T)) - e^{-\alpha\theta_{m_{i-1}}}\, v^*(B_{m_i-1}, S(\hat{\theta}_{m_{i-1}})) .$$

Equation (34) can be interpreted as the decomposition of the difference of (optimal) utility gained from the process (B(t),S(t)) **between** two times of intervention **into** a difference caused by the evolution of the non-controlled process between these times **and** a difference caused by actions at the second intervention time. By summing up right and left sides of (34) from 1 to j, we get

$$e^{-\alpha\hat{\theta}_{m_j}}\, v^*(B(\hat{\theta}_{m_j}), S(\hat{\theta}_{m_j})) - v^*(B, S)$$

$$= \sum_{i=1}^{j} \left(e^{-\alpha\hat{\theta}_{m_i}}\, v^*\left(B_{m_i-1}, S\left(\hat{\theta}_{m_i}-\right)\right) - e^{-\alpha\hat{\theta}_{m_i}}\, v^*\left(B_{m_i-1}, S\left(\hat{\theta}_{m_{i-1}}\right)\right)\right)$$

$$+ \sum_{k=1}^{m_{j+1}-1} 1_{\{\theta_k < T\}} \left(e^{-\alpha\theta_k} \left(v^*(B_k, S_k) - v^*(B_{k-1}, S_k - \Delta S_k)\right)\right) \qquad (35)$$

Because v* is assumed to be a solution of the quasi-variational inequalities, we have the following two relations :

$$v^*(B_{k-1}, S_k - \Delta S_k) \geq v^*(B_k, S_k) + U(c_k)$$

$$v^*(B_{k-1}, S_k - \Delta S_k)\, e^{-\alpha\hat{\theta}_k} - v^*(B_{k-1}, S_{k-1})\, e^{-\alpha\hat{\theta}_{k-1}}$$

$$= \int_{\hat{\theta}_{k-1}}^{\hat{\theta}_k} e^{-\alpha s} Lv^*(B(s),S(s))ds + \int_{\hat{\theta}_{k-1}}^{\hat{\theta}_k} e^{-\alpha s} v^*_S(B(s),S(s))\sigma S(s)dW(s)$$

$$\leq \int_{\hat{\theta}_{k-1}}^{\hat{\theta}_k} e^{-\alpha s} v^*_S(B(s),S(s))\sigma S(s)dW(s).$$

Combining these relations with relation (35) results in

$$v^*(B, S) - e^{-\alpha\hat{\theta}_{m_j}} v^*(B(\hat{\theta}_{m_j}), S(\hat{\theta}_{m_j}))$$

$$\geq \sum_{k=1}^{m_{j+1}-1} 1_{\{\theta_k < T\}} U(c_k) e^{-\alpha\theta_k} - \int_0^{\hat{\theta}_{m_j}} e^{-\alpha s} v^*_S(B(s),S(s))\sigma S(s)dW(s). \quad (36)$$

Now, compute the expectation of the left side of relation (36), let j go to infinity, and use relation (32) to achieve

$$v^*(B, S) - E\left(e^{-\alpha\hat{\theta}_{m_j}} v^*\left(B\left(\hat{\theta}_{m_j}\right), S\left(\hat{\theta}_{m_j}\right)\right)\right)$$

$$\xrightarrow{j\to\infty} v^*(B, S) - E\left(e^{-\alpha T} v^*(B(T),S(T))\right). \quad (37)$$

On the right side of (36), we obtain (remember relations (31) and (37))

$$E\left(\sum_{k=1}^{m_{j+1}-1} \left(1_{\{\theta_k < T\}} U(c_k) e^{-\alpha\theta_k}\right) - \int_0^{\hat{\theta}_{m_j}} e^{-\alpha s} v^*_S(B(s),S(s))\sigma S(s)dW(s)\right)$$

$$\xrightarrow{j\to\infty} E\left(\sum_{k=1}^{\infty} \left(1_{\{\theta_k < T\}} U(c_k) e^{-\alpha\theta_k}\right)\right).$$

Letting T go to infinity and using assumption (32) results in

$$v^*(B, S) - E\left(e^{-\alpha T} v^*(B(T),S(T))\right) \xrightarrow{T\to\infty} v^*(B, S)$$

and in

$$E\left(\sum_{k=1}^{\infty} \left(1_{\{\theta_k < T\}} U(c_k) e^{-\alpha\theta_k}\right)\right) \xrightarrow{T\to\infty} E\left(\sum_{k=1}^{\infty} \left(1_{\{\theta_k < \infty\}} U(c_k) e^{-\alpha\theta_k}\right)\right).$$

Putting all these relations together, we have proved relation (33), because we have

$$v^*(B, S) \geq E\left(\sum_{k=1}^{\infty}\left(1_{\{\theta_k < T\}} U(c_k) e^{-\alpha\theta_k}\right)\right). \tag{38}$$

By observing that we get equality in relation (38) if we use an admissible qvi-control (to see this, note that all the inequalities yielding relation (38) are then in deed equalities), we have also shown the optimality of this control strategy and that v* coincides with the value function.

□

Remark 20

a) The above proof demonstrates that we could even weaken the requirement "v* satisfies Itô's formula in the sense of Remark 18 b)". Relation (33) can also be proved if instead of the equation (*) we only have an inequality of the form that f(t, S(t)) should be smaller or equal to the right hand side of equation (*). Examples for such relations can be found in (Krylov 1980). Further examples are implied by variants of the Itô-formula for concave functions (see e.g. Sections 3.6/7 of (Karatzas and Shreve 1988) where one has to do the obvious modifications for changing from convex to concave functions).

b) If there exists a solution of the quasi-variational inequalities that satisfies a polynomial growth condition (and Itô's formula) then it can easily be verified that it satisfies the assumptions of Theorem 19. Moreover, if there exists such a solution (and a corresponding qvi-control), it must be the unique such solution of the qvi due to Theorem 19 (the value function of an optimisation problem must be unique). Also in this case, it can be shown as in Section 4.4 of (Korn 1992) that every qvi-control must be admissible.

c) As in the optimal stopping problem in sub-section ii), the quasi-variational inequalities can be solved by an iteration procedure similar to the one presented there:

"Start with $v_0 \equiv 0$, and then define v_n as the unique, sufficiently regular (in the sense of Theorem 19) solution of the variational inequalities

$$\begin{aligned} &\text{i)} && L\, v_n(B, S) \leq 0 \\ &\text{ii)} && v_n(B, S) \geq Mv_{n-1}(B, S) \\ &\text{iii)} && L\, v_n(B, S)\,(v_n(B, S) - Mv_{n-1}(B, S)) = 0 \end{aligned} \tag{39}$$

(if such a solution exists)."

Once again, $v_n(B, S)$ is the value function of the problem where at most n interventions of the investor are allowed. The implicit inequality $v(B, S) \geq Mv(B, S)$ is now replaced by the explicit inequality (39) ii) where the right side is explicitly known at step n of the iteration. Therefore, we only have to deal with variational inequalities at every stage of the iteration. Convergence of the sequence $v_n(B, S)$ is guaranteed under the assumptions of Theorem 19.

d) Although we will not introduce the subject here, we report that under the assumption of continuity of the value function and the validity of a certain Bellman principle, it is possible to prove that the value function of our portfolio problem is a viscosity solution of the qvi. This might be important for the convergence of numerical discretisation schemes of the qvi.

iv) Extensions

Finite time horizon problems

If we consider problems with a finite time horizon, such as Problem 1 or Problem 3 in sub-section i), then our method of analysis presented in sub-sections ii) and iii) is applicable with only some minor changes. First of all, the time variable t will appear in the value function. So, in problem 1 the value function will be defined as

$$v(t, B, S) = \sup_{(\theta_i, \Delta S_i, c_i) \in Z} E_{t,B,S}(U(X(T))).$$

If we define the analogue to the maximum operator M as

$$Mv(t, B, S) = \max_{(c, \Delta S) \in A(B,S)} v(t, B-K-c-\Delta S-k|\Delta S|, S + \Delta S),$$

we can once again argue that we must have

$$v(t, B, S) \geq Mv(t, B, S),$$

and at the first time (after the start in (t, B, S)) when it is optimal to intervene, v and Mv must coincide. But because the time horizon T is finite, there may be situations when it is optimal not to intervene on the whole time interval [0, T] at all. Therefore, the Bellman principle (26) must have the form

$$v(t,B,S) = \max\left\{ v_0(t,B,S),\ \sup_{\tau \in \Sigma_{t,T}} E_{t,B,S}\left(Mv(\tau,B,S(\tau))\right)\right\} =: G\, v(t,B,S) \quad (26^*)$$

where $v_0(t, B, S)$ is the expected utility of the no intervention strategy starting in (t, B, S), i.e.

$$v_0(t, B, S) = E_{t,B,S}\left(U\left(B - K + (1-k)Se^{(b-\frac{1}{2}\sigma^2)(T-t)+\sigma(W(T)-W(t))} \right)\right).$$

Substituting this in Theorem 11 (note that the iteration procedure now has to start with $v_0(t, B, S)$ defined above !) will deliver analogous results for the finite time horizon case. Of course, the quasi-variational inequality approach has to be modified, too. One has to add a boundary condition corresponding to the finite time horizon to the quasi-variational inequalities. Further, the differential operator will now include a partial derivative with respect to the time variable t. So, for Problem 1 the qvi have the following form :

$$\begin{aligned}
&\text{i)} \quad Lw(t,B,S) := \tfrac{1}{2}\sigma^2 S^2 w_{SS}(t,B,S) + bSw_S(t,B,S) + w_t(t,B,S) \le 0 \\
&\text{ii)} \quad w(t,B,S) \ge Mw(t,B,S) \\
&\text{iii)} \quad (w(t,B,S)-Mw(t,B,S))\, Lw(t,B,S) = 0 \qquad (30^*) \\
&\text{iv)} \quad w(T,B,S) = U(B-K+(1-k)S)
\end{aligned}$$

$$\text{for } (t,B,S)\in[0,T)\text{x}[0,\infty)^2.$$

Due to the fact that we could end up with $B(T)-K+(1-k)S(T) < 0$, i.e. a negative terminal wealth after selling all of our holdings, we have to decide what to do in this case. We could still require that all shares must be sold. Then our utility functions must be defined (at least) on $[-K,\infty)$ with $U(-K) = 0$. We could also say that in this case an investor will not sell his securities. Then we have to replace condition (30*) iv) by

$$\text{iv*)} \quad w(T,B,S) = U((B-K+(1-k)S)^+)\,. \qquad (30^*)$$

With these modifications, we will obtain the same optimality results as in Theorem 19. As in the optimal stopping method, one has to start the iteration procedure described in Remark 20 d) with $v_0(t, B, S)$ defined as above.

More than one stock

If we consider a securities market with more than one stock then the foregoing methods of analysis will go through with some multi-dimensional modifications. Substituting the scalar process $S(t)$ by a vector process consisting of components $S_i(t)$, $i = 1,\ldots, n$ with dynamics given by

$$dS_i(t) = S_i(t)\left(b_i dt + \sum_{j=1}^{n} \sigma_{ij} dW_j(t)\right)$$

does not alter any result in sub-section iii) if we parallel the one-dimensional arguments. For example, the argument proving Lemma 13 is still valid but the notation becomes more complicated because we do not compare two-dimensional feasible regions any longer. Further, the operator L occurring in the qvi is now defined as

$$L\,w(B,S) = \tfrac{1}{2}\sum_{i=1}^{n}\sum_{j=1}^{n} a_{ik}(S)\, w_{S_i S_k}(B,S) + \sum_{i=1}^{n} b_i S_i w_{S_i}(B,S) - \alpha w(B,S)$$

with

$$a_{ik}(S) := \sum_{j=1}^{n} \sigma_{ij}\sigma_{kj} S_i S_k .$$

With this (and other minor changes of notation) all the arguments of sub-sections ii) and iii) will stay correct.

More general price processes

If we consider a stock price process of the form

$$dS(t) = b(S(t), t)\,dt + \sigma(S(t), t)\,dW(t), \quad S(0) = s \in [0,\infty)$$

where b and σ are linearly bounded, Lipschitz continuous functions such that 0 will be an absorbing point for (every component of) S(t) then the qvi-analysis of sub-section iii) will go through (one- or more-dimensional) if one substitutes bS(t) by b(S(t), t) and $\sigma S(t)$ by $\sigma(S(t), t)$. The only critical point in the optimal stopping approach lies in the proof of Lemma 12 under assumption (d) which has to be modified corresponding to the exact form of the functions b(S(t), t) and $\sigma(S(t), t)$.

v) Simple Examples

Up to now, one should have got the feeling that explicit analytical solutions to impulse control problems are much harder to obtain than in the case of stochastic control problems as in Section 3.3. But even there, such explicit analytical solutions were the exception rather than the rule. However, the use of impulse control methods in practise is directly connected with the possibility to solve such problems (at least approximately) and not only with finding nice characterisations for their value functions. Therefore, in this section, we give some examples to illustrate the structure of the optimal strategies and to show methods how to solve the optimisation problems. Two examples of maximising the utility of terminal wealth will be studied. To solve them, we use the optimal stopping method for the first and the qvi-characterisation for the second example.

However, there is still a big need for efficient numerical algorithms or approximation methods to solve "real problems" (i.e. ones with at least a moderate number of securities). This subject is currently undergoing a lot of research and future results will show if the highly realistic impulse control approach will gain success in practical applications.

In our examples, we first consider the following problem of

Optimal terminal wealth : Linear utility

We assume the market setting of sub-section i) and consider the problem

$$\max_{(\theta_i, \Delta S_i) \in Z} E(X(T)) = \max_{(\theta_i, \Delta S_i) \in Z} E(B(T) - K + (1-k)S(T))$$

where Z is the set of admissible strategies in the sense of relation (30*). This problem will be solved by applying the optimal stopping method of sub-section ii). For this, observe that we have

$$v_0(t, B, S) = B - K + (1-k)Se^{b(T-t)} \tag{40}$$

By computing $Mv_0(t, B, S)$ we will see that we have to distinguish between different cases. Note further that because of the prohibition of short selling, we know that v(t, B, S) is bounded from below by –K which corresponds to (B, S) = (0,0). Therefore, we set $Mv_i(t, B, S)$ equal to –K if the feasible region for the corresponding optimisation problem will be empty (This will cause an obvious modification of foregoing arguments in sub-section ii)). One can also work with the utility function U(x) = x + K ensuring a non-negative utility). We have to distinguish between three different cases for the value of the mean rate of stock return b.

<u>i) b < 0</u>

$$Mv_0(t, B, S) = \begin{cases} v_0(t, B+(1-k)S-K, 0), & \text{if } B+(1-k)S-K \geq 0 \\ -K & , \text{else} \end{cases} \tag{41}$$

This corresponds to an immediate selling of all stocks if this action does not involve short selling of bonds. Comparing (41) and (40) and using the definition of $v_1(t, B, S)$ leads to

$v_1(t, B, S)$

$$= \begin{cases} v_0(t, B+(1-k)S-K, 0), & \text{if } (1-k)S-K \geq 0 \text{ and } \frac{1}{t}\ln(1-\frac{K}{S(1-k)}) \geq b \\ E^{t,B,S}\left[v_0(J, B+(1-k)S_\vartheta - K, 0)1_{(\vartheta < T)} + (B-K+(1-k)S(T))1_{(\vartheta \geq T)}\right], & \text{else} \end{cases} \tag{42}$$

with

$$\vartheta := \inf\left\{ s > t \,\middle|\, \frac{1}{s}\ln\left(1 - \frac{K}{S_s(1-k)}\right) \geq b \right\}.$$

Hence, the optimal one-intervention strategy is to sell all stocks at the first time instant when the expected gains from not holding the stocks to the terminal time T will pay the transaction costs. This will also be the optimal impulse control strategy. To see this, consider the situation in the next iteration: v_2 is equal to v_1 on the first set mentioned in (42) because there, we have $v_2 = Mv_1 = Mv_0 = v_1$. On the second set, we will again have $v_2 = v_1$ due to the linearity of v_0, the supermartingale property of S(t), and the optional sampling theorem (see Theorem A.18). Thus, one has the same strategy as in the previous iteration, and therefore $v_1(t, B, S)$ is the limit of the iteration procedure. Hence, it coincides with the value function.

ii) $b \in \left[0, \frac{1}{T}\ln\left(\frac{1+k}{1-k}\right)\right)$

In this case, we have

$$Mv_0(t, B, S) = \begin{cases} v_0(t, B-K, S), & \text{if } B > K \\ -K & \text{, else} \end{cases},$$

i.e. the optimal immediate action is to do "a zero intervention" which implies

$$v_1(t, B, S) \quad = v_0(t, B, S).$$

Hence, it is optimal to do nothing until terminal time.

iii) $b \geq \frac{1}{T}\ln\left(\frac{1+k}{1-k}\right)$

Now, we have

$$Mv_0(t, B, S) = \begin{cases} v_0\left(t, 0, S + \frac{B-K}{1+k}\right), & \text{if } B > K \\ -K & \text{, else} \end{cases}$$

i.e. the optimal immediate action is to sell all bonds. Comparing this with $v_0(., ., .)$ given by equation (40) leads to

$$v_1(t, B, S)$$
$$= \begin{cases} v_0\left(t, 0, S + \frac{B-K}{1+k}\right), \text{ if } B > K \text{ and } b \geq \frac{1}{t}\ln\left(\frac{(1+k)B}{(1-k)(B-K)}\right) \\ E^{t,B,S}\left[v_0\left(\vartheta, 0, S_\vartheta + \frac{B-K}{1+k}\right)1_{(\vartheta < T)} + (B-K+(1-k)S(T))1_{(\vartheta \geq T)}\right], \text{ else} \end{cases}$$

Similar to case i), the optimal one-intervention strategy will now consist of selling all bonds at the first time instant where the gains from holding stocks instead of bonds will pay the transaction costs. Also, one can show (as in case i)) that this must be the optimal strategy for the whole problem.

Another, more interesting example will be the problem of

Optimal terminal wealth : Exponential utility
As in the preceding example, we consider the same market model as in sub-section i) (i.e. a market with a bond with a constant price and a single stock with a price dynamics given by a geometric Brownian motion) . There, we look at the problem

$$\max_{(\theta_i, \Delta S_i) \in Z} E\left(-e^{-\lambda X(T)}\right)$$

(where $\lambda > 0$ is the coefficient of absolute risk aversion) and assume for simplicity that there will be no transaction costs at the terminal time T (or equivalent: the investor is only interested in maximising his paper wealth). For notational convenience, we have dropped the constant term "1" in the utility function compared to Example 3.31. This will only affect the optimal utility and not the optimal strategy. The optimal strategy in the no transaction cost case was derived by Pliska (1986) and is already given in Section 3.5 as

$$\varphi^*(t) = X(t) - \frac{b}{\lambda\sigma^2}, \quad \psi^*(t) = \frac{b}{\lambda\sigma^2 P_1(t)}$$

where $\psi^*(t)$ is the number of shares of the stock held at time t, and $\varphi^*(t)$ is the amount of money invested in the bond at time t (recall that the bond price is always equal to one). The remarkable feature of this result is that the optimal amount of money invested in the stock depends only on the coefficient of risk aversion of the investor and not on his wealth. Note further that the corresponding wealth process is a Brownian motion with drift (see Section 3.5), in particular, we have

$$X(t) = x + \frac{1}{\lambda}\left(\frac{b}{\sigma}\right)^2 t + \frac{1}{\lambda}\frac{b}{\sigma} W(t) .$$

In particular, X(t) can attain negative values. To compare this result to the case including transaction costs, we will also drop the requirement of a non-negative wealth process in the corresponding impulse control formulation. Thus, the set Z consists of all impulse control strategies satisfying the conditions (24) i), ii), iii) and v). Note that c_i must be deleted in all these conditions as we do not consider consumption in this example. Due to the characteristics of the exponential function, we have the following relation for the value function v(t, B, S) of our problem:

$$v(t, B, S) = e^{-\lambda B} v(t, 0, S).$$

We would therefore expect that the optimal strategy for stock investment is independent of the total wealth of the investor. More precisely, we would expect the following form for the optimal strategy:

"Intervene only if the stock price $P_1(t)$ has moved in such a way that your current number of stocks owned is far away from $\psi^*(t)$."

The bond holdings are uniquely given by the self-financing condition. Thus the optimal strategy should be given by an interval $[\psi^*(t) + y_-(P_1(t),t), \psi^*(t) + y_+(P_1(t),t)]$ with boundaries depending on the stock price and the number of shares of the stock currently owned. We further have to specify the optimal restarting points after an intervention, $\psi^*(t) + \hat{y}_-(P_1(t),t)$ and $\psi^*(t) + \hat{y}_+(P_1(t))$. It will be convenient for our subsequent analysis to separate the stock price and the number of shares by introducing the function

$$q(t, p, y) := v(t, 0, py) .$$

Setting up the qvi for our original problem and expressing $v(.,.,.)$ in terms of $q(.,.,.)$ leads to the following form of the qvi:

$$\begin{aligned}
&\text{i)} \quad L\, q(t, p, y) := \tfrac{1}{2}\sigma^2 p^2\, q_{pp}(t, p, y) + bpq_p(t, p, y) + q_t(t, p, y) \le 0,\\
&\text{ii)} \quad q(t, p, y) \ge M\, q(t, p, y) := \max_{u \in R}\left\{e^{-\lambda(up+k|u|p+K)} q(t,p,y+u)\right\}, \qquad (43)\\
&\text{iii)} \quad (q(t, p, y) - M\, q(t, p, y))\, L\, q(t, p, y) = 0,\\
&\text{iv)} \quad q(T, p, y) = -e^{-\lambda py}
\end{aligned}$$

$$\text{for } (t, p, y) \in [0,T) \times (0,\infty) \times \mathbf{R}.$$

Since no bond component enters the qvi, the problem appears to be totally similar to the one-dimensional ones in (Harrison e.a. 1983), (Sulem 1986) or (Korn 1997d). But note that there is no closed form solution of the boundary value problem corresponding to the non-intervention strategy in this case. In particular, we would have no starting point v_0 for an optimal stopping iteration. Moreover, due to the fixed time horizon, we cannot expect the boundary of the non-intervention region having a shape as simple as in the examples presented in the above cited references.

For these reasons, we will solve the above qvi (43) (only) approximately by an asymptotic analysis. This approach was first presented in a portfolio optimisation context (but with a completely different cost structure) by (Atkinson and Wilmott 1995). We will look at this problem in the next section. Our analysis of the present example follows the idea and presentation of (Whalley and Wilmott 1994). They performed an asymptotic analysis on the results of (Davis, Panas, Zariphopolou 1993) for option pricing in the presence of transaction costs.

The basic idea behind this approach is that we assume the transaction costs being non-zero but small. They are assumed to be a function of a (small) parameter ε which will vanish for $\varepsilon = 0$. Further, (a suitably transformed version of) the value function will also be viewed as a function of this parameter ε. Thus, for $\varepsilon = 0$ the value function of our impulse control problem and the value function of the corresponding portfolio problem without transaction costs coincide (provided the value function is continuous in ε). In the no transaction region (i.e. the region where we have $Lq = 0$ and $q > Mq$), the value function will then (formally) be written as an expansion of powers of ε. This expression is put into the equation $Lq = 0$. The resulting terms are then ordered by different powers of ε. This will give us an infinite set of equations for the coefficient functions of the expansion as all coefficients have to vanish simultaneously. However, believing that at least for small values of ε, we could replace the infinite expansion of the value function by a finite one (i.e. by ignoring terms of higher order of ε) to get a good approximation, we will only consider the first three such equations which we can solve explicitly with the help of appropriate boundary conditions.

The natural point about which such an expansion should be made, is the optimal number of shares of the stock of the problem without transaction cost, i.e. the Pliska solution

$$y^*(p, t) := \psi^*(t) = b/(\sigma^2 \lambda P_1(t)) .$$

As in (Atkinson and Wilmott 1995) and in (Whalley and Wilmott 1994), we first rescale the y-variable and the cost function by introducing Y and K given by

$$y = y^*(p, t) + \varepsilon^{1/4} Y , \tag{44}$$

$$\varepsilon K(p, u) = k(p, |\varepsilon^{1/4} u|)$$

where k(p, u) represents the transaction costs (fixed and proportional) if buying of u shares of the stock for a price of p takes place (where a negative u corresponds to selling). It will prove to be very convenient for our subsequent analysis to assume the following form of the cost function :

$$k(p, u) = A\varepsilon^{3/4} |u| p + B\varepsilon \qquad (0<\varepsilon <<1,\ A, B > 0).$$

This form coincides with our usual form for the cost function if we take $k = A\varepsilon^{3/4}$ and $K = B\varepsilon$. As a consequence, we have

$$K(p, u) = A|u|p + B,$$

i.e. the scaled costs $K(p, u)$ are independent of ε. The above scaling by $\varepsilon^{1/4}$ seems to be a unmotivated but will prove to be the appropriate one in the end by delivering the magnitude of the width of the no transaction interval (for fixed (p, t)). Thus, it keeps the whole analysis in the interesting area. In deed, it is a crucial point to work out a scaling of the variables such that the asymptotic analysis finally leads to a solution (see (Nayfeh 1981) for an introduction to the method of asymptotic analysis and its underlying ideas). Another convenient notation is provided by

$$Q(t, p, Y) := q(t, p, y) \tag{45}$$

We further assume that the (p, y)-space (i.e. the space of stock holdings (expressed in numbers of shares) as functions of the stock price) can be decomposed in three regions (that of course could change over time) as indicated in the schematic Figure 9 below. NT is the no transaction region, i.e. the region where the optimal action of the investor consists of not changing his holdings. The region SS is the region where the optimal action of the investor is to sell sufficiently many shares of the stock to get to NT. BS is the region where he should buy shares to reach NT. The dashed lines represent the optimal restarting points of the (p, y) process after reaching the upper/lower boundary (or after starting above/below NT). We refer to them in a moment.

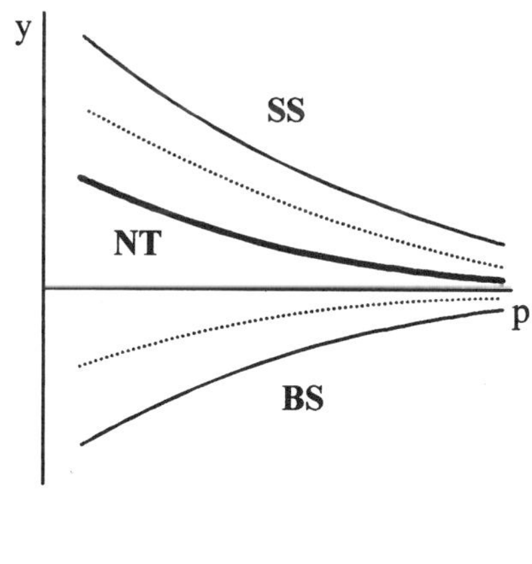

Figure 9: Schematic diagram of the (p, y)-space

We assume that the upper boundary of NT and the optimal upper restarting line (i.e. the upper dashed line) have the forms

$$y^*(p, t) + \varepsilon^{1/4}\hat{Y}^+, \quad y^*(p, t) + \varepsilon^{1/4} Y^+$$

where for notational convenience we have suppressed the dependence of $\hat{Y}^+$, Y^+ on p and t. Hence, if the process $(P_1(t), y(t))$ of stock prices and numbers of shares held by the investor reaches the upper boundary of NT, $y^*(P_1(t),t) + \varepsilon^{1/4}\hat{Y}^+$(due to both changings in the stock price and the boundary over time), after starting inside NT then the optimal action of the investor is to sell $\varepsilon^{1/4}(\hat{Y}^+ - Y^+)$ shares. Then, the investor's new pair $(P_1(t), y(t))$ is $(P_1(t), y^*(P_1(t),t) + \varepsilon^{1/4}Y^+)$, i.e. the corresponding point on the upper restarting line. Also, we assume a similar form for the lower boundary of NT and the lower restarting line

$$y^*(p, t) + \varepsilon^{1/4}\hat{Y}^-, \quad y^*(p, t) + \varepsilon^{1/4} Y^-$$

(but where the signs of $\hat{Y}^-, Y^-$ are negative). These assumptions will be justified in the end when we obtain explicit expressions for $\hat{Y}^\pm, Y^\pm$. Further, in SS we must have equality in relation (43) ii). Hence :

$$Q(t, p, Y) = q(t, p, y) = e^{\lambda\left(K + k\varepsilon^{1/4}\left|Y - Y^+\right|p + \varepsilon^{1/4}(Y - Y^+)p\right)} q(t, p, y^*(t, p) + \varepsilon^{1/4} Y^+)$$

$$= Q(t, p, Y^+) e^{\lambda\left(\varepsilon K(p, Y - Y^+) + \varepsilon^{1/4}(Y - Y^+)p\right)}, \qquad (p, y) \in SS. \qquad (46)$$

Similarly, in BS we have

$$Q(t, p, Y) = Q(t, p, Y^-) e^{\lambda\left(\varepsilon K(p, Y - Y^-) + \varepsilon^{1/4}(Y - Y^-)p\right)}, \qquad (p, y) \in SB.$$

In NT we (formally) expand Q(t, p, Y) in powers of $\varepsilon^{1/4}$ around the Pliska solution $y^* = y^*(t, p)$ in the following way :

$$Q(t, p, Y) =$$

$$-e^{-\left[\lambda p y^* + H_0(p,t) + \varepsilon^{1/4}\lambda p Y + \varepsilon^{1/2} H_2(p,t) + \varepsilon^{3/4} H_3(p,t) + \varepsilon H_4(p,t,Y) + \varepsilon^{5/4} H_5(p,t,Y) + ..\right]} \tag{47}$$

This choice of expansion will again be justified by the final result and is inspired by the papers of (Atkinson and Wilmott 1995) and (Whalley and Wilmott 1994). As in (Whalley and Wilmott 1994), we could also include a further coefficient function $H_1(p, t)$ for $\varepsilon^{1/4}$, but our analysis would show that it would be identically zero.

Therefore, we omit it. For the following computations, note that due to the definition (44), Y is also a function of y*. From the form (47) of Q and from (45), we obtain :

$$q_t = -Q\cdot[H_{0_t} + \varepsilon^{1/2} H_{2_t} + \varepsilon^{3/4} H_{3_t} + \ldots.] \tag{48}$$

$$q_p = -Q\cdot[\lambda y^* + H_{0_p} + \varepsilon^{1/4}\lambda Y + \varepsilon^{1/2} H_{2_p} + \varepsilon^{3/4}(H_{3_p} - y^*_p H_{4_Y}) + \varepsilon H_{4_p} + \ldots] \tag{49}$$

$$\begin{aligned} q_{pp} = {} & Q\cdot[\lambda y^* + H_{0_p} + \varepsilon^{1/4}\lambda Y + \varepsilon^{1/2} H_{2_p} + \varepsilon^{3/4}(H_{3_p} - y^*_p H_{4_Y}) + \varepsilon H_{4_p} + \ldots]^2 \\ & - Q\cdot[H_{0_{pp}} + \varepsilon^{1/2}(H_{2_{pp}} + (y^*_p)^2 H_{4_{YY}}) + \varepsilon^{3/4}(H_{3_{pp}} - y^*_{pp} H_{4_Y} - 2y^*_p H_{4_{Yp}}) \\ & + \varepsilon H_{4_{pp}} + \ldots] \end{aligned} \tag{50}$$

where the (lowest) subscripts denote partial derivatives with respect to the corresponding variables, and the dots indicate that all the remaining terms include powers of ε of higher (or at least equal) order. Substituting the representations (48–50) into the partial differential equation Lq = 0 (recall that we assume to be in NT), and ordering the resulting terms by powers of ε results in

$$\begin{aligned} 0 & = q_t + bpq_p + \tfrac{1}{2}\sigma^2 p^2 q_{pp} \\ & = -Q\cdot[\{H_{0_t} + bp(\lambda y^* + H_{0_p}) + \tfrac{1}{2}\sigma^2 p^2 (H_{0_{pp}} - (\lambda y^* + H_{0_p})^2)\} \\ & \quad + \varepsilon^{1/4}\{bp\lambda Y - \sigma^2 p^2(\lambda^2 y^* Y + H_{0_p}\lambda Y)\} \\ & \quad + \varepsilon^{1/2}\{H_{2_t} + bpH_{2_p} + \tfrac{1}{2}\sigma^2 p^2 (H_{2_{pp}} + (y^*_p)^2 H_{4_{YY}}) \\ & \quad - \tfrac{1}{2}\sigma^2 p^2(\lambda^2 Y^2 + 2\lambda y^* H_{2_p} + 2H_{0_p} H_{2_p})\} + \ldots] \end{aligned}$$

which in particular yields the three equations

$$0 = H_{0_t} + bp(\lambda y^* + H_{0_p}) + \tfrac{1}{2}\sigma^2 p^2 (H_{0_{pp}} - (\lambda y^* + H_{0_p})^2), \tag{O(1)}$$

$$0 = bp\lambda Y - \sigma^2 p^2(\lambda^2 y^* Y + H_{0_p}\lambda Y), \tag{$O(\varepsilon^{1/4})$}$$

$$\begin{aligned} 0 = {} & H_{2_t} + bpH_{2_p} + \tfrac{1}{2}\sigma^2 p^2 (H_{2_{pp}} + (y^*_p)^2 H_{4_{YY}}) \\ & - \tfrac{1}{2}\sigma^2 p^2(\lambda^2 Y^2 + 2\lambda y^* H_{2_p} + 2H_{0_p} H_{2_p}), \end{aligned} \tag{$O(\varepsilon^{1/2})$}$$

because all coefficients of different powers of ε have to vanish simultaneously such that Lq = 0 can be satisfied (if the region NT of the assumed form has non-empty interior). By solving $(O(\varepsilon^{1/4}))$ for H_{0_p} and substituting the result in (O(1)), we obtain

$$H_{0_t} + \tfrac{1}{2}\sigma^2 p^2 H_{0_{pp}} = -\frac{b^2}{2\sigma^2}. \tag{51}$$

Note that for ε = 0, Q(t, p, Y) of the form (47) can only be equal to the (suitably transformed) value function of the problem without transaction costs if we have

$$H_0(p, t) = \frac{b^2(T-t)}{2\sigma^2}$$

which is the solution of the partial differential equation (51) with zero final data which is, of course, the natural final condition. Next, we look at the $(O(\varepsilon^{1/2}))$-equation. Using the just obtained form for H_0 results in a simpler form of this equation which becomes a differential equation for $H_{4_{YY}}$,

$$0 = H_{2_t} + \tfrac{1}{2}\sigma^2 p^2 (H_{2_{pp}} + (y^*_p)^2 H_{4_{YY}}) - \tfrac{1}{2}\sigma^2 p^2 \lambda^2 Y^2$$

with solution

$$H_4 = -\frac{Y^2 G(H_2)}{\left(\sigma p y^*_p\right)^2} + \frac{\lambda^2 Y^4}{12(y^*_p)^2} + aY + b\,, \quad G(H_2) := H_{2_t} + \tfrac{1}{2}\sigma^2 p^2 H_{2_{pp}}$$

where the "constants" a and b still have to be determined.

We have now reached the point where the boundary conditions at the border of NT enter the scene. We will require the usual smooth-pasting conditions for the value function in this case (compare also to the Davis and Norman method in Section 1), i.e. we will require that the value function is continuous and continuously differentiable in p (for fixed (y, t)). It is shown in (Korn 1992), Chapter 4, that there exists a suitable version of Itô's formula that can be applied in this case (Hence, the assumptions of Theorem 19 would be satisfied). As usual, these assumptions will be used to determine the explicit form of the value function by considering its behaviour at the borders between NT, SS, and BS.

We will first concentrate at the border between NT and SS. Using relation (46) with $Y = Y^+$ and the expansion (47) of Q, the requirement of continuity of the value function at the border between NT and SS leads to the equation

$$e^{-\left[\lambda p y^* + H_0(p,t) + \varepsilon^{1/4}\lambda p \hat{Y}^+ + \varepsilon^{1/2} H_2(p,t) + \varepsilon^{3/4} H_3(p,t) + \varepsilon H_4(p,t,\hat{Y}^+) + \ldots\right]}$$

$$= e^{-\left[\lambda p y^* + H_0(p,t) + \varepsilon^{1/4}\lambda p Y^+ + \varepsilon^{1/2} H_2(p,t) + \varepsilon^{3/4} H_3(p,t) + \varepsilon H_4(p,t,Y^+) + \ldots\right]}$$

$$e^{\lambda\left(\varepsilon K(p, Y^+ - \hat{Y}^+) + \varepsilon^{1/4}(Y^+ - \hat{Y}^+)p\right)}$$

Comparing the coefficients of ε yields the relation

$$H_4(p, Y^+, t) - H_4(p, \hat{Y}^+, t) = \lambda K(p, Y^+ - \hat{Y}^+) \tag{52}$$

for all $(p, \hat{Y}^+)$ such that the corresponding pair (p, y) is part of the border between NT and SS. Due to relation (46), equation (52) is also valid for all Y such that the corresponding pair (p, y) is in SS. Optimality of the transaction $\Delta y = Y^+ - \hat{Y}^+$ requires

$$\frac{d}{d(\Delta y)}\left(Q\left(t, p, \hat{Y}^+ - \Delta y\right) e^{\lambda\left(\varepsilon K(p, -\Delta y) + \varepsilon^{1/4}(-\Delta y)p\right)} \right)\Bigg|_{\Delta y = Y^+ - \hat{Y}^+} = 0$$

leading to

$$H_{4_Y}(p, Y^+, t) = \lambda K_u(p, Y^+ - \hat{Y}^+) \tag{53}$$

for all $(p, \hat{Y}^+)$ as above (here, K_u denotes the partial derivative of K with respect to the second component). Continuity of $H_{4_Y}(p,Y,t)$ at the border between NT and SS gives us

$$H_{4_Y}(p, \hat{Y}^+, t) = \lambda K_u(p, Y^+ - \hat{Y}^+). \tag{54}$$

We obtain similar equations at the border between NT and BS. So we (first) concentrate on the set (52–54), the equations for the "upper" border. By noting that we have

$$H_{4_Y}(p, Y, t) = -\frac{2YG(H_2(p,t))}{(\sigma p y^*_p)^2} + \frac{\lambda^2 Y^3}{3(y^*_p)^2} + a,$$

and using (the equality of the left sides of) equations (53) and (54), we acquire

$$G(H_2(p, t)) = \tfrac{1}{6}(\sigma\lambda p)^2\left((\hat{Y}^+)^2 + \hat{Y}^+ Y^+ + (Y^+)^2\right)$$

for all $(p, \hat{Y}^+)$ such that the corresponding pair (p, y) is part of the border between NT and SS at time t. Using this representation of $G(H_2(p, t))$, equation (54) implies

$$-A = \frac{a}{\lambda} - \frac{\lambda}{3(y^*_p)^2}\left(Y^+ + \hat{Y}^+\right)\hat{Y}^+ Y^+. \tag{*}$$

If we further put the above expression for $G(H_2(p, t))$ into the general form of H_4 then equation (52), together with the remark following it, tells us that we must have

$$\lambda K(p, Y^+ - Y) = \frac{\lambda^2((Y)^4 - (Y^+)^4)}{12(y^*_p)^2} + a(Y^+ - Y)$$

$$-\frac{\lambda^2}{6(y^*_p)^2}\left((Y)^2 - (Y^+)^2\right)\left((\hat{Y}^+)^2 + \hat{Y}^+ Y^+ + (Y^+)^2\right) \tag{55}$$

for all $(p, y) \in SS$. We have a similar equation for all $(p, y) \in SB$. The left hand sides of these two equations are even functions in $u = Y^+ - Y$, respectively in $u = Y^- - Y$. Further, we can also obtain the form

$$G(H_2(p, t)) = \tfrac{1}{6}(\sigma\lambda p)^2\left((\hat{Y}^-)^2 + \hat{Y}^- Y^- + (Y^-)^2\right).$$

Together with the preceding remarks, this implies that the (two) right hand side(s) of equation (55) (and its analogue at the border between NT and SB) can only be an even function in u if we have

$$Y^+ = -Y^- \quad \textbf{and} \quad a = 0. \tag{56}$$

Keeping this in mind, the equality between the two different representations of $G(H_2(p, t))$ also implies

$$\hat{Y}^+ = -\hat{Y}^-.$$

Hence, it is enough to find $(Y^+, \hat{Y}^+)$. Using equation (55) for $Y = \hat{Y}^+$ results in

$$K(p, Y^+ - \hat{Y}^+) = \frac{\lambda}{12(y^*_p)^2}\left(\hat{Y}^+ + Y^+\right)^3\left(\hat{Y}^+ - Y^+\right). \tag{57}$$

From the explicit form of $y^*(t,p)$, relations (*) and (56), we have

$$\left(Y^+ + \hat{Y}^+\right)\hat{Y}^+ Y^+ = \frac{3b^2 A}{\sigma^4(\lambda p)^3}. \tag{58}$$

Substituting this into the left side of equation (57) yields

$$\left(\hat{Y}^+ - Y^+\right)^3\left(\hat{Y}^+ + Y^+\right) = \frac{12Bb^2}{\lambda^3(\sigma p)^4}. \tag{59}$$

Note that due to the form of these equations, both $Y^+(p)$ and $\hat{Y}^+(p)$ must be linear in p^{-1}. These two equations have a unique solution which will be proved in

Proposition 21

a) For every choice of positive parameters λ, p, σ, b, A, B there exists a unique solution $(Y^+, \hat{Y}^+)$ to the system (58/59) of non-linear equations satisfying

$$0 < Y^+ < \hat{Y}^+.$$

b) In the case "b = 0" the only non-negative solution to the system (58/59) is given by

$$(Y^+, \hat{Y}^+) = (0, 0).$$

c) If all parameters but A are positive and A is equal to zero then the only non-negative solution to (58/59) with $Y^+ < \hat{Y}^+$ is given by

$$(Y^+, \hat{Y}^+) = \left(0, \sqrt[4]{12Bb^2\lambda^{-3}(\sigma p)^{-4}}\right).$$

d) If all parameters but B are positive and B is equal to zero then the only non-negative solution to (58/59) is given by

$$(Y^+, \hat{Y}^+) = \left(\sqrt[3]{3b^2A\sigma^{-4}(\lambda p)^{-3}}, \sqrt[3]{3b^2A\sigma^{-4}(\lambda p)^{-3}}\right).$$

Remark 22

a) The form of the solutions correspond to the forms in the different cases obtained in (Whalley and Wilmott 1994). However, their explicit forms of the solution pairs differ from ours due to the different optimal strategies in the underlying problems without transaction costs. Further, they have not made any comment on existence or uniqueness of the solutions in the "general" case a) of the above proposition, but our proof also applies to their case. Hence, there is existence and uniqueness in their results, too.

b) The interpretation of cases c) and d) of Proposition 21 is similar to the one in (Whalley and Wilmott 1994): In the case of only fixed costs (i.e. A = 0), it is always optimal to transact to the optimal solution without transaction costs, y*. In the case of only proportional transaction costs (i.e. B = 0), the optimal action is to keep the y-process at the border of NT, i.e. to have a strategy similar to the one of Davis and Norman in Section 1. The case "b = 0" is the degenerate one. In this case, the stock holdings should be zero as in the corresponding setting without transaction costs.

c) One could also compute $H_2(p, t)$ from one of the representations of $G(H_2(p, t))$ by solving the relevant partial differential equation. This would give us an indicator for the impact of the transaction costs on the optimal utility. However, since this can only be done numerically, and, more important, since the change in the optimal utility has no clear implications we will not do this here.

Proof (of Proposition 21):
Assertions b), c) and d) are fairly obvious. We only prove part a). For notational convenience we rewrite equations (2.36/37) as

$$(Z+Y)YZ = C \tag{60}$$

$$(Z-Y)^3(Y+Z) = D \tag{61}$$

with positive constants C, D and $0 < Y < Z$. Due to the (specific) quadratic form of (60), for every $Y \in (0, \sqrt{C/2})$ there exists a unique value $Z(Y) \in (\sqrt{C/2}, \infty)$ such that (Y, Z(Y)) solves equation (60). Further, Z decreases continuously in Y. For Y close to 0, Z(Y) must be large to satisfy equation (60). Hence, for sufficiently small Y and Z(Y) thus sufficiently large, we have $(Z(Y)-Y)^3(Y+Z(Y)) > D$. On the other hand, for Y close to $\sqrt{C/2}$ the corresponding Z(Y) must be close to $\sqrt{C/2}$. Consequently, for Y being sufficiently close to $\sqrt{C/2}$, we must have $(Z(Y)-Y)^3$ $(Y+Z(Y)) < D$. By continuity, there exists a pair (Y, Z(Y)) solving (60) and (61). To see the uniqueness, assume that there is a further pair (Y', Z') satisfying equations (60) and (61). Equalling the left sides of equations (60) and (61) for the pairs (Y, Z) and (Y', Z') results in

$$(Z+Y)YZ = (Z'+Y')Y'Z',$$

$$(Z-Y)^3(Y+Z) = (Z'-Y')^3(Y'+Z').$$

Solving the second equation for (Y+Z) and substituting the resulting expression into the first one leads to

$$\frac{(Z'-Y')^3}{(Z-Y)^3}\, YZ = Y'Z'.$$

If we assume that the above quotient is bigger than 1 then the assumption $Z' > Z$ leads to $Y > Y'$ and to $Y + Z > Y' + Z'$ which contradicts $YZ < Y'Z'$ (i.e. the validity of the above equation under our assumption on the quotient). In a similar way, the assumption $Y' < Y$ leads to a contradiction. Hence, the quotient cannot be bigger than 1. By symmetry it can also not be smaller than 1. Hence, we have proved our desired uniqueness result.

□

We have thus completed the computations for the asymptotic analysis. Figure 10 shows a typical example of an approximately optimal strategy resulting from an asymptotic analysis. In this case, we have chosen $b = 0.1$, $\sigma = 0.2$, $\lambda = 0.01$. Our cost function is given by $k(p,\Delta y) = 0.005\ |\Delta y| p + 0.01$ with $A = 1$, i.e $\varepsilon = 0.000855$. The inner line in Figure 10 marks the optimal strategy without transaction costs, the Pliska solution. The outer dotted lines give the boundaries of the no transaction region while the inner dotted lines mark the optimal restarting points.

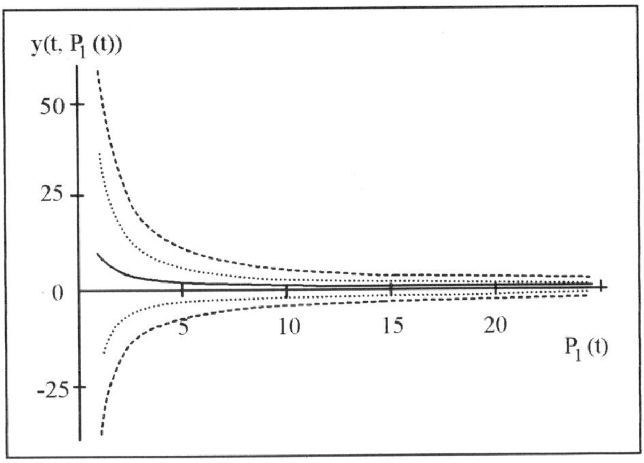

Figure 10: (Asymptotically) optimal strategy for terminal wealth with exponential utility

It can be read off from this figure that starting with a positive initial number of shares, one should sell an optimal number of shares if the share price increases sufficiently. In contrast it is never optimal to purchase additional shares. The gain of utility from the additional shares is not worth the transaction costs. Therefore, if one starts with an initially positive stock position, the optimal wealth process is bounded from below. If also the initial amount invested in the bond is positive then the optimal wealth process is non-negative. Thus, in this case we could also require to have a non-negative wealth process without changing the (approximately) optimal solution. We have indicated the possible path of the number of shares held over time by the "step function" in Figure 10. Another fact that can be seen in Figure 10 is that even if one starts with a negative amount of shares of the stock (i.e. if one has sold shares short), it could be possible not to intervene. In this case the gain of utility from doing an intervention is not worth the transaction costs. Only if the stock prices increases by a certain amount it will be optimal to buy some shares, i.e. to decrease the number of shares sold short.

5.3 Maximising the Growth Rate under Fixed Transaction Costs

In this section, we take up an approach given in (Morton and Pliska 1995) who maximise the (expected) growth rate of the wealth process of an investor in the presence of (specially structured) transaction costs. In their setting, at every time instant where the investor rebalances his holdings, he has to pay a fixed fraction $1-\alpha$ of his actual wealth (this is what Morton and Pliska call "fixed transaction costs"). However, note that on one hand, the amount of transaction costs is independent of the type and volume of the transaction, but on the other hand, it varies with the wealth of the current holdings of the investor (and thus these transaction costs are not fixed cost in the sense of absolute amounts). Therefore, this transaction cost structure is no special case of the one of the previous section.

Although this kind of transaction costs lacks some realism, we will present the results of (Morton and Pliska 1995) in sub-section i) below in an informal manner. The main reason for presenting it is that there is an interesting method of approximate solution of the problem for a model with a moderate number of stocks (say at least around 20–30), an asymptotic analysis of the problem given in (Atkinson and Wilmott 1995) and in (Atkinson, Pliska and Wilmott 1995). The method given in sub-section ii) below is similar to the treatment of the exponential utility problem in the previous section, but will be extended to a multi-dimensional setting. It would therefore be interesting to apply it to multi-dimensional problems with a more realistic transaction cost structure. Some numerical examples close this section.

i) Optimal Growth Rate under Transaction Costs

We consider a market consisting of a bond and n stocks with prices given by

$$dP_0(t) = P_0(t)r\,dt,\ \ P_0(0) = 1,$$

$$dP_i(t) = P_i(t)\left(b_i dt + \sum_{j=1}^{n} \sigma_{ij} dW_j(t)\right),\ \ P_i(0) = p_i\,.$$

The goal of an investor is to maximise the (expected) growth rate of his wealth X(t), i.e. to maximise

$$\lim_{T \to \infty} \frac{1}{T} E\,(\ln(X(T)))$$

by choosing the best portfolio process $\pi(t)$, $t \geq 0$. In the case of no transaction costs, the optimal portfolio process (the one that maximises the growth rate) and the maximum growth rate can be directly deduced from Example 3.19 as

$$\tilde{\pi}(t) \equiv \tilde{\pi} = (\sigma\sigma')^{-1}(b - r\underline{1}),$$
$$\tilde{R} = (1 - \tilde{\pi}'\underline{1})r + \tilde{\pi}'b - \tfrac{1}{2}\tilde{\pi}'\sigma\sigma'\tilde{\pi}\ .$$

Following such a portfolio process requires trading at every time instant which is not advantageous in the presence of (any reasonable form of) transaction costs. In (Morton and Pliska 1995) the investor is required to pay a constant fraction (not a constant amount !) $1-\alpha$ of the wealth of his current holdings at every time he rebalances these holdings (independent of the type and volume of the actual transaction !). Further, in the Morton–Pliska setting short-selling is prohibited. This ensures a positive wealth process $X(t)$ of the investor at every time instant. Hence, $\ln(X(t))$ is always defined. In this context, an admissible policy is defined in the following way:

Definition 23
An **admissible policy** is a sequence $\{(t_i, \pi(t_i)),\ i = 0, 1, 2, \ldots\}$ where each t_n is a stopping time with respect to the Brownian filtration $\{F_t\}_{t \geq 0}$ with $t_n \leq t_{n+1}$, $t_0 = 0$. Each (n-dimensional) portfolio vector $\pi(t_i)$ should be F_{t_i}-measurable with

$$\pi_j(t_i) > 0 \quad \forall j = 1, \ldots, n \ \ \textbf{and} \ \ \underline{1}'\pi(t_i) < 0\ .$$

If an n-dimensional vector π satisfies the last two inequalities we will write "$\pi \in \Pi$". Z will denote the set of admissible policies.

Remark 24
The requirement "$\pi(t) \in \Pi$" in Definition 23 simplifies the analysis in (Morton and Pliska 1995). To be consistent with their presentation, we also assume from now on that $\tilde{\pi}$, the optimal portfolio process without transaction costs, lies in Π (of course, this is an assumption on the market coefficients r, b, σ).

If we denote by $X(t)$ the wealth process at time t of an investor following an admissible policy (suppressing the dependence on the policy $\{(t_i, \pi(t_i))\}$) then our optimisation problem under transaction costs reads

$$\max_{(t_i,\pi(t_i))\in Z} \lim_{T\to\infty} \frac{1}{T} E(\ln(X(T))). \tag{62}$$

If an investor follows an admissible policy $(t_i, \pi(t_i))$ then at time t_1- (i.e. strictly before he rebalances his holdings for the first time) his wealth will be given by

$$X(t_1-) = x\left[(1-\pi(0)'\underline{1})e^{rt_1} + \sum_{i=1}^{n} \pi_i(0)\frac{P_i(t_1)}{p_i}\right]. \tag{63}$$

To see this, note that $x\pi_i(0)$ represents the money invested in security i at time $t = 0$ which will have "grown" by a factor $P_i(t)/p_i$ by the time t_1. As under our transaction cost structure we always have $X(t_i) = \alpha X(t_i-)$, from equation (63), by an easy induction argument, we obtain the representation

$$\begin{aligned} X(t_k) &= \alpha^k x \prod_{i=1}^{k}\left[(1-\pi(t_{i-1})'\underline{1})e^{r(t_i-t_{i-1})} + \sum_{j=1}^{n} \pi_j(t_{i-1})\frac{P_j(t_i)}{P_j(t_{i-1})}\right] \\ &= \alpha^k x \prod_{i=1}^{k}\Big[(1-\pi(t_{i-1})'\underline{1})e^{r(t_i-t_{i-1})} + \\ &\qquad \sum_{j=1}^{n} \pi_j(t_{i-1})e^{(b_j-\frac{1}{2}\|\sigma j\|^2)(t_j-t_{j-1})+\sigma_j(W(t_i)-W(t_{i-1}))}\Big]. \end{aligned}$$

(where σ_j denotes the j th row of the matrix σ). Hence, if an investor starts at time t with a portfolio of π and an initial wealth of $X(t)$ and stops at time $t + \tau$ to choose a new portfolio then his gain in log wealth over the period $[t, t + \tau]$ (after paying transaction costs at time $t + \tau$), $g(\tau, \pi) := \ln(X(\tau)/X(t))$, is given by

$$g(\tau,\pi) = \ln(\alpha) + \ln\left((1-\pi'\underline{1})e^{r\tau} + \sum_{j=1}^{n} \pi_j e^{(b_j-\frac{1}{2}\|\sigma j\|^2)\tau+\sigma_j W(\tau)}\right).$$

With the help of the theory of semi-Markov decision processes, it is shown in (Morton and Pliska 1995) that to solve the optimisation problem (62), we can indeed concentrate on a single period starting with an already chosen portfolio vector and ending with the first rebalancing after this initial choice:

Proposition 25

Assume that R is the maximum growth rate for our optimisation problem (62). Further, assume that (τ^*, π^*) maximises the quotient $E(g(\tau, \pi))/E(\tau)$ over all finite stopping times τ and $\pi \in \Pi$ (where τ^* is a finite stopping time and π^* lies in Π). Then, we have

$$R = \sup_{\tau,\pi} \frac{E(g(\tau,\pi))}{E(\tau)}, \qquad (64)$$

and the optimal policy is given by $t_{n+1} = t_n + \tau^*$, $\pi_n = \pi^*$ for all $n \in \mathbf{N}$ (with $t_0 = 0$)

Remark 26

a) Proposition 25 states that we have a stationary policy in the sense that it does not change over time which is in that way an analogue to the optimal portfolio process without transaction costs being a constant. From equation (64), we can also see that the optimal growth rate R is given as the maximum quotient of the expected gain in log-wealth of the period starting with the portfolio vector π up to the first rebalancing and the expected duration of this period. Of course, a high gain per period is something desirable. But as we consider an average gain criterion, it is the gain per unit time that counts.

b) For our subsequent analysis we often use the following form of equation (64) :

$$0 = \sup_{\tau,\pi} \left(E(g(\tau, \pi)) - R\, E(\tau) \right).$$

To solve problem (62), we need some further considerations. Note first that between two transaction times, the portfolio process $\pi(t)$ evolves freely according to the stochastic differential equation

$$d\pi(t) = \mathrm{diag}(\pi(t))[I - \underline{1}\pi(t)'][(b - r\underline{1} - \sigma\sigma'\pi(t))dt + \sigma dW(t)], \quad t \in [t_i, t_{i+1}), \qquad (65)$$

where I is the n-dimensional identity matrix and diag(x) is an n-dimensional diagonal matrix where the entries in the diagonal are the components of the vector x. To see equation (65), note that for $t \in [t_i, t_{i+1})$ we have

$$\pi(t) = \varphi'\, \mathrm{diag}(P(t)) / X(t) \qquad (66)$$

with $P(t) = (P_1(t), \ldots, P_n(t))'$, $\varphi \in \mathbf{R}^n$ given by $\varphi_j = \pi_j(t_i)X(t_i)/P(t_i)$, i.e. φ_j denotes the number of shares of stock i held between the two transaction times t_i and t_{i+1}. Then, application of Itô's formula to representation (66) yields the stochastic differential equation (65) for the portfolio process. If we further note that we have

$$(1-\underline{1}'\pi(0))\, X(0)\, e^{rt_1} = (1-\underline{1}'\pi(t_1-))\, X(t_1-)$$

(to see this equality, note that both sides of it are equal to the amount of money invested into the bond immediately before the first transaction) then we acquire

$$X(t_1) = \alpha X(t_1-) = \alpha \frac{(1-\underline{1}'\pi(0))X(0)}{1-\underline{1}'\pi(t_1-)} e^{rt_1},$$

$$\ln(X(t_1)) = \ln(\alpha) + \ln(1-\underline{1}'\pi(0)) + \ln(X(0)) + rt_1 - \ln(1-\underline{1}'\pi(t_1-))$$

which yield

$$g(\tau, \pi) = \ln(\alpha) + \ln(1-\underline{1}'\pi)) + r\tau - \ln(1-\underline{1}'\pi(\tau-)),$$

$$R = r + \sup_{\tau,\pi} \frac{\ln(\alpha) + \ln(1-\underline{1}'\pi) - E_\pi(\ln(1-\underline{1}'\pi(\tau-)))}{E_\pi(\tau)}.$$

Therefore, optimisation problem (62) is equivalent to the equation

$$0 = \sup_{\pi \in \Pi} \{\ln(\alpha) + \ln(1-\underline{1}'\pi) + f_R(\pi)\}, \tag{67}$$

$$f_R(\pi) := \sup_{\tau} [E_\pi(-\ln(1-\underline{1}'\pi(\tau-))) - (R-r)E_\pi(\tau)]. \tag{OS}$$

Assuming concavity and sufficient regularity of f_R, we obtain the optimal portfolio process π^* as solution of the first order conditions

$$0 = \frac{\partial}{\partial \pi_i}(\ln(1-\underline{1}'\pi) + f_R(\pi)), \quad i = 1,..., n$$

(still assuming that we have $\pi^* \in \Pi$). Substituting π^* into Equation (67) results in

$$0 = \ln(\alpha) + \ln(1-\underline{1}'\pi^*) + f_R(\pi^*)$$

from which we could determine R if $f_R(\pi^*)$ would be known (as a function of R). Thus, everything boils down to the computation of $f_R(\pi)$. But this can be done by identifying it as the value function of an optimal stopping problem where the process to stop is the portfolio process $\pi(t)$. As its dynamics are explicitly given by (65), we can directly set up the corresponding variational inequalities that characterise the value function (compare also to Proposition 3.2 in (Morton and Pliska 1995)):

Proposition 27

Let the operator A be defined by

$$(A\,h)(\pi) = \tfrac{1}{2}\sum_{i,j=1}^{n} h_{ij}(\pi)\pi_i\pi_j(e_i'-\pi')\sigma\sigma'(e_j-\pi) + \sum_{i=1}^{n} h_i(\pi)\pi_i(e_i'-\pi')(b-r\underline{1}-\sigma\sigma'\pi)$$

where the subscripts on h: $\mathbf{R}^n \to \mathbf{R}$ denote partial derivatives. Then f_R, the value function of the optimal stopping problem (OS), is the smallest function satisfying Itô's formula on Π and the variational inequalities

$$(A\,f)(\pi) \le R - r\,, \tag{68}$$

$$f(\pi) \ge -\ln(1-1'\pi), \tag{69}$$

$$((A\,f)(\pi) - (R-r))(f(\pi) + \ln(1-1'\pi)) = 0 \tag{70}$$

for $\pi \in \Pi$. Further, the optimal stopping time τ is given by the first time when the second inequality holds as an equality, i.e. we have

$$\tau = \inf\{\, t \ge 0 \mid f_R(\pi(t)) = -\ln(1-1'\pi(t)) \,\}.$$

Remark 28

a) Note that the inequality (69) is an explicit one and not an implicit one as the corresponding part of the qvi in Section 5.2. The main reason for this is that for the above optimal stopping problem, we know the reward function for stopping immediately. The two parts of it can be interpreted in the following way: If we stop immediately then the gain is $-\ln(1-1'\pi)$. If we let the π-process continue then this gain changes according to $-\ln(1-1'\pi(t))$, but there is also a continuation fee of $-(R-r)t$ we have to pay which is expressed by inequality (68). Thus, it is a balance problem between watching the gain process (and waiting for the best moment to stop) and paying fees for having the opportunity to watch.

b) As is shown in (Morton and Pliska 1995), the continuation set for the optimal stopping problem (i.e. the sub-set of Π where it is optimal not to stop the process) is non-empty and contains $\tilde{\pi}$, the optimal portfolio process in the absence of transaction costs. Further, we have $(A\,f)(\pi) = R - r$ in the continuation set.

ii) Asymptotic Analysis of the Problem

Although at first sight, the problem of solving the variational inequalities of Proposition 27 seems to be a much easier task than to solve the qvi of Section 2, there is in general only a possibility to solve them numerically for a small number of stocks (typically for n = 1, 2 as is done in the numerical examples of (Morton and Pliska 1995). To overcome this problem, Atkinson and Wilmott performed an asymptotic analysis of the problem (see (Atkinson and Wilmott 1995)) which is centred around $\tilde{\pi}$, the optimal portfolio without transaction costs. As in the asymptotic expansion of Section 2, it is necessary that the transaction costs are small (but non-zero) to have hope for obtaining a reasonable approximation to the optimal solution by using asymptotic methods. We therefore rename the "transaction cost fraction" $1-\alpha$ of sub-section i) by ε and assume it to be small.

Following (Atkinson and Wilmott 1995), we start presenting the case of a single risky security. Introduce the function $G(\pi)$ by

$$f_R(\pi) = -\ln(1-\pi) + G(\pi).$$

This definition of $G(\pi)$ leads to the following form of inequality (68)

$$R-r \geq (A\, f_R)(\pi) = (AG)(\pi) + \tfrac{1}{2}\sigma^2\left(\frac{2(b-r)\pi}{\sigma^2} - \pi^2\right)$$

$$= \tfrac{1}{2}\sigma^2\, \pi^2(1-\pi)^2 G'' + \sigma^2\left(\frac{b-r}{\sigma^2} - \pi\right)\pi(1-\pi)G' + \tfrac{1}{2}\sigma^2\left(\frac{2(b-r)\pi}{\sigma^2} - \pi^2\right)$$

with equality for $\pi_l \leq \pi \leq \pi_u$ where the borders π_l, π_u of the continuation region are determined by the smoothness requirements on f_R (compare to the Davis–Norman approach of Section 1 and the calculations in Section 6.3) which, expressed in terms of G, read

$$G(\pi_l) = G(\pi_u) = G'(\pi_l) = G'(\pi_u) = 0\,.$$

Note that we have implicitly assumed that the continuation region is given by an interval. However, by Remark 28 b), we have good reasons to do so. The next step is to perform a rescaling of the variables π, R by introducing the new variables $\overline{\pi}$, $\overline{R}$ and the function $\overline{G}$ as

$$\pi = \tilde{\pi} + \varepsilon^{1/4}\overline{\pi}, \quad R = \tilde{R} + \varepsilon^{1/2}\overline{R} + o(\varepsilon^{1/2}), \quad G(\pi) = \varepsilon\overline{G}(\overline{\pi}) + o(\varepsilon)\,.$$

We will comment on the form of that rescaling later when it becomes obvious that it is an appropriate one. Inside the continuation region we replace R, G and π by the right hand sides of the corresponding equations above, and, after dividing by $\varepsilon^{1/2}$, end up with the following differential equation:

$$\left(\tfrac{1}{2}\sigma^2\tilde{\pi}^2(1-\tilde{\pi})^2 + O\left(\varepsilon^{1/4}\right)\right)(\overline{G}'' + o(1)) = \overline{R} + \tfrac{1}{2}\sigma^2\overline{\pi}^2 + o(1) \tag{71}$$

where $\overline{G}''$ denotes second derivative of $\overline{G}$ with respect to $\overline{\pi}$ and o(1) refers to $\varepsilon \to 0$. Neglecting all terms of order (at least) of o(1) in equation (71) results in the equation

$$\tfrac{1}{2}\sigma^2\,\tilde{\pi}^2(1-\tilde{\pi})^2\overline{G}'' = \overline{R} + \tfrac{1}{2}\sigma^2\overline{\pi}^2 \tag{72}$$

together with the boundary conditions

$$\overline{G}(\overline{\pi}_l) = \overline{G}(\overline{\pi}_u) = \overline{G}'(\overline{\pi}_l) = \overline{G}'(\overline{\pi}_u) = 0 \tag{73}$$

where still $\overline{R}$, $\overline{\pi}_l$, $\overline{\pi}_u$ have to be determined. Next, we look at equation (67) which expressed in terms of $\overline{\pi}$ and $\overline{G}$ has the form

$$0 = \sup_{\overline{\pi}}\{\ln(1-\varepsilon) + \varepsilon\overline{G}(\overline{\pi}) + o(\varepsilon)\}\ .$$

Expanding $\ln(1-\varepsilon)$ for small values of ε as $\ln(1-\varepsilon) = \ln(1) - \varepsilon + o(\varepsilon)$, dividing by ε, and neglecting the o(1) terms, we obtain

$$\sup_{\overline{\pi}}\{\overline{G}(\overline{\pi})\} = 1. \tag{74}$$

Integrating equation (72) yields that $\overline{G}(\overline{\pi})$ is a polynomial of the form

$$\overline{G}(\overline{\pi}) = \left(X + Y\overline{\pi} + \tfrac{1}{2}\overline{\pi}^2\overline{R} + \tfrac{1}{24}\sigma^2\overline{\pi}^4\right)\left(\tfrac{1}{2}\sigma^2\tilde{\pi}^2(1-\tilde{\pi})^2\right)^{-1} \tag{75}$$

with suitable constants X, Y. Using equations (73) and (75) leads to the form

$$\overline{G}(\overline{\pi}) = D(\overline{\pi} - \overline{\pi}_u)^2(\overline{\pi} - \overline{\pi}_l)^2 \tag{76}$$

with a suitable constant D. By carrying out the multiplications in equation (76) and comparing the coefficients of the two different representations for $\overline{G}(\overline{\pi})$, we obtain

$$\overline{\pi}_u = -\overline{\pi}_l, \quad Y = 0, \quad \overline{R} = -\tfrac{1}{6}\sigma^2\overline{\pi}_u^{\,2}\,. \tag{77}$$

Using the first order condition for a maximum together with equation (74) yields the optimal rebalance point (to leading order) $\overline{\pi}^*$ (from the final form of $\overline{G}(\overline{\pi})$ it can be checked that the maximum in relation (74) will be attained in the interior of the region where $\tilde{\pi} + \varepsilon^{1/4}\overline{\pi} \in (0, 1)$, i.e. $-\tilde{\pi}\varepsilon^{-1/4} < \overline{\pi} < (1-\tilde{\pi})\varepsilon^{-1/4}$, and not at the boundary) and coefficient X in equation (75) as

$$\overline{\pi}^* = 0, \quad X = \tfrac{1}{2}\sigma^2\,\tilde{\pi}^2(1-\tilde{\pi})^2. \tag{78}$$

Inserting the representation for $\overline{R}$ in the equation $\overline{G}(\overline{\pi}_u) = 0$ (where we can now use $Y = 0$ and $X = \tfrac{1}{2}\sigma^2\tilde{\pi}^2(1-\tilde{\pi})^2$) yields the final results of

$$\overline{G}(\overline{\pi}) = \frac{\overline{\pi}^4 - \beta^4}{12\tilde{\pi}^2(1-\tilde{\pi})^2}, \quad \overline{R} = -\tfrac{1}{6}\lambda^2\beta^2,$$

$$\overline{\pi}_u = -\overline{\pi}_l = \beta, \quad \beta = 3^{1/4}\sqrt{2\tilde{\pi}(1-\tilde{\pi})}\ .$$

Summing up all the previous considerations, we get the approximate solutions (in the above sense) to the optimisation problem (62) as

$$[\pi_l, \pi_u] = \left[\tilde{\pi} - (3\varepsilon)^{1/4}\sqrt{2\tilde{\pi}(1-\tilde{\pi})}, \tilde{\pi} + (3\varepsilon)^{1/4}\sqrt{2\tilde{\pi}(1-\tilde{\pi})}\right],$$

$$\pi^* = \tilde{\pi}, \quad R = \tilde{R} - \tfrac{1}{6}\lambda^2\beta^2\varepsilon^{1/2}$$

where π^* denotes the optimal rebalance point to leading order (i.e. the optimal rebalance point obtained via asymptotic analysis).

Remark 29

a) Note that the above approximate solutions have some remarkable features:

- The no-transaction region $[\pi_l, \pi_u]$ is symmetric about the Merton point $\tilde{\pi}$ with a length of order $\varepsilon^{1/4}$. In particular, for $\varepsilon = 0$, it reduces to the Merton point.
- The impact of transaction costs on the (approximately) optimal growth rate is a decrease of order $\varepsilon^{1/2}$.
- The investor following the above strategy will always rebalance back to the Merton point.

Thus, our rescaling kept the whole analysis in the area of interest. The symmetry around the Merton point would probably get lost if we would take into account higher order terms of ε in our equations (in particular in the considerations leading to equation (71)). However, this could turn the solution of the problem into a non-tractable task.

b) As reported in (Morton and Pliska 1995), the optimal rebalance point will in general not be equal to the Merton point and the no-transaction region will not be symmetric around $\tilde{\pi}$. Thus, one can only hope for a good approximation of the solution obtained by asymptotic analysis if the optimal rebalance point is close to $\tilde{\pi}$ and if the no-transaction region is not too asymmetric. However, this is usually the case in the examples given in (Atkinson and Wilmott 1995).

We will now turn to the case of an arbitrary number of stocks n. There, the function G is given via

$$f_R(\pi) = -\ln(1 - 1'\pi) + G(\pi).$$

We then get

$$(A\, f_R)(\pi) = (AG)(\pi) + \tfrac{1}{2}\tilde{\pi}'\sigma\sigma'\tilde{\pi} - \tfrac{1}{2}(\pi-\tilde{\pi})'\sigma\sigma'(\pi-\tilde{\pi}).$$

The n-stock analogue of the rescaling procedure has the form

$$\pi = \tilde{\pi} + \varepsilon^{1/4}\bar{\pi}, \quad R = \tilde{R} + \varepsilon^{1/2}\bar{R} + o(\varepsilon^{1/2}), \quad G(\pi) = \varepsilon\bar{G}(\bar{\pi}) + o(\varepsilon)$$

with

$$\tilde{\pi} = (\sigma\sigma')^{-1}(b - r\underline{1}), \quad \tilde{R} = r + \tfrac{1}{2}\tilde{\pi}'\sigma\sigma'\tilde{\pi}\ .$$

In a similar way as in the case of a single risky security we obtain the "leading order system":

$$\tfrac{1}{2}\sum_{i,j=1}^{n}\bar{G}_{ij}\tilde{\pi}_i\tilde{\pi}_j(e_i' - \tilde{\pi}')\sigma'\sigma(e_j' - \tilde{\pi}') = \bar{R} + \tfrac{1}{2}\bar{\pi}'\sigma\sigma'\bar{\pi}, \qquad (79)$$

$$\bar{G} = 0 \ \text{ and } \ \nabla\bar{G} = 0 \text{ on } \partial C, \qquad (80)$$

where C denotes the free boundary (which has to be determined as part of the solution of problem (79), (80)). We also arrive at the "leading order form"

$$\sup_{\bar{\pi}}\{\bar{G}(\bar{\pi})\} = 1 \qquad (81)$$

of equation (67) where again the supremum has to be taken over the region such that $\tilde{\pi} + \varepsilon^{1/4}\bar{\pi}$ is in the interior of the unit simplex. Fortunately, the free boundary problem (79), (80) has a surprisingly simple solution,

$$\bar{G}(\bar{\pi}) = (\bar{\pi}'M\bar{\pi} - 1)^2 \qquad (82)$$

with a suitable positive definite matrix M (we use positive definite in the sense that it also includes the requirement on M to be symmetric), and the set C is simply the interior of the ellipsoid given by the equation

$$\bar{\pi}'M\bar{\pi}-1=0.$$

Inside this set, the supremum in equation (81) is attained for $\bar{\pi}=0$. Hence, the optimal rebalance point for the leading order problem is the Merton point $\tilde{\pi}$. The ellipsoid, describing C, is uniquely determined by its semi-axes. Their directions are given by the n eigenvectors of M, and their lengths coincide with the corresponding eigenvalues. The continuation set for our original problem (up to leading order) is thus given by $\tilde{\pi}+\varepsilon^{1/4}C$. Hence, we have totally specified the no-transaction region and its boundary in arbitrary dimensions.

What remains is to determine the matrix M. By introducing the matrix H via

$$H_{ij}=\tilde{\pi}_i\tilde{\pi}_j(e_i-\tilde{\pi})'\sigma\sigma'(e_j-\tilde{\pi}),\quad i,j=1,\ldots,n$$

and using the symbol ∇ for the gradient, we can rewrite equation (79) as

$$\tfrac{1}{2}\nabla' H\,\nabla\bar{G}(\bar{\pi})=\bar{R}+\tfrac{1}{2}\bar{\pi}'\sigma\sigma'\bar{\pi}\,. \tag{83}$$

If we insert $\bar{G}(\bar{\pi})$ given by equation (82) into equation (83), use the relations

$$\nabla\bar{\pi}'M\bar{\pi}=2M\bar{\pi},\quad \nabla'HM\bar{\pi}=\mathrm{Tr}(HM)$$

(where Tr(A) denotes the trace of the matrix A), we obtain that $\bar{G}(\bar{\pi})$ of the form (82) is a solution to the free boundary problem (79), (80) if and only if M satisfies the equation

$$8\,MHM+4\,\mathrm{Tr}(HM)\,M=\sigma\sigma'\,. \tag{84}$$

Then, $\bar{R}$ is given as

$$\bar{R}=-2\,\mathrm{Tr}(HM).$$

To show that equation (84) has a unique, positive definite solution M, we assume that we have H = I (we could do so as this can be achieved by a suitable coordinate transformation as H is positive definite). This leads to a simpler form of equation (84):

$$8\,M^2+4\,\mathrm{Tr}(M)\,M=\sigma\sigma'. \tag{85}$$

As a positive definite matrix is uniquely determined by its eigenvalues and the corresponding eigenvectors, we will compute these using equation (85), thereby showing that they are uniquely determined by the requirement of a positive definite M. Let v_i, λ_i be the set of eigenvectors and corresponding eigenvalues of M. Applying both sides of equation (85) to v_i leads to

$$\left(8M^2 + 4\mathrm{Tr}(M)M\right)v_i = \left(8\lambda_i^2 + 4C\lambda_i\right)v_i = \sigma\sigma' v_i$$

where we have set $C = \mathrm{Tr}(M)$. In particular, the eigenvectors of M and $\sigma\sigma'$ coincide. Even more, if δ_i, $i = 1,\dots, n$, are the eigenvalues of $\sigma\sigma'$ (which are all positive) then we have

$$\delta_i = 8\lambda_i^2 + 4C\lambda_i .$$

This equation for λ_i has exactly one positive solution (given C), namely

$$\lambda_i = \tfrac{1}{4}\left(-C + \sqrt{C^2 + 2\delta_i}\right).$$

Adding up all these n equations for λ_i leads to the following equation for C,

$$f(C) = (n+4)C - \sum_{i=1}^{n} \sqrt{C^2 + 2\delta_i} = 0$$

which has a unique positive solution as we have $f(0) < 0$ and $f'(C)$ is strictly positive and bounded away from zero. Hence, we have shown that there exists a unique matrix M satisfying equation (85) and having n eigenvectors (which coincide with those of the positive definite matrix $\sigma\sigma'$) with corresponding positive eigenvalues, and so the leading order problem is solved. Note in particular that our whole analysis was greatly simplified by the fact that the free boundary problem (79), (80) had an explicit solution of a simple form.

iii) Some Numerical Examples

To illustrate the dependence of the "leading order solutions" computed via asymptotic analysis on the magnitude of the transaction costs, we will give some simple numerical examples. We start with the case of a single risky security. For the choice of $r = 0.07$, $b = 0.182$, $\sigma = 0.4$, we have plotted the upper and lower boundary of the continuation region for the portfolio process (together with $\tilde{\pi} = 0.7$) as a function of the transaction cost fraction ε in Figure 11. It is reported in (Atkinson and

Wilmott 1995) that there is excellent agreement with the numerical solution obtained in (Morton and Pliska 1995). Note that transaction costs of one percent cause a very wide continuation interval. This clearly demonstrates that the losses from holding a non-optimal position are not as dramatic as the loss of money caused by transaction fees.

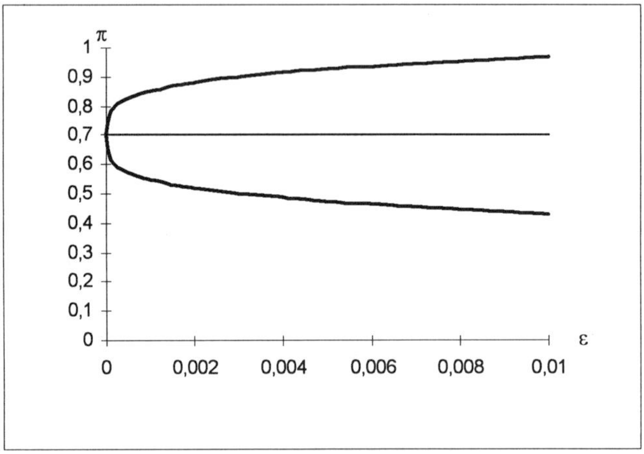

Figure 11: Continuation region and optimal rebalance point to leading order in dependence on the transaction costs

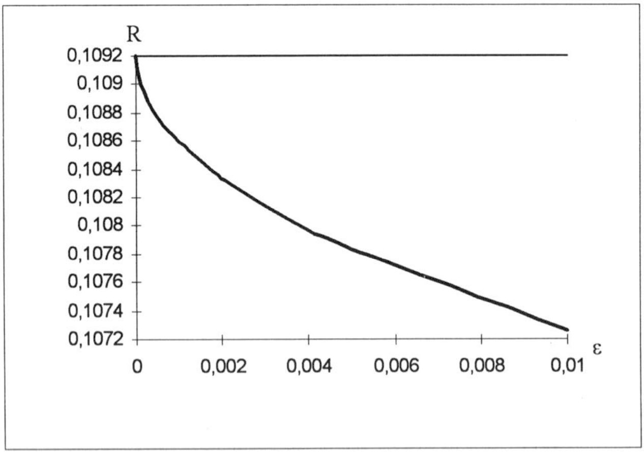

Figure 12: Optimal growth rate to leading order as a function of transaction costs

In Figure 12, we have plotted the optimal growth rate to leading order, R, as a function of ε. The impact of the transaction costs on the asymptotic growth rate are not too dramatic. Transaction costs of one percent only result in a decrease of around 0,2 percent in the optimal growth rate to leading order.

Also, the mean time between two trades, $E(\tau)$, is an interesting quantity. Figure 13 contains this quantity plotted as a function of ε. To produce this figure, note that we have

$$E(\tau_\pi) = \frac{y\left(\exp(-2\nu z/\sigma^2)-1\right) - z\left(\exp(-2\nu y/\sigma^2)-1\right)}{\nu\left(\exp(-2\nu z/\sigma^2) - \exp(-2\nu y/\sigma^2)\right)} \tag{86}$$

with

$$\tau_\pi = \inf\{t > 0 \mid \pi(t) \in \{\pi_l, \pi_u\}, \pi(0) = \pi\}, \quad \pi \in [\pi_l, \pi_u],$$

$$y = \ln(\pi_u(1-\pi)) - \ln(\pi(1-\pi_u)), \quad z = \ln(\pi_l(1-\pi)) - \ln(\pi(1-\pi_l)),$$

$$\nu = b - r - \tfrac{1}{2}\sigma^2.$$

From equation (86), $E(\tau)$ can be obtained by setting $\pi = \bar{\pi}$. To prove equation (86), choose the unique solution f(x) of the boundary value problem

$$f'(x)\, x(1-x)(b-r-\sigma^2 x) + \tfrac{1}{2} f''(x)\, x^2(1-x)^2\sigma^2 = -1, \quad x \in (\pi_l, \pi_u),$$

$$f(\pi_l) = f(\pi_u) = 0,$$

and apply Itô's formula to $f(\pi(\tau_\pi))$. Taking expectation on both sides of the resulting equation yields

$$E(\tau_\pi) = f(\pi(0))$$

where $f(\pi)$, as the unique solution of the above boundary value problem, coincides with the right hand side of Equation (86).

As can be read off from Figure 13, the expected time between two trades increases rapidly as a function of ε which demonstrates the significant impact of the transaction costs on the trading behaviour of the investor. In particular, the expected time between two trades seems to increase linearly in the transaction costs for $\varepsilon >$ 0,001. It is also worth looking at the actual values of the mean time between two trades. For example, if the transaction cost fraction is $\varepsilon = 0{,}001$ then the mean time between two trades is around three and a half years. This somewhat strange behaviour is a consequence of the non-realistic transaction cost structure. Note that the better your performance up to the next trade the higher will be your transaction costs! In particular, the conclusions drawn from Figure 13 are totally in line with the comments made on the form of Figure 11.

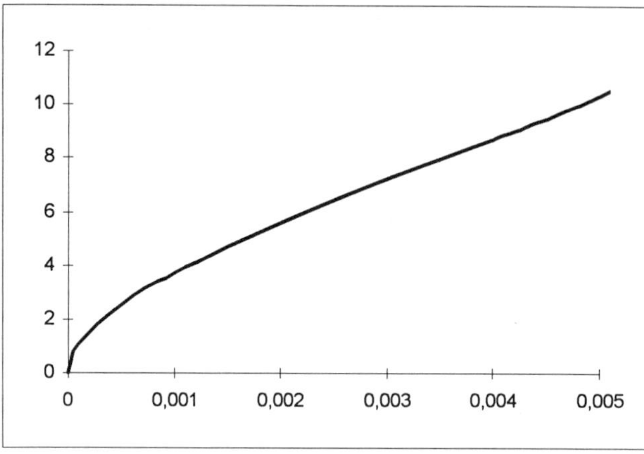

Figure 13: Mean time between two transactions for the optimal strategy to leading order as a function of transaction costs

Figure 14 displays an ellipsoid describing the no transaction region (to leading order obtained by asymptotic analysis) for the case of n = 2. We have chosen the parameters $r = 0.05$, $b_1 = 0.08$, $b_2 = 0.09$, $\varepsilon = 0.001$, $\sigma_{11} = 0.3$, $\sigma_{22} = \sqrt{0.12}$, $\sigma_{12} = \sigma_{21} = 0$. The ellipsoid is centred around the Merton point which is in this case given by

$$\tilde{\pi} = (1/3,\ 1/3)'\ .$$

It can directly be verified that the matrix H is a diagonal matrix in our setting. Solving the matrix equation (84) for M, we obtain the eigenvalues of M as

$$\lambda_1 = 0.76731, \quad \lambda_2 = 0.094152$$

where the corresponding eigenvectors are the two unit vectors. With these facts we can compute the semi-axes of the ellipsoid that describes the continuation region as indicated in sub-section ii).

Remark 30

In a recent paper, (Atkinson, Pliska and Wilmott 1995), the authors report to have even better agreement of the solutions computed by a slightly different variant of an asymptotic analysis and the ones computed numerically for the case n = 2. The difference to the method given in sub-section ii) above lies in another transformation of the coordinates before performing the asymptotic analysis. The power of their method is demonstrated by the solution of the above portfolio problem under transaction costs for all the 30 Dow Jones industrials, at task which is definitely impossible without asymptotic analysis.

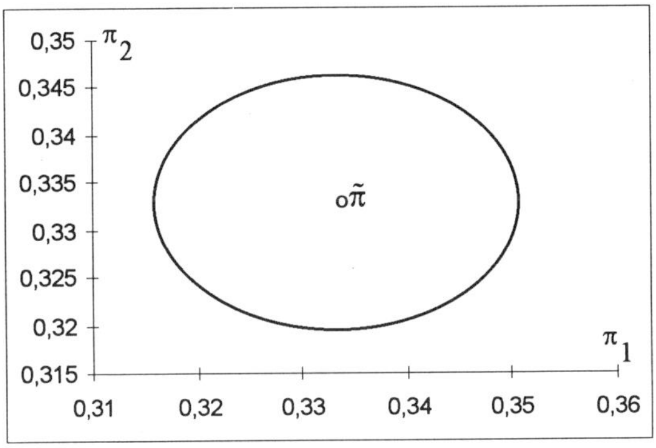

Figure 14: Continuation set (to leading order)

CHAPTER 6

NON-UTILITY BASED PORTFOLIO SELECTION MODELS

This chapter will be a collection of some approaches that differ from the usual expected utility maximising approach presented in the foregoing chapters. We do not claim to give a complete choice of such alternative approaches. Instead, we have just picked out some interesting ones, and we will highlight their main ideas and features.

6.1 Universal Portfolios : The Discrete-Time Model

This approach was developed by Cover in (Cover 1991) in a discrete-time setting. The continuous-time version was given in (Jamshidian 1992). In contrast to the expected utility maximisation approach of the preceding chapters, the goal in this approach is to obtain the **universal portfolio** which is optimal in the following sense: If the time horizon is large then the universal portfolio should perform (approximately) as well as the best constantly rebalanced portfolio (in a strictly pathwise sense). Of course, the best constantly rebalanced portfolio could not be determined a priori without knowledge of the future security prices which makes it an impossible goal to reach. Further, we will see that a very attractive feature of the universal portfolio is that for setting it up, no knowledge or statistical information on market parameters, such as interest rates or volatility, is required. Indeed, we do not even need to specify a particular model for the price dynamics in the discrete-time setting. Remark 4 below sums up the philosophy behind this approach and gives a simple, descriptive explanation why it works.

In this section, we will describe the discrete-time approach in detail and continue with the continuous-time version in the next section.

i) The Discrete-Time Model and Some Basic Definitions

We consider a securities market consisting of m securities. Let

$$x_i = (x_{i1}, \ldots, x_{im})'$$

be the so-called **market vector** at day i (we will measure time in days here, but the time scale can of course be adjusted to an arbitrary discrete setting). Its components are the quotients

$$x_{ij} = \frac{\text{closing price of security j at day i}}{\text{opening price of security j at day i}} \tag{1}$$

i.e. $(x_{ij}-1)$ is the relative return of security j at day i where we assume that the closing prices at day i coincide with the opening prices at the following day.

Following the notation in (Cover 1991), we denote by b_i the portfolio of an investor at day i, and we do not allow for short selling, i.e. we have

$$b_i \in B = \left\{ b \,\middle|\, b = (b_{i1},\dots,b_{im})',\ b_{ij} \geq 0,\ \sum_{j=1}^{m} b_{ij} = 1 \right\}.$$

Thus, the use of the portfolio strategy b_i at day i results in a change of the investor's wealth by a factor $b_i'x_i$. Further, we assume that dividends are immediately taken out of the investor's portfolio. This could have the interpretation that they are paid in a separate account, are consumed immediately, or simply do not occur. Therefore, for the rest of this section, we assume that there are no dividends.

Definition 1
A portfolio strategy $\mathbf{b} = (b_1, b_2, \dots)$ (where b_i are the portfolios at day i) will be called **self-financing** if the whole wealth at period i will be reinvested. A self-financing portfolio strategy will be called a **constant portfolio strategy** if there exists a vector $b \in B$ with $b_i = b$ for all $i \in \mathbf{N}$.

Starting with an initial wealth of $S_0(\mathbf{b}) = 1$, the wealth after n days of an investor following a constant portfolio strategy $\mathbf{b}$ will be

$$S_n(\mathbf{b}) = \prod_{i=1}^{n} b'x_i .$$

$$S_n^* := \max_{b \in B} S_n(\mathbf{b})$$

will denote the maximal attainable wealth by using a constant portfolio strategy. As we have seen in the previous chapters, there are many situations where optimal portfolios are constant ones. We will give some further reasons for considering S_n^*, the wealth of the best constant portfolio after n days, as an attractive goal.

Proposition 2
At time n, S_n^* exceeds the wealth of
a) the best **buy-and-hold** strategy (i.e. a strategy that consists of making an initial investment (in possibly different assets) and leaving the holdings untouched in subsequent periods) which can be expressed by

$$S_n^* \geq \max_{j=1,\ldots,n} S_n(\mathbf{e_j}) .$$

$\mathbf{e_j}$ is the constant portfolio strategy of the form $b = e_j$ for and some $1 \leq j \leq n$ where e_j is the j th unit vector.
b) the geometric mean of securities prices (the "**Value line**"), i.e.

$$S_n^* \geq \left(\prod_{j=1}^{n} S_n(e_j) \right)^{1/m} .$$

c) the weighted average of the securities prices (the "**Dow Jones index**"), i.e.

$$S_n^* \geq \sum_{j=1}^{m} \alpha_j S_n(e_j) \qquad \forall\ \alpha \in B.$$

Proof:
Part a) follows from the fact that the maximal wealth over all elements of B, S_n^*, must be at least as big as the maximum over the vertices of B. Further, every buy-and hold strategy performs at most as good as the best stock. Therefore, the inequality of part a) indeed implies that S_n^* exceeds the wealth of the best buy-and-hold strategy. As the geometric and the weighted arithmetic mean are smaller than the maximum of the values entering these means, parts b) and c) of the proposition are a consequence of part a).

□

Note further that $S_n^* = S_n^*(x_1,\ldots, x_n)$ is invariant under permutations of the market vectors x_i. This has the interpretation that a market crash over a sequence of days has the same impact on S_n^* as if all the "bad days" were distributed arbitrarily over the whole time set $i = 1,\ldots, n$. However, there is a major drawback regarding the best constant strategy $b^*(n)$ over the first n days. In a "non-degenerate" market, we will in general have

$$b^*(n) \neq b^*(n+1),$$

i.e. the best constant portfolio over the first n days will **not** coincide with that over the first n+1 days. But this will imply

$$S_{n+1}(b^*(n)) < S_{n+1}^* = S_{n+1}(b^*(n+1)) .$$

Hence, it is not possible to change from $b^*(n)$ to $b^*(n+1)$ at day $(n+1)$ in a self-financing way **and** at the same time obtain a wealth of S_{n+1}^*. Without knowledge of the future securities prices up to the time horizon, it is impossible to know the best constant strategy a priori. Thus, S_n^* is an impossible goal to reach via an a priori chosen constant portfolio strategy. To achieve this goal approximately (in a sense which will be defined shortly), Cover introduces the **universal portfolio** $\hat{b}$ defined in the following way:

$$\hat{b}_1 = \left(\frac{1}{m},\dots,\frac{1}{m}\right), \qquad \hat{b}_{k+1} = \frac{\int_B b S_k(\mathbf{b})db}{\int_B S_k(\mathbf{b})db}, \quad k \geq 1.$$

The corresponding wealth process, the **universal wealth** $\hat{S}_n$, is then given by

$$\hat{S}_n = \prod_{k=1}^{n} \hat{b}_k{}' x_k.$$

Note that the universal portfolio is not a constant portfolio. It starts with equal fractions of wealth in all securities and is a weighted average over all constant portfolios in later periods. Here, the weights are determined by the performance of the particular securities. Our goal will now be to show that this is a "good" strategy. To do so, we introduce the following notations

$$W(\mathbf{b}, F_n) := \int \ln(b'x)dF_n(x),$$

$$W^*(F_n) := \max_{b \in \mathbf{R}^m} W(\mathbf{b}, F_n)$$

where F_n denotes the empirical distribution function of the given vectors $x_1,\dots, x_n$ in $\mathbf{R}_+^m$ with mass $1/n$ in every x_i. The constant portfolio strategy $\mathbf{b}$ is given by $b \in \mathbf{R}^m$. Then, we have

$$S_n^* = \max_{b \in \mathbf{R}^m} \prod_{i=1}^{n} b'x_i = \max_{b \in \mathbf{R}^m} \exp\left(n\frac{1}{n}\sum_{i=1}^{n}\ln(b'x_i)\right)$$

$$= \max_{b \in \mathbf{R}^m} \exp\left(n\int \ln(\mathbf{b'x})dF_n(x)\right) = \exp(nW^*(F_n)) \tag{2}$$

and similarly

$$S_n(b) = \exp(nW(b, F_n)).$$

$W(b, F_n)$ will be called the **exponential growth rate** of $S_n(b)$.

The following proposition is a collection of some first properties of the universal portfolio.

Proposition 3

a) $\hat{S}_n$ is the (arithmetic) mean over the whole simplex B of the wealth $S_n(\mathbf{b})$ corresponding to all constant portfolio strategies, i.e. we have

$$\hat{S}_n = \frac{\int_B S_n(\mathbf{b})\,db}{\int_B db}.$$

b) The universal wealth exceeds the value line, i.e. we have

$$\hat{S}_n \geq \left(\prod_{j=1}^{n} S_n(e_j)\right)^{1/m}$$

c) $\hat{S}_n = \hat{S}_n(x_1,\ldots,x_n)$ is invariant under permutations of $x_1,\ldots,x_n$.

Proof:

a)

$$\hat{b}_k{}'x_k = \frac{\int_B b'x_k S_{k-1}(b)\,db}{\int_B S_{k-1}(b)\,db} = \frac{\int_B \prod_{i=1}^{k} b'x_i\,db}{\int_B \prod_{i=1}^{k-1} b'x_i\,db}$$

results in

$$\hat{S}_n = \prod_{i=1}^{n} \hat{b}_k{}'x_k = \frac{\int_B \prod_{i=1}^{n} b'x_i\,db}{\int_B db} = \frac{\int_B S_n(\mathbf{b})\,db}{\int_B db}.$$

b) Using the representation of $\hat{S}_n$ given in part a), we can interpret it as the expected value of the wealth $S_n(\mathbf{b})$ if we assume that the vector b describing the constant strategy **b** is uniformly distributed over B. Having this in mind, we obtain:

$$\hat{S}_n = E_{\mathbf{b}}(S_n(\mathbf{b})) = E_{\mathbf{b}}(\exp(nW(\mathbf{b}, F_n))) \geq \exp(E_{\mathbf{b}}(nW(\mathbf{b}, F_n)))$$

$$= \exp\left(E_{\mathbf{b}}\left(n\int \ln(\mathbf{b}'\mathbf{x})\,dF_n(x)\right)\right) \geq \exp\left(nE_{\mathbf{b}}\left(\sum_{j=1}^{m} b_j \int \ln(\mathbf{e}_j{}'\mathbf{x})\,dF_n(x)\right)\right)$$

$$= \exp\left(\frac{1}{m}\sum_{j=1}^{m} n\int \ln(\mathbf{e}_j{}'\mathbf{x})\,dF_n(x)\right) = \left(\prod_{j=1}^{n} S_n(e_j)\right)^{1/m}.$$

To obtain the first inequality, we have used Jensen's inequality. The second inequality follows from the fact that we have

$$\ln\left(\sum_{i=1}^{n} b_i x_i\right) \geq \sum_{i=1}^{n} b_i \ln(x_i)$$

for $x_i > 0$, $i = 1,\ldots, n$, $b \in B$, $n \in \mathbf{N}$ which can be proved by an induction argument using the concavity of the logarithm. Finally, we have used $E_{\mathbf{b}}(\mathbf{b}_j) = 1/m$ for the second but last equality.

c) The assertion in part c) follows directly from part a) in the form

$$\hat{S}_n = \frac{\int_B \prod_{i=1}^n b'x_i \, db}{\int_B db}$$

□

Remark 4 (not to be skipped !!!)

a) The representation of the universal wealth in part a) of Proposition 3 allows for an interpretation of the universal portfolio. By following the universal portfolio, we formally assign an amount of $\mathbf{db}/\int_B \mathbf{db}$ to each constant portfolio strategy given by b $\in B$, invest according to that strategy over the period $\{1,\ldots, n\}$, receive a wealth of $S_n(\mathbf{b})\mathbf{db}/\int_B \mathbf{db}$ at time n, and pool these gains at day n resulting in a total wealth of $\hat{S}_n$. In this sense, the universal portfolio seems to be a strategy that "plays safe" by following every (constant) possibility.

b) In deed, the interpretation given in a) can be underlined by a very simple example. Assume that we have got the possibility to invest money in n different bank accounts with different interest rates r_i. After the initial allocation of money to these different accounts there is no rebalancing. For ease of exposition, we assume that we have continuous compounding of interest, i.e. one unit of money invested in bank account i today leads to an amount of money of $e^{r_i t}$ at time t. However, the rates r_i are not known to us at the initial time. Let r* be the biggest interest rate. Then, we have that compared to an investment in the corresponding account all other investments in just a single account are outperformed exponentially in the sense of

$$e^{r_i t} / e^{r^* t} = e^{(r_i - r^*)t},$$

and this quotient converges to zero exponentially fast for t large if r_i is not equal to r^*. In our situation, the only way to ensure that the quotient of the sum of our wealth at time t (i.e. the sum of the money that we have in the different bank accounts) and $e^{r^* t}$ (the wealth corresponding to investing all money in the best bank account) does not converge to zero exponentially fast for large t, is to allocate a positive amount of money to **each** bank account today. Then, our wealth at time t, X(t), is given by

$$X(t) = \sum_{i=1}^{n} \alpha_i e^{r_i t}$$

with the consequence that we have

$$X(t) / e^{r^*t} \xrightarrow{t\to\infty} \alpha^*$$

where α^* is the amount of money initially invested in the (then unknown) best account. We can rephrase this fact as "the average of exponential functions $\exp(r_i t)$ has the same exponential growth rate r^* as the maximum $\exp(r^*t)$ of these n exponential functions". By noting that we have $S_n(b) = \exp(nW(b, F_n)))$ and looking at the interpretation of the universal wealth in part a) of this remark, we conjecture that the universal wealth will asymptotically grow as S_n^* in the sense that it has the same asymptotic growth rate. We will make this more precise below.

To compare $\hat{S}_n$ with S_n^*, we have to introduce some notations. We will write $b^* = b^*(F_n)$ for the optimal constant portfolio $b^*(n) = b^*(n; x_1,..., x_n)$ of the first n days corresponding to the market vectors $x_1,..., x_n$. Due to relation (2), maximising $S_n(\mathbf{b})$ over B is equivalent to maximising $W(\mathbf{b}, F_n)$ over B.

Definition 5
We say that all securities are
a) "**active**" (at time n) if there exists some $\mathbf{b^*}$ attaining $W^*(F_n)$ (of equation (2)) with $(b^*(F_n))_i > 0$ for all i = 1,..., m.
b) "**strictly active**" (at time n) if for all $\mathbf{b^*}$ attaining $W^*(F_n)$ we have $(b^*(F_n))_i > 0$ for all i = 1,..., m.

As usual, the vectors $x_1,..., x_n$ $(n \geq m)$ are said to have full rank if they span $\mathbf{R}^m$. In financial terms, this is equivalent to the assumption of absence of dependent securities in the market.

For presenting the main ideas and examining the performance of the universal portfolio relative to the (non-attainable) target S_n^*, we will restrict ourselves to the case of a market consisting of only two securities. We will state some results for the general case of an arbitrary number of securities in sub-section iii) without proving them. The analysis of the case of an arbitrary number of assets is notationally more complicated, but in principal the same (although it can be quite cumbersome to do the generalisation).

ii) The Case of two Securities

Our main goal will be to examine and present the announced asymptotic behaviour of the universal wealth in a rigorous way. More precisely, we will show that the quotient $\hat{S}_n / S_n{}^*$ is asymptotically equivalent to $\sqrt{2\pi/(nJ_n)}$ where the "volatility index" J_n still has to be defined. In particular, we will be able to prove that in a very general setting $\sqrt{2\pi/(nJ_n)}$ is a lower bound for $\hat{S}_n / S_n{}^*$ for all $n\in\mathbf{N}$. To prove that it is also an asymptotic upper bound, we have to make some strong assumptions. However, the lower bound is the one which is relevant for practical purposes.

Let us first simplify our notation for the two-dimensional case. Consider a sequence of arbitrary market vectors $x_i = (x_{i1}, x_{i2})'$, $i = 1,\ldots, n$. In this setting, a constant portfolio $\mathbf{b}$ has the form $(b, 1-b)'$ for a real number $b\in[0, 1]$. Therefore, in the following, we will identify a portfolio with its first component. Some further simplifying notations are given by

$$S_n(b) := S_n(\mathbf{b}) = \prod_{i=1}^{n}(bx_{i1} + (1-b)x_{i2}),$$

$$S_n{}^* = \max_{b\in[0,1]} S_n(b) = S_n(b_n{}^*),$$

$$W_n(b) := \frac{1}{n}\ln(S_n(b)) = \frac{1}{n}\sum_{i=1}^{n}\ln(bx_{i1} + (1-b)x_{i2}),$$

$$=: \int_{\mathbf{R}^2}\ln(bx_1 + (1-b)x_2)dF_n(x),$$

$$W_n{}^* = \max_{b\in[0,1]} W_n(b)$$

for $b\in[0, 1]$ (where again F_n is the empirical distribution function of the x_i, $i = 1,\ldots, n$). We can also simplify the representation of the universal wealth and the universal portfolio by using the diffeomorphism $(b, 1-b) \to b$ from B to $[0, 1]$. From the integral transformation formula we obtain

$$\hat{S}_n = \int_0^1 S_n(b)db \Big/ \int_0^1 db = \int_0^1 S_n(b)db = \int_0^1 \exp\{nW_n(b)\}db,$$

and that $\hat{\mathbf{b}}_k = (\hat{b}_k, 1-\hat{b}_k)$ is now given by

$$\hat{b}_k = \int_0^1 bS_{k-1}(b)db \Big/ \int_0^1 S_{k-1}(b)db .$$

To analyse the performance of $\hat{S}_n$, we define the **relative range** τ_n of the sequence $x_1,\ldots, x_n$ as

$$\tau_n = \sqrt[3]{2}\left(\frac{\max\limits_{i=1,\ldots,n,j=1,2}(x_{ij}) - \min\limits_{i=1,\ldots,n,j=1,2}(x_{ij})}{\min\limits_{i=1,\ldots,n,j=1,2}(x_{ij})}\right).$$

Further, denote by J_n the "volatility index" which measures the curvature of $S_n(b)$ about its maximum S_n^* and which is defined by

$$J_n = \frac{1}{n}\sum_{i=1}^{n}\frac{(x_{i1}-x_{i2})^2}{\left(b_n^* x_{i1}+(1-b_n^*)x_{i2}\right)^2}.$$

It is straight forward to show that we have $J_n = -W_n''(b_n^*)$ if both stocks are active at time n. We will assume that this is the case and that the (component wise) minimum of the market vectors x_i is strictly positive which simply means that all security prices should stay positive. In particular, this will guarantee a finite relative range.

The following theorem provides a lower bound for the quotient $\hat{S}_n / S_n^*$.

Theorem 6

Let $x_1, x_2, \dots$ be an arbitrary sequence of market vectors (in $(0, \infty)^2$) such that we have $\min_{i,j}\{x_{ij}\} > 0$, and let

$$a_n = \min\left\{ b_n^*, (1-b_n^*), \frac{3J_n}{\tau_n^3} \right\} > 0 \quad \text{for all } n \in \mathbf{N}.$$

Then for n sufficiently large (i.e. n must be big enough that there exists at least one index $i \le n$ with $x_{i1} \neq x_{i2}$) and arbitrary $\varepsilon \in (0,1)$ we have

$$\frac{\hat{S}_n}{S_n^*} \ge \sqrt{\frac{2\pi}{nJ_n(1+\varepsilon)}} - \frac{2}{\varepsilon(1+\varepsilon)a_nJ_nn}\, e^{-\frac{\varepsilon^2(1+\varepsilon)a_nJ_nn}{2}}.$$

Proof :

By performing a Taylor expansion of $W_n(b)$ in b_n^*, we obtain

$$W_n(b) = W_n(b_n^*) + (b-b_n^*)W_n'(b_n^*) + \tfrac{1}{2}(b-b_n^*)^2 W_n''(b_n^*) + \tfrac{1}{6}(b-b_n^*)^3 W_n'''(\bar{b}_n)$$

with a suitable number $\bar{b}_n$ between b and b_n^*. Due to the optimality of b_n^* (which is assumed to be an inner point, i.e. $0 < b_n^* < 1$), the different terms in this expansion satisfy

$$W_n(b_n^*) = \frac{1}{n}\ln(S_n^*) =: W_n^*,$$

$$W_n'(b_n^*) = 0,$$

$$W_n''(b_n^*) = -J_n < 0,$$

$$W_n'''(\bar{b}_n) = 2\int \frac{(x_1 - x_2)^3}{(\bar{b}_n x_1 - (1-\bar{b}_n)x_2)^3} dF_n(x) .$$

Because of

$$|\bar{b}_n x_1 + (1-\bar{b}_n)x_2| \geq \bar{b}_n \min_{i,j}\{x_{ij}\} + (1-\bar{b}_n)\min_{i,j}\{x_{ij}\} = \min_{i,j}\{x_{ij}\},$$

we have got the estimate

$$|W_n'''(\bar{b}_n)| \leq \tau_n^3 \quad \forall \bar{b}_n \in [0,1]$$

which enables us to give a lower bound for $S_n(b)$:

$$S_n(b) = \exp\{nW_n(b)\} \geq \exp\left(nW_n^* - \tfrac{n}{2}(b - b_n^*)^2 - \tfrac{n}{6}|b - b_n^*|^3 \tau_n^3 \right).$$

Using this estimate and the change of variable $u = \sqrt{n}(b - b_n^*)$ yields

$$\hat{S}_n = \int_0^1 S_n(b)db \geq \frac{S_n^*}{\sqrt{n}} \int_{-\sqrt{n}b_n^*}^{\sqrt{n}(1-b_n^*)} \exp\left(-\tfrac{1}{2}u^2 J_n - \tfrac{1}{6\sqrt{n}}|u|^3 \tau_n^3\right) du.$$

The last integral will decrease if we further restrict the region of integration. By noting that we have

$$-\tfrac{1}{2}u^2 J_n - \tfrac{1}{6\sqrt{n}}|u|^3 \tau_n^3 \geq -\tfrac{1}{2}u^2 J_n(1+\varepsilon) \quad \text{for } |u| \leq 3\varepsilon\sqrt{n}\frac{J_n}{\tau_n^3},$$

and using the definition of a_n, we obtain

$$\begin{aligned} \frac{\sqrt{n}\hat{S}_n}{S_n^*} &\geq \int_{-\sqrt{n}b_n^*}^{\sqrt{n}(1-b_n^*)} \exp\left(-\tfrac{1}{2}u^2 J_n - \tfrac{1}{6\sqrt{n}}|u|^3 \tau_n^3\right) du \\ &\geq \int_{-\sqrt{n}a_n\varepsilon}^{\sqrt{n}a_n\varepsilon} \exp\left(-\tfrac{1}{2}u^2 J_n(1-\varepsilon)\right) du \\ &= \int_{-\infty}^{\infty} \exp\left(-\tfrac{1}{2}u^2 J_n(1-\varepsilon)\right) du - 2\int_{-\infty}^{-\sqrt{n}a_n\varepsilon} \exp\left(-\tfrac{1}{2}u^2 J_n(1-\varepsilon)\right) du \\ &= \sqrt{\frac{2\pi}{J_n(1+\varepsilon)}} - \sqrt{\frac{8\pi}{J_n(1+\varepsilon)}}\,\Phi\left(-\varepsilon a_n \sqrt{nJ_n(1+\varepsilon)}\right) \end{aligned}$$

where $\Phi(x)$ denotes the distribution function of the standard normal. Using an inequality relating $\Phi(x)$ to the density function of the standard normal (see Remark A2 c)) with $x = \varepsilon a_n \sqrt{nJ_n(1+\varepsilon)}$, we obtain

$$\Phi\left(-\varepsilon a_n\sqrt{nJ_n(1+\varepsilon)}\right) < \frac{1}{\sqrt{2\pi(\varepsilon a_n)^2(1+\varepsilon)nJ_n}} e^{-\frac{\varepsilon^2(1+\varepsilon)a_nJ_nn}{2}}$$

which completes the proof.

□

The theorem more or less states that we have $\hat{S}_n / S_n^* \geq \sqrt{2\pi/(nJ_n)}$. If J_n would be bounded from below by a positive constant J then we would acquire that the quotient $\hat{S}_n / S_n^*$ is bounded from below by a multiple of $1/\sqrt{n}$, i.e. the exponential growth rates of $\hat{S}_n$ and S_n^* coincide. This will be made rigorous in the following theorem.

Theorem 7
Let $x_1, x_2, \ldots$ be an arbitrary sequence of market vectors (in $(0, \infty)^2$). Assume that there exist constants δ, τ and J satisfying $\delta \leq b_n^* \leq 1-\delta$, $\tau_n \leq \tau < \infty$, and $J_n \geq J > 0$ along a subsequence of times $n_1, n_2, \ldots$. Then, along this subsequence we have

$$\liminf_{i\to\infty} \frac{\hat{S}_{n_i} / S_{n_i}^*}{\sqrt{2\pi/(n_iJ_{n_i})}} \geq 1 .$$

Proof :
The boundedness assumptions on b_n^*, J_n and τ_n together with Theorem 6 imply

$$\frac{\hat{S}_n / S_n^*}{\sqrt{2\pi/(nJ_n)}} \geq \sqrt{\frac{1}{1+\varepsilon_n}} - \frac{2}{\varepsilon_n(1+\varepsilon_n)a_n\sqrt{2\pi J_n n}}$$

$$\geq \sqrt{\frac{1}{1+\varepsilon_n}} - \frac{2}{\varepsilon_n(1+\varepsilon_n)\min(\delta, 3J/(\tau^3))\sqrt{2\pi Jn}}$$

for all n belonging to the subsequence $n_1, n_2, \ldots$ where we can choose $\varepsilon_n \in (0,1)$ arbitrarily. For the choice of $\varepsilon_n = n^{-1/4}$, we have convergence of the right hand side of this inequality against 1 along the subsequence n_i.

□

The two inequalities of Theorems 6 and 7 yield lower bounds on the performance of the universal portfolio relative to the sequence of the best constant portfolios at the different time instants, i.e. they are of the form "the universal portfolio is asymptotically at least as good as $S_n^*\sqrt{2\pi/(nJ_n)}$" (which is the relevant direction for practical use). To show that the universal portfolio does indeed not perform better, we are going to demonstrate that the lower bound of Theorem 7 will also provide an asymptotic upper bound for the quotient $\hat{S}_n / S_n^*$. However, to do so, we have to impose some drastic constraints on the market vectors and the sequence of optimal constant portfolios and corresponding optimal wealths (see also (Cover 1991)). We will comment on the possibility to prove that these assumptions can be satisfied after the following theorem.

Theorem 8 "Asymptotic behaviour of the universal portfolio"
Assume that there exists a subsequence of times $n_1, n_2, \ldots$ and a twice continuously differentiable function $W:[0,1] \to \mathbf{R}$ with $W_{n_i}(b) \to W(b)$ as $i \to \infty$ for all $0 \le b \le 1$ such that

(i) $W(b)$ attains its unique maximum in $b^* \in (0, 1)$,

(ii) $W_n(b) \le W(b)$ for $0 \le b \le 1$,

(iii) $b_n^* \to b^*$ (where b_n^* is the maximiser of $W_n(b)$),

(iv) $W_n''(b_n^*) \to W''(b^*)$, $W_n''(b_n^*) \le -J$ for some $J > 0$ and all $0 \le b \le 1$.

for all n which are part of the above subsequence. Assume further that the assumptions of Theorem 7 are satisfied. Then along the subsequence n_i we have

$$\frac{\hat{S}_n}{S_n^*} \sim \sqrt{2\pi / (nJ_n)}$$

(where $a_n \sim b_n$ means that the sequence (a_n / b_n) converges to 1).

Proof:
Due to Laplace's method of integration (see Part D of the Appendix), for a twice continuously differentiable function g with $g''(x) \le -J$ with a unique maximum $u^* \in (0, 1)$, we have

$$\int_0^1 e^{ng(u)} du \sim e^{ng(u^*)} \sqrt{\frac{2\pi}{n|g''(u^*)|}} .$$

Thus, due to our assumptions on W(b), we obtain

$$\hat{S}_n = \int_0^1 \exp\{nW_n(b)\} db \overset{(ii)}{\le} \int_0^1 \exp\{nW(b)\} db \sim e^{nW(b^*)} \sqrt{\frac{2\pi}{n|W''(b^*)|}} \sim S_n^* \sqrt{\frac{2\pi}{nJ_n}} .$$

Note that the corresponding lower bound is already given in Theorem 7 which completes the proof.

□

Remark 9

If we assume that the market vectors $x_i \in [0, \infty)^2$ satisfy $A_* \le x_{i1}, x_{i2} \le A^*$ (for suitable positive constants A_*, A^*) and that we have $W_n''(b) \le -J < 0$ for all $n \in \mathbf{N}$ and all $b \in [0, 1]$ then one can prove the existence of a function W: $[0,1] \to R$ with

(i) $W \in C^2([0,1])$,

(ii) $W(b) = \lim_{i \to \infty} W_{n_i}(b) \quad \forall\, 0 \le b \le 1$ (along a subsequence n_i),

(iii) $W''(b) \le -J < 0 \quad \forall\, 0 \le b \le 1$.

The proof consists of showing that the sets $\{W_n, n \in N\}$, $\{W_n', n \in N\}$, $\{W_n'', n \in N\}$ are uniformly bounded and uniformly continuous and of multiple application of the Arzéla-Ascoli Theorem. This is done in (Müller 1995).

iii) The Case of an Arbitrary Number of Securities

In the general case of m securities the notation becomes more complicated but the above results will in principle stay the same as long as all securities are active, of full rank and satisfy $b_n^* \to b^* \in \text{int}(B)$. To state these results, we first have to adjust the definition of the volatility index J_n.

Definition 10

Let F(x) be a distribution function on $[0, \infty)^m$ such that for all $1 \le i, j \le m-1$ the integrals

$$J_{ij}(b) = \int \frac{(x_i - x_m)(x_j - x_m)}{\left(\sum_{j=1}^{m-1} b_j x_j + (1 - \sum_{j=1}^{m-1} b_j) x_m\right)^2} dF(x), \quad 1 \le i, j \le m-1$$

exist. The corresponding matrix-valued function J(b) = J(b; F) is called the **sensitivity matrix function** (as a function of $b \in B$) with respect to F. We further set $J^* := J(b^*(F))$ where $b^*(F)$ maximises $W(\mathbf{b}, F)$ and $J_n(b) := J(b; F_n)$.

As in the two-security case, it is straight-forward to show that if all securities are active at time n then for $b^* \in \text{int}(B)$ we have

$$J^*_{ij} = J_{ij}(b^*(F)) = -\frac{\partial^2 W(\mathbf{b^*}, F)}{\partial b_i \partial b_j}$$

(i.e. $J_n(b_n^*)$ represents the curvature of $S_n(b)$ around the maximiser b_n^*). With this definition in mind, we can state an m-dimensional analogue to Theorem 8 (see (Cover 1991) for a (sketch of a) proof).

Theorem 11 "Asymptotic behaviour of the universal portfolio"
Suppose that we have a sequence of market vectors $x_1, x_2, \ldots$ with $x_i \in [\alpha, \gamma]^m$ for two constants $0 < \alpha \le \gamma < \infty$. Further, along a subsequence of times n_i, we assume the existence of a function W such that

(i) $W_n(b) \uparrow W(b) \quad \forall\, b \in B$ and some $W \in C^2(B)$,

(ii) $b_n^* \to b^* \quad$ for some $b^* \in \text{int}(B)$,

(iii) $J_n(b_n^*) \to J^* := -\left(\dfrac{\partial^2 W(\mathbf{b}^*, F)}{\partial b_i \partial b_j}\right)_{1 \le ij \le m-1}$.

Moreover, W(b) should be strictly concave with bounded third partial derivatives on B, and b_n^* should be in int(B). Then, we have

$$\frac{\hat{S}_n}{S_n^*} \sim \left(\sqrt{\frac{2\pi}{n}}\right)^{m-1} \frac{(m-1)!}{\sqrt{\det(J^*)}}.$$

Remark 12
a) Comparing Theorems 8 and 11, we see that with every additional security in the market, the asymptotic performance of the universal portfolio (relative to S_n^*) is getting worse by a factor of $1/\sqrt{n}$. However, this does not change the fact that $\hat{S}_n$ and S_n^* have the same asymptotic exponential growth rate.
b) So far, we have always assumed that every security in the market has an impact on S_n^* for all n (by assuming $b_n^*, b^* \in \text{int}(B)$) which is a natural assumption if there are no redundant securities. However, in real securities markets there are often redundant (or nearly redundant) securities. This can result in a sequence of optimal portfolios b_n^* which lie in a k-dimensional sub-simplex of B. Consequently, the universal portfolio restricted to the corresponding subset of k securities would asymptotically perform as $(1/\sqrt{n})^{k-1} S_n^*$. Of course, this k-subset of all m securities is a priori unknown. Therefore, Cover proposes to follow a mixture of universal portfolios each for one of the (2^m-1) subsets of $\{1, \ldots, m\}$ corresponding to k securities with $1 \le k \le m$. This mixture of universal portfolios is called the **generalised universal portfolio**. More precisely, let $V \subseteq \{1, \ldots, m\}$ and let μ_V be the measure on

$$B(V) := \{b \in B \mid b_i = 0 \quad \forall\, i \notin V\}$$

corresponding to the uniform distribution. With the definition of the measure

$$\mu := \frac{1}{2^m - 1} \sum_{\emptyset \neq V \subseteq \{1,..,m\}} \mu_V$$

the generalised universal portfolio is defined in (Cover 1991) by

$$\hat{b}_{n+1} := \frac{\int_B b S_n(b)\mu(db)}{\int_B S_n(b)\mu(db)}, \quad S_n(b) = \prod_{i=1}^{n} b' x_i, \quad S_0(b) = 1$$

By defining $J_n^{(k)}(F_n)$ to be the (k–1, k–1) sensitivity matrix corresponding to the k securities in V where V is the smallest subset of {1,..., m} such that all optimal portfolios $b_n^*(F_n)$ are in the interior of B(V) (in fact it is enough that all but finitely many $b_n^*(F_n)$ are in int(B(V))), the following asymptotic behaviour for the generalised universal wealth is reported in (Cover 1991):

$$\frac{\hat{S}_n}{S_n^*} \sim \frac{1}{2^m - 1} \left(\sqrt{\frac{2\pi}{n}}\right)^{k-1} \frac{(k-1)!}{\sqrt{\det(J_n^{(k)}(F_n))}}$$

(where the factor $(2^m-1)^{-1}$ has its origin in the fact that only this mass is assigned to the "correct" universal portfolio, i.e. to the one that uses the "correct" subset of the securities). Of course, this formula is only valid for k > 1. In the case of k = 1 (i.e. there is one security that totally dominates the other securities), it is obvious to have

$$\frac{\hat{S}_n}{S_n^*} \sim \frac{1}{2^m - 1}.$$

iv) Stochastic Markets

One striking feature of all the considerations in the previous sub-sections is that there is no need for an (explicit) assumption on the stochastic nature of the market vectors (i.e. on the securities prices). In this sub-section, we will show the naturalness of our goal to "track" S_n^* and demonstrate once more the desirable properties of the universal portfolio in a more specialised stochastic setting.

Let us therefore assume that the market vectors are modelled as independent identically distributed (for short: iid) random variables $X_1, X_2, \ldots$ which are $[0,\infty)^m$-valued, and each is distributed according to the distribution function F(x). We also assume that $H(b) := E(\ln(b'X_i))$ exists for all $b \in B$ and is continuous as a function in b. As in the setting of the preceding subsections we denote by

$$S_n(b) = \prod_{i=1}^{n} b'X_i, \qquad S_0(b) = 1$$

the wealth process of a constant portfolio strategy after n days starting with unit initial capital. Using the strong law of large numbers, we have

$$S_n(b) = \exp\left(\sum_{i=1}^{n} \ln(b'X_i)\right) = \exp\big(n\big(E(\ln(b'X_1)) + o_p(1)\big)\big)$$
$$=: \exp\big(n\big(W(b,F) + o_p(1)\big)\big) \tag{3}$$

with

$$W(b, F) := E(\ln(b'X_1) = \int \ln(b'x)dF(x),$$

and where $o_p(1)$ is a random variable satisfying $o_p(1) \xrightarrow{n\to\infty} 0$ almost surely.

Remark 13
Representation (3) explains the name "exponential growth rate" for W(b, F) due to the fact that W(b, F) describes the growth of the exponent of $S_n(b)$ up to first order.

Further, we write

$$W^*(F) := \max_{b\in\mathbf{R}^m} \int \ln(b'x)dF(x)$$

and assume that $b^*(F)$ attains the corresponding maximum. Then, denote by $S_n{}^*$ the wealth $S_n(b^*(F))$ corresponding to the constant strategy $b^*(F)$ which is the analogue (up to $o_p(1)$) of the optimal constant portfolio $S_n{}^*$ of the previous sub-sections. In this setting, a classical result of (Breiman 1961) is the following:

Theorem 14
With the above notations and assumptions, for all constant portfolios $b\in\mathbf{R}^m$, we have

$$\limsup_{n\to\infty} \frac{1}{n} \ln\left(\frac{S_n(b)}{S_n{}^*}\right) \le 0.$$

Proof :
By using representation (3) for both $S_n(b)$ and $S_n{}^*$, we obtain

$$\frac{S_n(b)}{S_n{}^*} = \exp\big(n\big((W(b,F) - W^*(F)) + (o_p(1) - \tilde{o}_p(1))\big)\big) \le \exp\big(n\hat{o}_p(1)\big)$$

with $o_p(1)$, $\tilde{o}_p(1)$, $\hat{o}_p(1):=o_p(1)-\tilde{o}_p(1)$ all converging to 0 for n going to infinity. Thus

$$\frac{1}{n}\ln\left(\frac{S_n(b)}{S_n{}^*}\right) \le \frac{1}{n}\ln\left(\exp\left(n\hat{o}_p(1)\right)\right) = \hat{o}_p(1) \xrightarrow{n\to\infty} 0 \quad \text{a.s..}$$

□

This theorem states that W*(F) is the biggest possible exponential growth rate that can be attained by following a constant strategy which is exactly the sense in which b*(F) — and therefore $S_n{}^*$ — is optimal. As W(b, F) is smaller than W*(F) for all b that do not deliver the maximum corresponding to W*(F), these constant portfolios b are exponentially outperformed by $S_n{}^*$ (in the long run).

The next result will show that the universal portfolio has the same (asymptotic) growth rate as $S_n{}^*$. Hence, the universal portfolio will also outperform all constant portfolios exponentially that do not have a growth rate of W*(F).

To prove this result, we make again use of the diffeomorphism $(b_1,\ldots,b_m) \to (b_1,\ldots,b_{m-1})$ from B to C where C is given by

$$C := \left\{ (c_1,\ldots,c_{m-1}) \,\middle|\, c_i \ge 0,\ \sum_{i=1}^{m-1} c_i \le 1 \right\} .$$

We will slightly abuse our notation below by writing W(c, F), $S_n(c)$, c*(F) for c∈C with the convention that we mean W(b(c), F), $S_n(b(c))$ by this where we have set

$$b(c) := \left(c_1,\ldots,c_{m-1}, 1-\sum_{i=1}^{m-1} c_i \right).$$

Theorem 15

Let all market vectors $X_i \ge 0$ be iid random variables with distribution function F. Assume that the maximum c*(F)∈C of W(c, F) is unique with c*(F)∈int(C). Further, let W(c, F) be twice continuously differentiable and uniformly concave as a function in c. Then, the (asymptotic) exponential growth rate, $\lim_{n\to\infty}\left(\ln(\hat{S}_n)/n\right)$, satisfies

$$\lim_{n\to\infty}\left(\frac{1}{n}\ln(\hat{S}_n)\right) = W^*(F) \quad \text{a.s. .}$$

Proof :

Using the above diffeomorphism and the definition of $\hat{S}_n$, we obtain

$$\hat{S}_n = (m-1)!\int_C S_n(c)dc = (m-1)!\int_C \exp(n(W(c,F)+o_p(1)))dc$$
$$= (m-1)!\ e^{no_p(1)}\int_C e^{n(W(c,F)}dc$$

Due to our assumptions, we can apply (a multi-dimensional variant of) Laplace's method of integration (see Part D of the Appendix) which results in

$$\int_C e^{n(W(c,F)}dc \sim e^{nW(c^*(F))}\left(\frac{2\pi}{n}\right)^{\frac{m-1}{2}} \frac{1}{\sqrt{\det(-W''(c^*(F),F))}} = K\left(\frac{1}{n}\right)^{\frac{m-1}{2}} e^{nW(c^*(F))}$$

(where W"(c, F) denotes the Hessian with respect to c, K is a suitable constant) with the consequence of

$$\hat{S}_n \sim (m-1)!K\left(\frac{1}{n}\right)^{\frac{m-1}{2}} e^{n(W(c^*(F))+o_p(1))} = (m-1)!K\left(\frac{1}{n}\right)^{\frac{m-1}{2}} S_n{}^* e^{no_p(1)}$$

where $o_p(1)$ does not necessarily coincide with the variable of the same name used in the first relation of the proof. Using the fact that the relation "~" is preserved under continuous transformations, we conclude

$$\left|\frac{1}{n}\ln(\hat{S}_n) - W^*(F)\right| \sim \left|\frac{1}{n}\ln(\hat{S}_n / S_n{}^*)\right| \sim \left|\frac{1}{n}\ln((m-1)!K\left(\frac{1}{n}\right)^{\frac{m-1}{2}} e^{no_p(1)})\right|$$
$$= \left|\frac{1}{n}\ln((m-1)!K) + o_p(1) - \frac{1}{n}\frac{m-1}{2}\ln(n)\right| \xrightarrow{n\to\infty} 0 \quad \text{a.s.}\ .$$

□

Remark 16

Some numerical examples using real market data are presented in (Cover 1991). The main conclusion that can be drawn from them is that the universal portfolio performs extremely well if the stock prices in the underlying market segment are not too similar. This fact will be underlined by our theoretical results for the continuous-time setting given in the next section (in particular, see Remark 33).

6.2 Asymptotically Optimal Portfolios and Universal Portfolios in Continuous Time

This section is concerned with the presentation of a continuous-time version of Cover's universal portfolio approach given in (Jamshidian 1992). As it will turn out, the gain, obtained by restriction to a continuous-time model of the type similar to those given in Chapter 2, lies in explicit formulae for the wealth of constant rebalanced portfolios. This will make the whole analysis more transparent than in the discrete-time case. As we now deal with a continuous market model, we will use the notation of Chapter 2 which does not coincide with the notation of Section 1, the presentation of the universal portfolio approach in discrete time.

i) Some Results on Constant Rebalanced Portfolios in Continuous Time

Our continuous-time market will be made up of n traded securities with prices $P_i(t)$ given as the unique (strong) solutions of the equations

$$\frac{dP_i(t)}{P_i(t)} = b_i(t)\,dt + \sum_{j=1}^{n} \sigma_{ij}(t)dW_j(t), \qquad i = 1,\dots, n, \tag{4}$$

where W(t) is an n-dimensional Brownian motion; b(t), σ(t) are assumed to be F_t-adapted (where $\{F_t\}_{t\geq 0}$ is the usual Brownian filtration) with

$$\int_0^t |b_i(s)|\,ds < \infty \quad , i = 1,\dots, n, \ \forall\, t \geq 0 \ \text{ a.s.},$$

$$\int_0^t \left(\sigma_{ij}(s)\right)^2 ds < \infty \quad , i, j = 1,\dots, n, \ \forall\, t \geq 0 \ \text{ a.s.}.$$

Remark 16

As we have not assumed regularity of σ(t), we have also included the case in our model that the first security could be a riskless bond.

We will not allow for consumption and short-selling of stocks. Thus the portfolio processes π(t) (defined as the vector of fractions of wealth invested in the different securities at time t) are constrained to satisfy

$$\pi(t) \in \Pi = \left\{ \pi \in \mathbf{R}^n \,\middle|\, \pi_i \geq 0, \sum_{i=1}^{n} \pi_i = 1 \right\} \quad \forall\, t \geq 0 \ \text{ a.s.}\,.$$

As usual, we only consider self-financing portfolios (and "ignore" dividends in the sense of the remark made in the previous section), i.e. we assume that the wealth process X(t) corresponding to the portfolio process π(t) satisfies

$$\frac{dX(t)}{X(t)} = \sum_{i=1}^{n} \pi_i(t) \frac{dP_i(t)}{P_i(t)} \quad , \quad X(0) = x \tag{5}$$

We will define a **constant rebalanced portfolio** (for brevity: a constant portfolio) to be a self-financing portfolio process $\pi(t)$ that is constant over time, i.e. $\pi(t) \equiv \pi$ for all t. Denote its wealth process by $X(t, \pi)$. As already stated in Chapter 2, we can calculate $X(t, \pi)$ as the explicit solution of the stochastic differential equation (5). We will give some convenient forms of this explicit solution below. To state them, we need some abbreviations. Let

$$\Sigma(t) := \sigma(t)\sigma(t)', \qquad \Lambda(t) := \int_0^t \Sigma(s)ds .$$

Proposition 17

$$X(t,\pi) = x\left(\prod_{i=1}^{n}\left[\frac{P_i(t)}{P_i(0)}\right]^{\pi_i}\right)e^{-\frac{1}{2}\pi'\Lambda(t)\pi+\frac{1}{2}\sum_{i=1}^{n}\Lambda_{ii}(t)\pi_i} \tag{6}$$

Proof :

First, note that from the explicit form of the stock prices and Itô's formula, we obtain

$$\ln\left(\frac{P_i(t)}{P_i(0)}\right) = \int_0^t \frac{dP_i(s)}{P_i(s)} - \frac{1}{2}\Lambda_{ii}(t)$$
$$= \int_0^t b_i(s)ds - \frac{1}{2}\Lambda_{ii}(t) + \sum_{j=1}^{n}\int_0^t \sigma_{ij}(s)dW_j(s).$$

Having this in mind and using equation (5), we obtain

$$\ln\left(\frac{X(t,\pi)}{X(0,\pi)}\right) = \int_0^t \frac{1}{X(s,\pi)}dX(s,\pi) - \frac{1}{2}\int_0^t \frac{1}{X(s,\pi)^2}d\langle X(.,\pi)\rangle_s$$
$$= \sum_{i=1}^{n}\int_0^t \pi_i \frac{dP_i(s)}{P_i(s)} - \frac{1}{2}\int_0^t \frac{1}{X(s,\pi)^2}d\langle X(.,\pi)\rangle_s$$
$$= \sum_{i=1}^{n}\pi_i\left(\ln\left(\frac{P_i(t)}{P_i(t)}\right) + \frac{1}{2}\Lambda_{ii}(t)\right) - \frac{1}{2}\pi'\Lambda(t)\pi .$$

Application of the exponential function to both sides of equation (7) yields representation (6).

□

As in the discrete-time setting, our target portfolio (i.e. the portfolio that will serve as a benchmark for the performance of all other portfolios) at every time instant t will be the best constant rebalanced portfolio up to this time which is defined via its wealth process

$$X^*(t) := X(t, \pi^*(t)) := \max_{\pi \in \Pi} X(t, \pi).$$

As Π is compact and $X(t,.)$ is a continuous function in π, it is clear that the maximum will be attained. Due to Proposition 17, it is given as the solution of the quadratic programming problem

$$\max_{\pi \in \Pi} \left(\sum_{i=1}^{n} \pi_i \left(\ln\left(\frac{P_i(t)}{P_i(t)} \right) + \tfrac{1}{2} \Lambda_{ii}(t) \right) - \tfrac{1}{2} \pi' \Lambda(t) \pi \right) \tag{8}$$

By construction, $\Lambda(t)$ is positive semi-definite. It is even positive definite if the instantaneous covariance matrix of the security prices, $\Sigma(s)$, is positive definite for all $s \le t$ (which is implied by the regularity of $\sigma(s)$). In this case, we would have a unique solution to the maximisation problem (8). Since the feasible set Π is a closed subset of an $(n-1)$-dimensional manifold, we can expect to weaken the condition of a positive definite $\Lambda(t)$. In fact, we only need that $\Lambda(t)$ is positive definite on Π. This implies that the objective function of our problem is strictly concave on Π, and thus attains a unique maximum over this compact set. An equivalent formulation of this condition which will prove to be useful in the sequel is given in the following proposition.

Proposition 18
Let $\Lambda \in \mathbf{R}^{n,n}$ be symmetric and positive semi-definite. Then the following two statements are equivalent :
a) The convex function $f(\pi) = \pi'\Lambda\pi$ restricted to Π is strictly convex.
b) The $(n-1, n-1)$-matrix V with

$$V_{ij} = \Lambda_{ij} - \Lambda_{in} - \Lambda_{jn} - \Lambda_{nn} \quad , \quad 1 \le i, j \le n-1$$

is positive definite.

Proof:
Set $\overline{\pi} = (\pi_1, \ldots, \pi_{n-1})$ for $\pi \in \Pi$. Due to $\pi_n = 1 - \pi_1 - \ldots - \pi_{n-1}$, we obtain

$$f(\pi) = \overline{\pi}' V \overline{\pi} + a'\overline{\pi} + c$$

for suitable $a \in \mathbf{R}^{n-1}$, $c \in \mathbf{R}$ which yields the equivalence of a) and b).

□

Condition b) is not only satisfied in the case of a positive definite Λ, but also in the important case of security n being a riskless bond and of a non-degenerate covariance matrix of the remaining securities, i.e. the case of

$$\sigma(t) = \begin{pmatrix} \tilde{\sigma}(t) \\ 0 \ldots 0 \end{pmatrix}$$

with an (n–1, n)-matrix $\tilde{\sigma}(t)$ such that $\tilde{\sigma}(t)\tilde{\sigma}(t)'$ is positive definite.

In the following, we will also use the terminology of an active, inactive and optimal stock in the same way as in the discrete-time model. To ensure that the "asymptotically optimal constant rebalanced portfolio" exists (in a sense that still has to be defined !), we have to make some assumptions. Jamshidian does this by introducing certain notions of regularity of stochastic processes and security market models.

Definition 19

a) A (vector-valued) stochastic process X(t) is called **weakly regular** if for all components $X_i(t)$ we have :

i) $E(|X_i(t)|) < \infty \quad \forall\, t \geq 0$,

ii) $\lim\limits_{t\to\infty} \dfrac{E(X_i(t))}{t} =: \eta_i$ exists ,

iii) $\dfrac{X_i(t)}{t} \xrightarrow{t\to\infty} \eta_i$ in probability .

b) A security market model is called **weakly regular** if the corresponding matrix $\Lambda(t)$ and the (logarithm of the) price vector $\left(\ln(P_1(t)),\ldots,\ln(P_n(t))\right)'$ are weakly regular in the sense of part a) above and if further the conditions of Proposition 17 are satisfied by $\Lambda(t)$ and $\lim\limits_{t\to\infty} (\Lambda(t)/t)$.

Remark 20

a) A simple example of a weakly regular model is given by the Black-Scholes model, i.e. by the case of constant market coefficients r, b, σ where the volatility matrix of the stock prices has maximum rank.

b) In a weakly regular model, we set (component wise)

$$\Sigma^{\infty} := \lim_{t\to\infty} E(\Lambda(t)/t) \quad \eta^{\infty} := \lim_{t\to\infty} E(\ln(P(t))/t)$$

If we note that the weak regularity of $\Lambda(t)$ implies

$$E(\Lambda_{ii}(t)) = E\left(\int_0^t \sum_{j=1}^{n} \sigma_{ij}(s)^2 \, ds \right) < \infty$$

then we obtain

$$E\left(\int_0^t b_i(s)ds \right) = E(\ln(P_i(t))) + \tfrac{1}{2} E(\Lambda_{ii}(t)) - E(\ln(P_i(0))),$$

and the weak regularity of the model implies the existence and form of the **long time average rate of return** of security i,

$$b_i^{\infty} := \lim_{t\to\infty} \tfrac{1}{t} E\left(\int_0^t b_i(s)ds \right) = \eta_i^{\infty} + \tfrac{1}{2} \Sigma_{ii}^{\infty} .$$

Via representation (6), $\ln(X(t,\pi))$ is weakly regular with **rate of return**

$$r(\pi) := \lim_{t\to\infty} \tfrac{1}{t} E(\ln(X(t,\pi))) = -\tfrac{1}{2} \pi'\Sigma^{\infty}\pi + \pi'b^{\infty}.$$

With the help of this remark, we are able to make the notion of an **asymptotically optimal constant rebalanced portfolio** rigorous.

Definition 21
In a weakly regular model, we define the **asymptotically optimal constant rebalanced portfolio** to be the unique portfolio $\pi^{\infty} \in \Pi$ with the highest rate of return

$$r(\pi^{\infty}) := \max_{\pi \in \Pi} r(\pi) .$$

Note that due to the weak regularity of the model, there exists exactly one constant portfolio attaining the highest rate of return. It is therefore justified to talk of "the" asymptotically optimal constant rebalanced portfolio. To justify that this portfolio is a reasonable target to reach, we first show that it outperforms all other constant portfolios exponentially.

Theorem 22
In a weakly regular model, every constant rebalanced portfolio $\pi \in \Pi \setminus \{\pi^{\infty}\}$ is outperformed exponentially by the asymptotically optimal constant portfolio, i.e. for all such π there exists a constant $\delta > 0$ with

$$\frac{X(t,\pi)e^{\delta t}}{X(t,\pi^{\infty})} \xrightarrow{t\to\infty} 0 \quad \text{in probability.}$$

For brevity, we will write "$X(t,\pi)/X(t,\pi^{\infty}) \approx 0$" to express this relation.

Proof:
Set $Z(t) = \ln(X(t, \pi^{\infty})/X(t, \pi))$, $c = r(\pi^{\infty}) - r(\pi) > 0$. Then the weak regularity of both $X(t, \pi^{\infty})$ and $X(t, \pi)$ implies

$$\tfrac{1}{t}Z(t) = \tfrac{1}{t}\ln(X(t, \pi^{\infty})) - \tfrac{1}{t}\ln(X(t, \pi)) \xrightarrow{t\to\infty} c \quad \text{in probability,}$$

and for $0 < \delta < \varepsilon$ and $0 < \varepsilon < c-\delta$ the above convergence yields

$$P(\,\delta t - Z(t) < -\varepsilon t\,) \xrightarrow{t\to\infty} 1\,.$$

Consequently, we have $\exp(\delta t - Z(t)) \xrightarrow{t\to\infty} 0$ in probability which, by the definition of $Z(t)$, implies $X(t, \pi)/X(t, \pi^{\infty}) \approx 0$.

□

Our next goal is to prove that the sequence of optimal constant portfolios $\pi^*(t)$ at time instant t converges in probability towards π^{∞} (which is a result one would expect in view of Theorem 22). To ensure this, we need some additional notation. In a weakly regular model, we therefore denote by J^{∞} the limit in probability of $V(t)/t$ which is given by

$$J_{ij}^{\infty} = \Sigma_{ij}^{\infty} - \Sigma_{in}^{\infty} - \Sigma_{jn}^{\infty} - \Sigma_{nn}^{\infty}\,.$$

Further, define the vectors $\lambda(t)$, $\beta^*(t)$, β^{∞}, γ^{∞} by

$$\lambda_i(t) := \ln\left(\frac{P_n(t)}{P_n(0)}\right) - \ln\left(\frac{P_n(t)}{P_n(0)}\right) - \tfrac{1}{2}V_{ii}(t)\,, \quad i = 1,\dots, n-1$$

$$\beta^*(t) := \left(V(t)\right)^{-1}\lambda(t), \quad \beta^{\infty} := -\left(J^{\infty}\right)^{-1}\gamma^{\infty} \;\text{ with}$$

$$\gamma_i^{\infty} := \eta_n^{\infty} - \eta_i^{\infty} - \tfrac{1}{2}J_{ii}^{\infty}.$$

Proposition 24 will show that $\beta^*(t)$ and β^{∞}coincide with the first n–1 components of $\pi^*(t)$ and π^{∞}, respectively.

Remark 23 (not to be skipped !)
Before continuing with our analysis, we will restrict our presentation to a special case of the results of (Jamshidian 1992). In (Jamshidian 1992), a security market model is called **asymptotically active** if every component of the asymptotically optimal constant rebalanced portfolio π^∞ is positive, i.e. every security enters π^∞. We will only concentrate on this case to avoid too much technicalities. Jamshidian also considers the general case of less than n securities being asymptotically active (i.e. not all securities have a corresponding positive component of π^∞), but this makes the presentation much more technically and notationally involved. As a compromise, we therefore restrict ourselves to the case of an asymptotically active model and refer the reader to the results of (Jamshidian 1992) for the remaining cases.

Proposition 24
Assume that V(t) is regular (as a matrix !).
a) Then, we have the following representations:

$$X(t,\pi) = x\,\frac{P_n(t)}{P_n(0)}\,e^{-\frac{1}{2}\bar{\pi}'V(t)\bar{\pi}-\lambda(t)'\bar{\pi}}, \tag{9}$$

$$X(t,\pi) = x\,\frac{P_n(t)}{P_n(0)}\,e^{\frac{1}{2}\beta^*(t)'V(t)\beta^*(t)-(\bar{\pi}-\beta^*(t))'V(t)(\bar{\pi}-\beta^*(t))}, \tag{10}$$

$$X(t,\pi^*(t)) = x\,\frac{P_n(t)}{P_n(0)}\,e^{\frac{1}{2}\overline{\pi^*(t)}'V(t)\overline{\pi^*(t)}}. \tag{11}$$

b) All stocks are active at time t if and only if $\beta^*(t)\in\overline{\Pi}:=\{(\pi_1,\dots,\pi_{n-1})\mid\pi\in\mathrm{int}(\Pi)\}$. In this case, we have

$$\beta^*(t) = \overline{\pi^*(t)}. \tag{12}$$

c) A weakly regular model is asymptotically active if and only if $\beta^\infty\in\overline{\Pi}$. Then, we have

$$\beta^\infty = \overline{\pi^\infty}.$$

Proof:
a, b) Using the relation $\pi_n = 1-\pi_1-\dots-\pi_{n-1}$, representation (6) for the wealth process $X(t,\pi)$, and the definition of $\lambda(t)$, we obtain representation (9) in the following way:

$$X(t,\pi) = x\,\frac{P_n(t)}{P_n(0)}\left(\prod_{i=1}^{n}\left[\frac{P_n(0)}{P_n(t)}\frac{P_i(t)}{P_i(0)}\right]^{\pi_i}\right)e^{\frac{1}{2}\sum_{i=1}^{n-1}\Lambda_{ii}(t)\bar{\pi}_i+\frac{1}{2}\Lambda_{nn}(t)(1-\sum_{i=1}^{n-1}\bar{\pi}_i)}.$$

$$\cdot e^{-\frac{1}{2}\left[\overline{\pi}'V(t)\overline{\pi}+2\sum_{i=1}^{n-1}(\Lambda_{jn}(t)-\Lambda_{nn}(t))\overline{\pi}_i+\Lambda_{nn}(t)\right]}$$

$$= x\,\frac{P_n(t)}{P_n(0)}\;e^{-\frac{1}{2}\overline{\pi}'V(t)\overline{\pi}-\lambda(t)'\overline{\pi}}.$$

Completing the square in the exponent of this representation yields representation (10). But from (10), it is obvious that the global maximum of $X(t, \pi)$ occurs at $\overline{\pi} = \beta^*(t)$. Hence, we have $\beta^*(t)\in\overline{\Pi}$ if and only if all stocks will be active at time t. Then, we have $\beta^*(t) = \overline{\pi^*(t)}$ as a consequence of representation (10). This yields representation (11) and the claims in part b). Equation (12) follows directly from representation (10) and the identity of $\beta^*(t)$ and $\overline{\pi^*(t)}$.

c) To prove c), note that the maximisation problem now reads

$$\max_{\overline{\pi}\in\overline{\Pi}}\left(\lim_{t\to\infty}\frac{1}{t}E\left(-\overline{\pi}'V(t)\overline{\pi}-\lambda(t)'\overline{\pi}\right)\right) = \max_{\overline{\pi}\in\overline{\Pi}}\left(-\frac{1}{2}\overline{\pi}'J^{\infty}\overline{\pi}-\gamma^{\infty\prime}\overline{\pi}\right) = r(\overline{\pi}^{\infty}).$$

Having this in mind, the assertions of part c) follow exactly in the same way as the corresponding ones of parts a) and b).

□

With the help of this proposition, we can now prove the already announced

Theorem 25

In a weakly regular model which is asymptotically active, we have

$$\pi^*(t)\xrightarrow{t\to\infty}\pi^{\infty}\text{ in probability.}$$

Further, the model is asymptotically active if and only if there is an $\varepsilon > 0$ with

$$P(\pi_i^*(t) > \varepsilon\;\;\forall\, i = 1,\ldots, n)\xrightarrow{t\to\infty} 1 \tag{13}$$

Proof :

a) As the model is weakly regular, we have

$$\beta^*(t) = -\left(\frac{V(t)}{t}\right)^{-1}\left(\frac{\lambda(t)}{t}\right)\xrightarrow{t\to\infty} -\left(J^{\infty}\right)^{-1}\gamma^{\infty} = \beta^{\infty}\text{ in probability}$$

which implies the desired convergence of $\pi^*(t)$ against π^{∞} (using part b) and c) of Proposition 24).

b) Assume that the model is asymptotically active. By part c) of Proposition 24, we have $\pi^{\infty} = \beta^{\infty} > 0$ (component wise) which implies the desired convergence in relation (13) due to the already proved convergence of $\pi^*(t)$ against π^{∞}. The converse is similar.

□

We have already shown that the asymptotically optimal constant rebalanced portfolio π^{∞} outperforms each other constant portfolio exponentially. We are now going to show that it is even optimal in a stronger sense. To do so, we have to recall some notions of optimality from (Jamshidian 1992) which are all based on the comparison of the wealth process of a portfolio at time t to that of the best constant portfolio at time t, $X^*(t)$.

Definition 26

a) A self-financing portfolio with wealth process $X(t)$ is called

- **asymptotically suboptimal** if $X(t) / X^*(t) \approx 0$ (see Theorem 22).
- **asymptotically weakly optimal** if there exists an $\varepsilon > 0$ with

$$P(\, X(t) / X^*(t) > \varepsilon t^{-c}) \xrightarrow{t \to \infty} 1$$

for some $c > 0$. We will then write "$X(t) / X^*(t) \succ t^{-c}$".

- **asymptotically optimal** if $X(t) / X^*(t)$ is bounded in distribution, i.e. if for every $\varepsilon > 0$ there exists a constant $M > 0$ and a $t_0 > 0$ with

$$P(|X(t) / X^*(t)| > M) < \varepsilon \quad \forall\, t > t_0.$$

- **asymptotically strongly optimal** if $X(t) / X^*(t)$ is asymptotic to 1, i.e. if we have

$$X(t) > 0,\ X^*(t) > 0, \text{ and } (X(t)/X^*(t)) \xrightarrow{t \to \infty} 1 \text{ in probability.}$$

We will then write "$X(t) / X^*(t) \sim 1$".

b) A weakly regular process $X(t)$ with $\eta := \lim (X(t)/t)$ is called **regular** if

$$Z(t) := \frac{X(t)}{\sqrt{t}} - \sqrt{t}\eta$$

is bounded in distribution, and it is called **strongly regular** if $Z(t)$ converges against zero in probability as t goes to infinity.

Under some further regularity assumptions, we will now prove that the asymptotically optimal constant rebalanced portfolio π^{∞} is asymptotically (strongly) optimal.

Theorem 27
If in an asymptotically active model both $\left(\ln(P_1(t)),\ldots,\ln(P_n(t))\right)$ and $\Lambda(t)$ are (strongly) regular then the constant portfolio π^{∞} is asymptotically (strongly) optimal.

Proof:
In view of representation (10), it suffices to show that $(\pi^{\infty}-\pi^*(t))'V(t)(\pi^{\infty}-\pi^*(t))$ is bounded in distribution (converges to zero in probability). For this, it is enough if the square root $V(t)^{1/2}(\pi^{\infty}-\pi^*(t))$ is so. Using the equation

$$V(t)^{1/2}(\pi^{\infty}-\pi^*(t)) = V(t)^{1/2}\left[V(t)^{-1}\lambda(t)-\left(J^{\infty}\right)^{-1}\gamma^{\infty}\right]$$

$$=\left(\frac{V(t)}{t}\right)^{-1/2}\left(\frac{\lambda(t)}{t}\right)\left[\left(\frac{\lambda(t)}{\sqrt{t}}-\sqrt{t}\gamma^{\infty}\right)-\left(\frac{V(t)}{\sqrt{t}}-\sqrt{t}J^{\infty}\right)\left(J^{\infty}\right)^{-1}\gamma^{\infty}\right]$$

yields the required result, because the (strong) regularity of the model implies the (strong) regularity of $\lambda(t)$, and therefore the right hand side of this equation is bounded in distribution (converges to zero in probability).

□

ii) The Universal Portfolio in Continuous Time

It would of course be nice to figure out the asymptotically optimal constant rebalanced portfolio π^{∞} to follow it. However, this would require the perfect knowledge of the long-term averages of the future instantaneous returns and covariances which rarely seems to be the case in practical situations. Also, following every other constant portfolio strategy would lead to an "exponential underperformance" compared to the use of π^{∞} (see Theorem 22). To overcome this danger of choosing the "wrong" constant portfolio, Jamshidian proposes the use of a continuous-time variant of the universal portfolio developed by Cover in discrete-time (see Section 1). The advantage of the universal portfolio is that it "tracks" the path of the optimal constant portfolios X*(t) via a performance weighted adjustment of the weights for investing in the different securities. As in the discrete-time case, the universal portfolio $\hat{\pi}$ is given by

$$\hat{\pi}_i(t) = \frac{\int_\Pi \pi_i X(t,\pi)d\pi}{\int_\Pi X(t,\pi)d\pi}, \quad i = 1, \ldots, n.$$

It follows directly from this definition that we start with $\hat{\pi}_i(0) = 1/n$ for all i. Further, we have the same representation of the "universal wealth" $\hat{X}(t)$ (i.e. the wealth process corresponding to the universal portfolio) as in the discrete-time case:

Proposition 28
For every $t \geq 0$ the universal wealth $\hat{X}(t)$ has the representation

$$\hat{X}(t) = \frac{\int_\Pi X(t,\pi)d\pi}{\int_\Pi d\pi}. \tag{14}$$

Proof :
Let Z(t) denote the right hand side of equation (14). Using Itô's formula, the self-financing condition and interchanging the Itô and the Lebesgue integral twice (see Lemmas 3.3 and 3.4 in (Kallianpur and Striebel 1969) for justifying this), we obtain

$$\int_0^t \frac{dZ(s)}{Z(s)} = \int_0^t \frac{d\left(\int_\Pi X(s,\pi)d\pi\right)}{\int_\Pi X(s,\pi)d\pi} = \int_\Pi \int_0^t \frac{1}{\int_\Pi X(s,\pi)d\pi} dX(s,\pi)d\pi$$

$$= \int_\Pi \int_0^t \sum_{i=1}^n \frac{\pi_i X(s,\pi)}{\int_\Pi X(s,\pi)d\pi} \frac{dP_i(s)}{P_i(s)} d\pi = \sum_{i=1}^n \int_0^t \frac{\int_\Pi \pi_i X(s,\pi)d\pi}{\int_\Pi X(s,\pi)d\pi} \frac{dP_i(s)}{P_i(s)}$$

$$= \sum_{i=1}^n \int_0^t \hat{\pi}_i(s) \frac{dP_i(s)}{P_i(s)} = \int_0^t \frac{d\hat{X}(s)}{\hat{X}(s)}$$

Hence, we have proved that both Z(t) and $\hat{X}(t)$ satisfy the same (linear) stochastic differential equation. Due to the fact that we also have $Z(0) = \hat{X}(0)$, application of the variation of constants formula, Theorem B15, yields that Z(t) and $\hat{X}(t)$ coincide. Thus, representation (14) is valid.

□

As in the discrete-time case, we can thus interpret following the universal portfolio as dividing all the initial capital into infinitesimally small parts and investing one of them into each possible constant strategy.

We will also give a result on the asymptotic performance of the universal portfolio (in the asymptotically active case) that is similar to Theorem 8 in Section 1.

Theorem 29 "Weak optimality of the universal portfolio"
If a weakly regular model is asymptotically active then we have

$$\frac{\hat{X}(t)}{X^*(t)} \sim \frac{(n-1)!}{\left|J^{\infty}\right|^{\frac{1}{2}}} \left(\frac{2\pi}{t}\right)^{(n-1)/2}. \tag{15}$$

In particular, the universal portfolio is weakly optimal in this case.

Remark 30
With the help of Theorems 29 and 31, we can now conclude that the universal portfolio is outperformed only polynomially by the asymptotically optimal constant rebalanced portfolio π^{∞}. Hence, it outperforms all other constant portfolios exponentially in the long run. Note further, that the universal portfolio does not require any prior information on market parameters to achieve this performance which clearly demonstrates its practical relevance.

Proof (of Theorem 30)**:**
We will decompose the proof into four different steps.
1. As in the discrete-time case, we have

$$\hat{X}(t) = \frac{\int_{\Pi} X(t,\pi)d\pi}{\int_{\Pi} d\pi} = \frac{\int_{\overline{\Pi}} X(t,\overline{\pi})d\overline{\pi}}{\int_{\overline{\Pi}} d\overline{\pi}} = (n-1)! \int_{\overline{\Pi}} X(t,\overline{\pi})d\overline{\pi}.$$

If we use this equality, the representations of $X(t, \pi)$, $X^*(t)$ given in Proposition 24, and the substitution $x = V^{\frac{1}{2}}(t)\left[\overline{\pi} - \overline{\pi^*(t)}\right]$ then we obtain

$$\begin{aligned}\frac{\hat{X}(t)}{X^*(t)} &= (n-1)! \int_{\overline{\Pi}} e^{-\frac{1}{2}\left(\overline{\pi}-\overline{\pi^*(t)}\right)' V(t)\left(\overline{\pi}-\overline{\pi^*(t)}\right)} d\overline{\pi} \\ &= \frac{(n-1)!}{\det\left(V^{\frac{1}{2}}(t)\right)} \int_{\Delta_t} e^{-\frac{1}{2}x'x} dx \end{aligned} \tag{16}$$

with $\Delta_t = V^{1/2}(t)\left[\overline{\Pi} - \overline{\pi^*(t)}\right]$. To apply Proposition 24, we can w.l.o.g. assume the model to be active at time t as it is asymptotically active which will by Theorem 25 imply that the model is at least active for large t. But this is all we need, since relation (15) to prove only depends on large values of t.

2. The set Δ_t will be examined further. Let therefore

$$U := \{ x \in \mathbf{R}^{n-1} \mid x'x < 1 \}.$$

As the model is asymptotically active, we have the existence of an $\varepsilon^* > 0$ with

$$P(\pi_i^*(t) > \varepsilon^*) \xrightarrow{t\to\infty} 1.$$

Choosing $\varepsilon = (\varepsilon^*/(n-1))^2$, we can check that we have

$$P(\overline{\Pi} - \overline{\pi^*(t)} \supset \varepsilon U) \xrightarrow{t\to\infty} 1 \tag{17}$$

3. With the usual matrix norm $\|A\|$ defined as

$$\|A\| = \sup_{0 \neq y \in \mathbf{R}^{n-1}} \frac{\|Ay\|}{\|y\|},$$

we have $\|A\|^2 \le \text{trace}(A'A)$ and $\|y\| \le \|A^{-1}\|\,\|Ay\|$ (where we have also used $\|.\|$ for the Euclidean vector norm) which together yield

$$\|Ay\|^2 \ge \frac{\|y\|^2}{\text{trace}\left((J^\infty)^{-1}\right)}.$$

We apply this to $y = \overline{\pi^*(t)} - \overline{\pi}$ and $A = V^{1/2}(t)$ and arrive at

$$\left\| V^{1/2}(t)\left(\overline{\pi^*(t)} - \overline{\pi}\right)\right\|^2 \ge \frac{\left\|\overline{\pi^*(t)} - \overline{\pi}\right\|^2}{\text{trace}\left(V^{-1}(t)\right)} \sim \frac{\left\|\overline{\pi^*(t)} - \overline{\pi}\right\|^2 t}{\text{trace}\left((J^\infty)^{-1}\right)} \tag{18}$$

where to obtain the last part of this relation, we have made use of the weak regularity of the model.

4. Due to relations (17) and (18), we have

$$P(\Delta_t \supset cU) \xrightarrow{t\to\infty} 1$$

for arbitrary positive c, hence

$$\int_{\Delta_t} e^{\frac{1}{2}x'x}dx \sim \int_{\mathbf{R}^{n-1}} e^{\frac{1}{2}x'x}dx = (2\pi)^{\frac{n-1}{2}}. \tag{19}$$

If we further note that we have $\det(V(t)) \sim t^{n-1}\det(J^{\infty})$ and that "~" is transitive (in fact, it is an equivalence relation) then we have shown the theorem by putting together the above relations, in particular relations (16) and (19).

□

Remark 31
As already mentioned, Jamshidian does not limit his considerations to the asymptotically active case. He also considers the case of only k asymptotically active securities (with $1 \leq k < n$) and shows that (under some technical requirements) even in this case, the asymptotically optimal constant rebalanced portfolio is strongly optimal and the universal portfolio is weakly optimal. However, the asymptotic expressions for the quotient $\hat{X}(t)/X^*(t)$ differ from the one in Theorem 29 (which is of course what to expect !). We refer the reader to (Jamshidian 1992) for the details of the general case. Also, Jamshidian discusses the concept of the generalised universal portfolio in the case of only k ($< n$) asymptotically active stocks. This concept is totally similar to the one in discrete time (see sub-section 1 iii)). We will not go into further detail of this topic but will instead look more closely at the case of a market consisting of only two stocks in the next section.

iii) An Example : A Market with only two Stocks

So far, we have seen that the universal portfolio has some very desirable features, but to gain some more insight into its structure and performance, we will now specialise to the case of a market consisting only of two stocks. The following proposition gives explicit formulae for the universal portfolio and wealth and the optimal constant portfolio and wealth, respectively.

Proposition 32
Let

$$v(t)^2 := V(t) = \Lambda_{11}(t) + \Lambda_{22}(t) - 2\Lambda_{12}(t),$$

$$h(t) := \frac{1}{v(t)}\left(\ln(P_2(t)/P_2(0)) - \ln(P_1(t)/P_1(0))\right).$$

a) Then we have the following representations:

$$\hat{X}(t) = X\left(t;\left(\tfrac{1}{2},\tfrac{1}{2}\right)'\right)\frac{\sqrt{2\pi}}{v(t)}e^{\frac{1}{2}h(t)^2}\left[\Phi\left(h(t)+\tfrac{1}{2}v(t)\right)-\Phi\left(h(t)-\tfrac{1}{2}v(t)\right)\right], \quad (20)$$

$$\hat{\pi}(t) = \left(\tfrac{1}{2}-B(t),\tfrac{1}{2}+B(t)\right)' \quad (21)$$

with

$$X\left(t;\left(\tfrac{1}{2},\tfrac{1}{2}\right)'\right) = X(0)\sqrt{\frac{P_1(t)P_2(t)}{P_1(0)P_2(0)}}\, e^{\frac{1}{8}v(t)^2}, \tag{22}$$

$$B(t) = \frac{h(t)}{v(t)} + \frac{P_1(t)/P_1(0) - P_2(t)/P_2(0)}{v(t)^2 \hat{X}(t)}$$

$$= \frac{1}{v(t)}\left[h(t) + \frac{e^{-(h(t)+\frac{1}{2}v(t))^2} - e^{-(h(t)-\frac{1}{2}v(t))^2}}{\sqrt{2\pi}(\Phi(h(t)+\frac{1}{2}v(t)) - \Phi(h(t)-\frac{1}{2}v(t))}\right] \tag{23}$$

where Φ is the distribution function of the standard normal distribution.
b) Moreover, both stocks are active at time t if and only if $|h(t)| < \frac{1}{2} v(t)$ in which case we have

$$\pi^*(t) = \left(\tfrac{1}{2} - \frac{h(t)}{v(t)}, \tfrac{1}{2} + \frac{h(t)}{v(t)}\right) \tag{24}$$

$$X^*(t) = X\left(t;\left(\tfrac{1}{2},\tfrac{1}{2}\right)'\right) e^{\frac{1}{2}h(t)^2} =$$

$$= X(0)\frac{P_1(t)}{P_1(0)} e^{\frac{1}{2}(h(t)+\frac{1}{2}v(t))^2} = X(0)\frac{P_2(t)}{P_2(0)} e^{\frac{1}{2}(h(t)-\frac{1}{2}v(t))^2}. \tag{25}$$

Proof:
As the detailed proof would be rather lenghty, but only consists of a multiple application of Itô's formula, we only sketch it and leave the details to the reader. The explicit form of

$$X\left(t;\left(\tfrac{1}{2},\tfrac{1}{2}\right)'\right)$$

as given in equation (22) is a consequence of Proposition 17. Using Proposition 28 and doing some tedious manipulations lead to

$$\hat{X}(t) = \int_0^1 X(t;(c,1-c)')dc$$

$$= X\left(t;\left(\tfrac{1}{2},\tfrac{1}{2}\right)'\right)\frac{\sqrt{2\pi}}{v(t)} e^{\frac{1}{2}h(t)^2}\left[\Phi\left(h(t)+\tfrac{1}{2}v(t)\right) - \Phi\left(h(t)-\tfrac{1}{2}v(t)\right)\right].$$

Computing $d\left(\ln\left(\hat{X}(t)\right)\right)$ from this representation (a really messy exercise !) and comparing the resulting coefficient of dW(t) to that of the representation

$$d\left(\ln\left(\hat{X}(t)\right)\right) = \sum_{i=1}^{n} \hat{\pi}_i(t)\frac{dP_i(s)}{P_i(s)} - \tfrac{1}{2}\hat{\pi}(t)'\Lambda(t)\hat{\pi}(t) \tag{26}$$

leads to the explicit form of $\hat{\pi}(t)$ and — as a by-product — B(t) as given above. Note that representation (26) is valid for every self-financing portfolio process (see also the proof of Proposition 17). The representations of the optimal constant portfolio at time t and its wealth process, $\pi^*(t)$ and $X^*(t)$, are a consequence of Proposition 24.

□

Remark 33 "Properties and performance of the continuous-time universal portfolio" To examine the behaviour of the universal portfolio and its wealth process, we take a closer look at the representations of the universal wealth and the universal portfolio. Note first that B(t) is an odd function in h(t). It is equal to zero if and only if h(t) equals zero. Hence, the universal portfolio consists of investing equal fractions of wealth in the two stocks if and only if they have performed equally well up to time t (in the sense that the quotients $P_i(t) / P_i(0)$, i = 1,2, are equal). Further, from the representation of B(t), one can deduce that B(t) is positive if and only if h(t) is positive. Hence, if one stock has performed better than the other on [0,t] (expressed in a higher quotient $P_i(t) / P_i(0)$) then an investor, following the universal portfolio strategy, invests more than half of the current wealth in this stock at time t, i.e. we have

$$\hat{\pi}_1(t) > \hat{\pi}_2(t) \Leftrightarrow (P_1(t) / P_1(0)) > (P_2(t) / P_2(0)). \tag{27}$$

In this sense, the universal portfolio is a type of "play the winner" rule. As a contrast to this we look at the **normalised components** of the universal portfolio

$$n_i(t) = \hat{\pi}_i(t)(P_i(0)/P_i(t)).$$

To understand the significance of this term, note that if we would have followed a buy-and-hold strategy starting with

$$\hat{\pi}_i(0)X(0) = \tfrac{1}{2}x$$

invested in each security at time t then the amount of money in the different securities would have changed to

$$\pi_i(t)X(t) = \hat{\pi}_i(0)X(0)\frac{P_i(t)}{P_i(0)}.$$

As such a buy-and hold strategy can be interpreted as leaving all the gains earned from a security invested in this particular one, its normalised portfolio process satisfies

$$\pi_i(t)\frac{P_i(0)}{P_i(t)} = \frac{1}{2x}.$$

However, it is claimed in (Jamshidian 1992) that the normalised components of the universal portfolio has the following behaviour,

$$(P_i(t) / P_i(0)) < (P_j(t) / P_j(0)) \;\Rightarrow\; n_i(t) > n_j(t), \tag{28}$$

for i, $j \in \{1, 2\}$, $i \neq j$, or expressed in (Jamshidian 1992): "In the universal portfolio, there are fewer normalised shares of the stock that has appreciated more". I.e. parts of the gains of the better security are used to buy additional shares of the other security. This is clearly a hedging or diversification aspect. We could verify relation (27) only for certain values of h(t) and v(t) (see Lemma 34 b)).

To sum up, if one stock performs better than the other then part of the profit is taken away from this stock to purchase additional shares of the other one (this corresponds to the behaviour of the normalised components of the universal portfolio), but it is always ensured that there is more money invested in the "better" stock (which corresponds to the behaviour of the portfolio).

To get a feeling for the evolution of the universal wealth, note that due to the characteristics of the distribution function of the standard normal, we have the following relations between the universal wealth $\hat{X}(t)$ and that of the wealth corresponding to the equally weighted strategy, $X(t;(\frac{1}{2},\frac{1}{2})')$ (see Lemma 34 a)):

$$|\,h(t)\,| < 1 \Rightarrow \hat{X}(t) < X\left(t;\left(\tfrac{1}{2},\tfrac{1}{2}\right)'\right), \tag{29}$$

$$|\,h(t)\,| > 1 + \tfrac{1}{2}\,v(t) \Rightarrow \hat{X}(t) > X\left(t;\left(\tfrac{1}{2},\tfrac{1}{2}\right)'\right), \tag{30}$$

i.e. if the two stocks perform nearly identical then the equally weighted strategy outperforms the universal portfolio, while if there is a significant difference in the performance of the stocks then the universal portfolio can use this to perform better. Due to our method of proof, we cannot give an inequality for the parameter setting not covered by relations (29) and (30).

Lemma 34

a) With the notations of Theorem 32 relations (29) and (30) are valid.

b) Assume that for h(t), v(t) as in Theorem 32 the relation

$$e^{v(t)h(t)}\left(\tfrac{1}{2}v(t)-h(t)\right) > \tfrac{1}{2}h(t)+v(t). \tag{31}$$

is satisfied. Then, the implication (28) is valid.

Proof:

a) Let $\varphi(x)$ be the density function of the standard normal distribution. To show relation (29), let $0 \le h(t) < 1$. Then, h(t) lies in the concave part of φ. Due to the form of $\varphi(x)$, for every pair (x, x') with

$$x < h(t) < x' \ \textbf{and} \ |h(t) - x| = |h(t) - x'|,$$

we have

$$\varphi'(x) > \varphi'(x').$$

From this relation, we obtain

$$\Phi\left(h(t)+\tfrac{1}{2}v(t)\right) - \Phi\left(h(t)-\tfrac{1}{2}v(t)\right) = \int_{h(t)-\frac{1}{2}v(t)}^{h(t)+\frac{1}{2}v(t)} \varphi(x)dx < \varphi(h(t))v(t),$$

and assertion (29) follows from representation (20) of the universal wealth. The argument for the case "$-1 < h(t) \le 0$" is identical. To show relation (30), note that then, h(t) lies in the convex part of φ. With this observation, relation (30) can be shown in a similar way as relation (29).

b) Note the equivalence

$$\hat{\pi}_1(t) < \hat{\pi}_2(t) \Leftrightarrow (P_1(t)/P_1(0)) < (P_2(t)/P_2(0)) \Leftrightarrow h(t) > 0. \tag{32}$$

W.l.o.g. we assume $P_1(0) = P_2(0) = 1$. Using the second representation of B(t) in equation (23), relation (21), and

$$\frac{P_2(t)}{P_1(t)} = e^{v(t)h(t)},$$

we obtain the equivalence:

$$n_1(t) > n_2(t)$$

$$\Leftrightarrow \quad e^{vh}\left(\left(\tfrac{1}{2}v-h\right)\left[\Phi\left(h+\tfrac{1}{2}v\right)-\Phi\left(h-\tfrac{1}{2}v\right)\right]-\left(\varphi\left(h+\tfrac{1}{2}v\right)-\varphi\left(h-\tfrac{1}{2}v\right)\right)\right)$$
$$> \left(\tfrac{1}{2}v+h\right)\left[\Phi\left(h+\tfrac{1}{2}v\right)-\Phi\left(h-\tfrac{1}{2}v\right)\right]+\left(\varphi\left(h+\tfrac{1}{2}v\right)-\varphi\left(h-\tfrac{1}{2}v\right)\right)$$

where for convenience we have omitted the argument t for h(t) and v(t). As from relation (32), we have $h(t) > 0$ under the assumption "$P_2(t) > P_1(t)$" and also v(t) is automatically positive, we have

$$\varphi\left(h(t) - \tfrac{1}{2}v(t)\right) > \varphi\left(h(t) + \tfrac{1}{2}v(t)\right),$$
$$\Phi\left(h(t) + \tfrac{1}{2}v(t)\right) > \Phi\left(h(t) - \tfrac{1}{2}v(t)\right).$$

Using these two inequalities, assumption (31) and the above equivalence for the relation between $n_1(t)$ and $n_2(t)$, we have shown the assertion of part b) in the case "$P_2(t) > P_1(t)$". The case "$P_1(t) > P_2(t)$" is similar.

□

6.3 Optimal Cash Management in Equity Index Tracking with Transaction Costs

i) Index Tracking : Idea and Practical Problems

A financial index (for brevity: an index) is made up of a basket of assets. The value of such an index is usually the weighted sum of the prices of the corresponding assets at a given time. According to their importance, some indices are updated minute by minute (such as the DAX or the FTSE). Their main purpose is to reflect the performance of the whole market, and therefore they often serves as benchmarks for investors, i.e. the performance of a particular investment is judged by comparing it with that of an appropriate index. Of course, the natural goal of an investor could be to "beat the market" (meaning that the value of his portfolio should increase faster than the corresponding market index). This idea also underlies the universal portfolio approach of the preceding two sections. Such trading behaviour is often referred to as "active trading". On the other hand, by following a so-called "passive trading strategy", the intention of an investor is not to beat the market but instead to replicate it. Hence, if the market is well represented by an index, the basic action of passive management is to hold the index portfolio, more precisely, to hold the assets that constitute the index in the right proportion. The philosophy behind such a strategy is that by holding an index the risk of being among the losers (compared to the whole market) is very low. Always hoping that the stock market grows faster than the riskless rate, a surplus in return should be obtained with a low risk by following a passive trading strategy.

This section is concerned with the practical problems of index tracking in the presence of random cash flows and is mainly taken from (Buckley and Korn 1997).

The paper that inspired this work is (Connor and Leland 1995) where the discrete- and continuous-time setting under a simpler transaction cost structure are treated (by methods that differ from ours).

We will consider a fund manager who is trying to passively track an equity index. In the ideal situation, the fund managers portfolio will always consist of holdings in all the securities that enter the index (in exactly the proportions in which they enter the index). So, after an initial investment in the exact proportions, all necessary trading would be done, and the corresponding portfolio would behave as a multiple of an exact copy of the index **if** the volume of the fund would not change. However, it is typical that the volume of the fund changes over time as there are irregular cash inflows/outflows to/from the fund over time which could originate in dividends of the shares, new subscriptions to the fund or fund redemptions. If all these inflows/outflows must be dealt with immediately by security transactions, this would result in large amounts of transaction costs. Therefore, it could be advantageous to the portfolio manager to hold a certain amount of cash to deal with minor inflows/outflows and thereby saving the transaction costs. If his cash account has reached a sufficiently high level then he will add parts of it to his portfolio by buying additional shares (in the appropriate proportions to track the index). If, in contrast, it has fallen below a critical level, he will increase it by selling some of his shares. However, a positive cash position will reduce the (expected) excess return of the portfolio over the riskless rate (as long as the expected rate of return of the portfolio lies above the riskless rate). Moreover, it will surely lead to a tracking error, i.e. a deviation of the performance of the investor's holdings (portfolio **and** the cash account) from the performance of the index, because there is no positive cash weight in an index.

In the following sub-sections, we will present a model for this situation and solve the resulting impulse control problem of the fund manager

ii) Some Results on Impulse Control Problems

Although we have presented some facts and results on impulse control problems in Section 5.2, we will give a short overview on the results and definitions that we will need to solve the cash management problem under transaction costs. This is necessary as the results of Section 5.2 do not fully cover the following setting.

Therefore, consider the situation where a "controller" is allowed to choose intervention times θ_i when he can shift the "fundamental process" $X(t)$ to another value $X(\theta_i) = X(\theta_i-) + \Delta x_i$ in an open interval I by choosing the "control action" Δx_i. Further, he is required to keep the process inside this interval, despite the costs that are incurred by each control action. We assume that between intervention times, our fundamental process is given as the solution of the stochastic differential equation

$$dX(t) = b(X(t))\,dt + \sigma(X(t))\,dW(t) \quad . \tag{29}$$

The functions b: $\mathbf{R} \to \mathbf{R}$ and σ: $\mathbf{R} \to \mathbf{R}$ should satisfy conditions guaranteeing the existence of a unique, non-exploding solution to the sde (29) (with finite second moment) for every initial condition X(0) = x (see Appendix B.3 for such conditions). As usual, W(t) is a one-dimensional standard Brownian motion with corresponding filtration $(F_t)_{t\geq 0}$ on some complete probability space (Ω, F, P).

Definition 35
An **admissible impulse control strategy** (with respect to the open interval $I \subseteq \mathbf{R}$) is a sequence $S=\{(\theta_i, \Delta x_i), i\in\mathbf{N}\}$ of intervention times θ_i and control actions Δx_i with

i) $0 \leq \theta_i \leq \theta_{i+1}$ a.s. $\forall\, i\in\mathbf{N}$

ii) θ_i is a stopping time with respect to the filtration
$\sigma\{\, X(s-),\ (\theta_n, \Delta X_n),\ n < i,\ s\leq t \,\},\ t \geq 0$

iii) $\theta_i \leq \inf\{\, t \geq \theta_i \mid X(t)\notin I \,\}$ (30)

iv) Δx_i is measurable with respect to $\sigma\{\, X(\theta_i-),\ \Delta X_n,\ n < i \,\}$ and such that $X(\theta_i) = X(\theta_i-) + \Delta x_i \in I$

v) $P(\lim\limits_{i\to\infty} \theta_i \leq T) = 0 \quad \forall\, T \geq 0$.

Let Z be the set of (admissible) impulse control strategies. Then we look at the following control problem:

$$\max_{\{(\theta_i, \Delta x_i), i\in\mathbf{N}\}\in Z} E^{x,S}\left(\int_0^\infty e^{-\alpha t} f(X(t))dt + \sum_{i=1}^{\infty} e^{-\alpha\theta_i}\left(K + k|\Delta x_i|\right) 1_{\{\theta_i < \infty\}} \right) \tag{31}$$

where α, K, k are positive constants, f: $\mathbf{R} \to [0,\infty)$ is a continuous function, and $E^{x,S}(.)$ denotes the expectation when the process X(t) starts in X(0) = x and when the impulse control strategy S is used by the controller. As in Section 5.2, we introduce the qvi corresponding to problem (31). The difference to Definition 5.10 lies in the additional state space constraint for the controlled process.

Definition 36

Let $\phi : \bar{I} \to \mathbf{R}$ be a continuous function (where $\bar{I}$ is the closure of the open interval I $\subseteq\mathbf{R}$). Define the operators L and M by

$$L\phi(x) = \tfrac{1}{2}\,\sigma^2(x)\,\phi''(x) + b(x)\phi'(x) - \alpha\phi(x) + f(x)$$

$$M\phi(x) = \max_{\Delta x | x+\Delta x \in I} [\,\phi(x) + \Delta x) - k\,|\Delta x| - K\,]$$

(where α, K, k, b, σ, f are already given in equation (29) and Definition 35). We say that φ solves the qvi (with respect to I) for problem (31) if it satisfies

$$\begin{aligned} &\text{i)} && L\phi(x) \le 0 \quad \forall\, x \in I, \\ &\text{ii)} && M\phi(x) \le \phi(x) \ \ \forall\, x \in I, \quad M\phi(x) = \phi(x) \ \ \forall\, x \in \bar{I} \setminus I, \\ &\text{iii)} && L\phi(x)(M\phi(x) - \phi(x)) = 0 \ \ \forall\, x \in I. \end{aligned} \tag{32}$$

Again, we introduce a qvi-control which will turn out to be an optimal one:

Definition 37

Let φ be a continuous solution of the qvi. Then the following admissible impulse control strategy is called a **qvi-control** (if it exists):

$$\begin{aligned} &\text{i)} && (\theta_0\,,\, \Delta x_0\,) := (0,0) \\ &\text{ii)} && \theta_i := \inf\,\{t \ge \theta_{i-1} \mid \phi(X(t-)) = M\phi(X(t-))\,\} \\ &\text{iii)} && \Delta\, x_i := \underset{\Delta x | X(\theta_i -)+\Delta x \in I}{\arg\max}\ [\,\phi(X(\theta_i -) + \Delta x) - k\,|\Delta x| - K\,] \end{aligned}$$

Hence, at every time instant when φ and Mφ coincide, a controller following a qvi-control intervenes and chooses an action that is a maximiser of the optimisation problem of Mφ. Note in particular that due to relation (32) ii), he has to intervene if the fundamental process has left I. In Definition 37, we assume implicitly that a qvi-control satisfies the requirements of Definition 36.

We cite a verification theorem relating quasi-variational inequalities, qvi-controls and impulse control problem (31). As its proof is similar to that of Theorem 5.11, we will omit it (see also (Korn 1997d) for a proof of a similar theorem).

Theorem 38

Let $\phi \in C^1$ be a solution to the qvi (32) such that φ'' exists in the distributional sense (see Theorem B24). Further, let φ satisfy the growth conditions

$$E^{x,S}\left(\int_0^{\infty} e^{-\alpha t}\sigma(X(t))\phi'(X(t)))^2\,dt\right) < \infty,$$

$$E^{x,S}\left(e^{-\alpha T}\phi(X(T))\right) \xrightarrow{T\to\infty} 0$$

for every process X(t) corresponding to an admissible impulse control S. Then we have

$$v(x) \le \phi(x) \quad \forall\, x \in \bar{I}.$$

Further, if there exists a qvi-control corresponding to ϕ then it is an optimal control strategy, and ϕ is identical to the value function v of our optimisation problem (31).

iii) Equity Index Tracking and Optimal Cash Management with Transaction Costs

To solve the index tracking problem in the presence of transaction costs (as described in sub-section i)), we will make use of the results and the method of impulse control given in the previous sub section, but first, we have to state a rigorous formulation of our problem.

Let the process C_t be the cash weight in the portfolio (i.e. the fraction of the wealth of the total holdings hold in cash at time t), and assume that it behaves like a Brownian motion with volatility σ and drift rate η between readjustment times θ_i and θ_{i+1}, i.e.

$$C_t = C_0 + \eta\, t + \sigma\, W(t) + \sum_{i=1}^{\infty}\left(C^{(i)} - C_{\theta_i -}\right)\cdot 1_{\{\theta_i \le t\}}$$

where $C^{(i)}$ is the value of the cash weight process that the portfolio manager has chosen at the readjustment time θ_i, C_0 is the initial cash weight and W(t) is a one-dimensional Brownian motion. Every readjustment causes costs that consist of a component which is proportional to the wealth $X(\theta_i-)$ of the portfolio immediately before the readjustment (one could imagine about provisions for the trader, see also Section 5.3 for the appearance of the same kind of costs) and of a component that is proportional to the absolute value of the difference

$$\left|C^{(i)} - C_{\theta_i -}\right| X(\theta_i -)$$

between the amount of cash hold after and before the readjustment, i.e. costs that are proportional to the value of the transaction. Hence, we assume transaction costs (relative to X(t)) of

$$K + k\left|C^{(i)} - C_{\theta_i -}\right|$$

(with k, $K \in (0, 1)$). Further, we assume that the index portfolio, we would like to track, has a relative rate of excess return π over the riskless rate — which is assumed to be zero — and a volatility of τ. For instructive reasons, one could think of the wealth process of the index portfolio, I(t), to follow a geometric Brownian motion, i.e.

$$dI(t) = I(t)\,[\pi dt + \tau dW(t)]\,,$$

but for our assumptions on π, τ to hold, it is only necessary to assume

$$\frac{1}{t}E\left(\int_0^t \frac{dI(s)}{I(s)}\right) = \pi, \quad \frac{1}{t}\mathrm{Var}\left(\int_0^t \frac{dI(s)}{I(s)}\right) = \tau^2 .$$

By holding a cash weight of C_t in our tracking portfolio, we achieve a relative rate of excess return of $(1-C_t)\pi$ at time t. Before we continue with our analysis, we must give a rigorous statement on what we mean by index tracking and by tracking error: We will say that X(t), the wealth process of our holdings, is tracking the index I(t) if we have

$$\int_0^t \frac{dX(s)}{X(s)} = \int_0^t \frac{dI(s)}{I(s)} \quad \forall\, t \geq 0,$$

i.e. if the "relative growth" of X(t) and I(t) coincide. If we define the tracking error as

$$\int_0^t \frac{dX(s)}{X(s)} - \int_0^t \frac{dI(s)}{I(s)}$$

and assume a riskless rate of zero then a constant fraction of c hold in cash results into a tracking error variance of

$$\mathrm{Var}\left(\int_0^t \frac{dX(s)}{X(s)} - \int_0^t \frac{dI(s)}{I(s)}\right) = \mathrm{Var}\left(\int_0^t c\,\frac{dI(s)}{I(s)}\right) = (c\tau)^2 \cdot t\ .$$

In this sense, we can say that a cash weight of C_t at time t leads to a tracking error variance of $(C_t\tau)^2$ (interpreted as a rate).

Our goal will be to have a **good excess return** and a **low tracking error variance** (over time) with not too exaggerated readjustment costs, i.e. we take on a mean-variance point of view. A further requirement is that the portfolio manager is not allowed to go negative in cash. Therefore, we look at the constrained optimisation problem

$$\psi(c) = \sup_{(\theta_i, C^{(i)}) \in Z} E^{c,S}\Bigg(\int_0^{\infty} e^{-\rho t}\left[(1 - C_t)\pi - \lambda \tau^2 C_t^{\,2}\right] dt - \sum_{i=1}^{\infty} \left(K + k\left|C^{(i)} - C_{\theta_i -}\right|\right) e^{-\theta_i t} 1_{\{\theta_i < \infty\}} \Bigg) \tag{33}$$

Here, ρ is a (positive) discount factor, λ the (positive) coefficient of absolute risk aversion of the utility function, c the (positive) initial cash weight, and Z is the space of admissible strategies with the additional requirement to use only strategies leading to a **non-negative cash weight** process C_t. Note also that all terms appearing in the objective function are formulated relative to the wealth of the current holdings. Thus, all terms agree in dimension.

The following proposition gives a sufficient condition characterising the value function, and it also describes the corresponding optimal control strategy.

Proposition 39

If there exists a continuous function V: $[0,\infty) \to \mathbf{R}$ and a triple (l, u, U), $0 < l < u < U$, with $V\big|_{(0,\infty)} \in C^1$, $V\big|_{(0,U]} \in C^2$ (where V''(U) is understood as the left limit of V''(c) in c = U) satisfying

$$\begin{aligned}
&\text{i)} && \tfrac{1}{2}\sigma^2 V''(c) + \eta V'(c) + [(1-c)\pi - \lambda(\tau c)^2] - \rho V(c) = 0 \qquad \forall\, c \in [0,U] \\
&\text{ii)} && V'(u) = V'(U) = -k\,, \quad V'(l) = k \\
&\text{iii)} && V(U) = V(u) - K - k(U-u), \quad V(0) = V(l) - K - k\,l \\
&\text{iv)} && V(c) = V(U) - k\,(c-U) \qquad \forall\, c \in [U, \infty)
\end{aligned} \tag{34}$$

then V coincides with the value function ψ given by (33). Moreover, in this case the strategy

> "Do nothing as long as C_t is in (0, U). If C_t reaches U then decrease the cash weight to u, if it reaches 0 then increase the cash weight to l. If C_0 is bigger than U then decrease the cash weight immediately to u"

is an optimal strategy.

Discussion of the form of the optimal strategy

Before proving this proposition, we will visualise and discuss the form of the (conjectured) optimal control strategy which is given in Figure 15. As long as the cash weight C_t is in (0, U), we let it evolve freely. A possible control action in (0, U) would not pay the transaction costs, i.e. it would not lead to a sufficient gain in our mean-variance sense. If it reaches U, it will be decreased to u (such as at times θ_1, θ_2 in Figure 15) via investing the difference $(U - u)X(\theta_i-)$ in the shares that form the index (note that also the transaction costs have to be paid !). If, in contrast, the cash weight process arrives at zero then the whole cash buffer is absorbed, and the fund manager has to sell some shares (of course in the appropriate proportions !) to increase the cash weight to l again (always remember that these are all **fractions** of the value of the total holdings and not absolute numbers !).

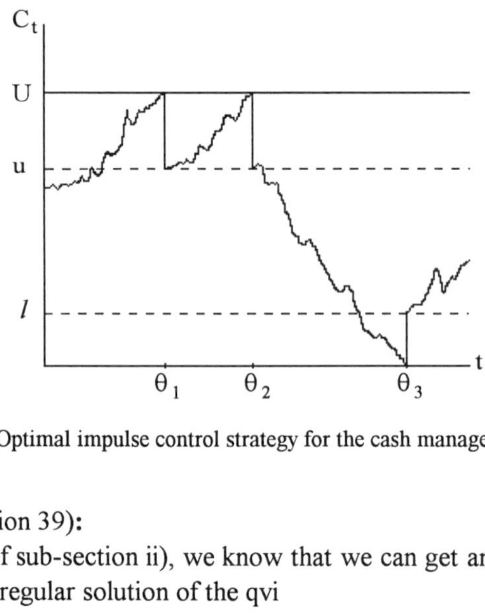

Figure 15 : Optimal impulse control strategy for the cash management problem

Proof (of Proposition 39)**:**
Using the results of sub-section ii), we know that we can get an optimal qvi-control from a sufficiently regular solution of the qvi

$$\text{i)}\quad LV(c) := \tfrac{1}{2}\sigma^2 V''(c) + \eta V'(c) + [(1-c)\pi - \lambda(\tau c)^2] - \rho V(c) \le 0 \quad \forall c \ge 0,$$

$$\text{ii)}\quad V(c) \ge MV(c) \quad \forall c \ge 0, \tag{35}$$

$$\text{iii)}\quad LV(c)(V(c) - MV(c)) = 0 \quad \forall c \ge 0,$$

corresponding to the impulse control problem (33). The definition of the operator M can also be written as

$$MV(c) = \sup_{C \ge 0} \{V(C) - K - k\,|C-c|\}.$$

Using Propositions 39, it suffices to show that a solution V of the equations (34) also satisfies the qvi (35). Then, the strategy described above is the corresponding qvi strategy and thus an optimal one. Note first that in the interval [0, U] we have

$$LV(c) = 0\,, \quad LV(c)(V(c) - MV(c)) = 0$$

by construction of V. The general form of V, satisfying equation (34) i), is given by

$$V(c) = a_1 e^{\lambda_1 c} + a_2 e^{\lambda_2 c} - \frac{\lambda\tau^2}{\rho} c^2 - \left(\frac{\pi + 2\lambda\tau^2\eta}{\rho}\right) c + d \quad \forall\, c \in [0,U]$$

where a_1, a_2, and d are real constants and λ_i, i = 1, 2, with $\lambda_1 < 0 < \lambda_2$ are given by

$$\lambda_i = -\frac{\eta}{\sigma^2} \mp \sqrt{\frac{\eta^2 + 2\sigma^2\rho}{\sigma^4}}\,.$$

It remains to demonstrate that V satisfies the qvi.

a) Demonstration that V satisfies (35) i) :

By construction of V, we only have to show this for c > U. Let therefore c > U.

$$\begin{aligned}
LV(c) &\overset{(34)\,iv)}{=} -\eta k + [(1-c)\pi - \lambda\,(\tau c)^2] - \rho\,[V(U) + k\,(c-U)] \\
&\overset{(34)\,i)}{=} -\eta k + [(1-c)\pi - \lambda(\tau c)^2] - k(c-U) - \tfrac{1}{2}\sigma^2 V''(U) - \eta V'(U) \\
&\qquad + [(1-U)\pi - \lambda(\tau U)^2] \\
&\overset{(34)\,iii)}{=} -\tfrac{1}{2}\sigma^2 V''(U) - [\pi(c-U) + \lambda\tau^2(c^2-U^2) + k(c-U)] < -\tfrac{1}{2}\sigma^2 V''(U)
\end{aligned}$$

Thus, it is sufficient to show that V"(U) is non-negative. We have to distinguish between different settings for the constants a_1, a_2.

Case 1 : $a_1 \cdot a_2 \le 0$

i) "$a_1 \le 0 \le a_2$"

It is easy to verify that in this case, we must have $V'''(x) \ge 0\ \forall\, x \in [0, U]$. This implies that V"(x) is monotonically increasing in x on [0, U]. Also, (34) ii) gives us the existence of a zero of V"(x) in [u, U]. These two facts together imply $V''(U) \ge 0$.

ii) "$a_1 \ge 0 \ge a_2$"

Here, one can verify that we must have $V'''(x) \le 0\ \forall\, x \in [0, U]$ which implies that V"(x) is monotonically decreasing on [0, U]. But this is a contradiction to the simultaneous existence of a zero of V"(x) in [u, U] **and** a negative value of V"(x) in [*l*, u].

Case 2 : $a_1 \cdot a_2 > 0$
i) "$a_1, a_2 < 0$"
We then have that V"(x) is strictly negative on [0, U] which contradicts the existence of a zero of V"(x) in [u, U] .
iia) "$a_1, a_2 > 0$": $V'''(0) \geq 0$
This implies that V'''(x) is non-negative on [0, U], and as in Case 1i), we can conclude $V''(U) \geq 0$
iib) "$a_1, a_2 > 0$": $V'''(0) \leq 0$
The existence of a zero of V"(x) at u* in [u, U] **and** a negative value of V"(x) in [l, u] together with $V'''(0) \leq 0$ imply that there is a $\hat{u} \leq u^*$ with $V'''(\hat{u}) = 0$. For all $x > \hat{u}$ we then have $V'''(x) > 0$, and as in Case 2iia) this implies $V''(U) \geq 0$.

b) Demonstration that V satisfies (35) ii) :

Case 1 : "Cases 1i) and 2iia) of the previous step"
i) "$c \in (0, l)$"
We have that V"(c) is non-positive on (0, l) which together with (34) ii) implies

$$V'(c) > k \quad \forall\, c \in (0, l) \tag{36}$$

Thus, the optimal action (note also $V'(l) = k$) in the maximisation problem of the operator M is to go to l, and, due to relations (36) and (34) iii), we have

$$V(c) > MV(c) \quad \forall\, c \in (0, l).$$

ii) "$c \in (l, u)$"

$$-k < V'(c) < k \quad \forall\, c \in (l, u)$$

implies that the "zero"-action (which does not move the process but result in fixed costs of K) is optimal in the maximisation problem of M. Hence,

$$V(c) > MV(c) \quad \forall\, c \in (l, u).$$

iii) "$c \in (u, U)$"
V"(c) is monotonically increasing on (u, U) with $V''(u) < 0 < V''(U)$. Hence, we have

$$V'(c) < -k \quad \forall\, c \in (u, U)$$

which, as for "$c \in (0, l)$", implies

$$V(c) > MV(c) \quad \forall\, c \in (u, U).$$

iv) "$c > U$"
Due to the affine linear form of V(c) on (U, ∞), the optimal action Δc (= c − C) must be bigger than c − U. But then the first order condition for a maximum implies

$$V'(c - \Delta c) = -k\,.$$

Hence, relation (34 ii) and the fact that V'(c) is bigger than -k on [0, U) yield $c - \Delta c = u$ and thus

$$MV(c) = V(c) - K - k\,(c - u) = V(c).$$

v) "$c \in \{0, U\}$"
Due to (34) ii) and iii), we have V(c) = MV(c) in both cases.
Case 2 : "Case 2iib) of the previous step"
As in Case 2iib) of the previous step, $V'''(0) < 0$ (and $V''(y) < 0$ for a $y \in (l, u)$) implies that we have $V''(x) \leq 0$ on $[0, u^*]$ and $V''(x) \geq 0$ on $[u^*, U]$ for a $u^* \in [0, U]$. Using this fact, the same proof as in Case 1 applies.

c) Finally, it follows immediately from the foregoing considerations that V also satisfies relation(35) iii).

□

In view of the proposition, we only have to show the existence of a function V satisfying the system of equations (34). We will construct it in several steps. As in related papers (see (Harrison e.a. 1983), (Jeanblanc Piqué 1993)), we will work with the derivative v of V and construct it on [0, U] only. The extension of V to (U, ∞) can then be easily obtained: due to its affine linear form and the continuity of V and V' in U, it is completely determined by the values V(U) and V'(U). This way of pasting together V constructed on [0, U] and its affine linear part with the help of smoothness requirements is usually referred to as **smooth pasting**.

The derivative v of V has to satisfy

$$\text{i)} \quad \tilde{L}v(c) := \tfrac{1}{2}\,\sigma^2 v''(c) + \eta\, v'(c) - \rho\, v(c) - 2\,\lambda\tau^2 c - \pi = 0 \qquad \forall\, c \in [0,U],$$

$$\text{ii)} \qquad \int_0^l (v(x) - k)\,dx = K, \quad \int_u^U (v(x) + k)\,dx = -K, \qquad (37)$$

$$\text{iii)} \qquad v(u) = v(U) = -k, \quad v(l) = k$$

In a first step to solve (37), we consider the boundary value problem

$$\tfrac{1}{2}\sigma^2 f''(c) + \eta\, f'(c) - \rho\, f(c) - 2\,\lambda\tau^2 c - (\pi + 2\,\lambda\tau^2 l) = 0 \qquad (38)$$

$$f(0) = k, \quad f(\Delta) = -k$$

with $\Delta = u - l$. The solution f(c) of this problem will also deliver v(c) via

$$v(c) = f\,(c - l)\,.$$

The general form of f(c) solving the boundary value problem (38) is given by

$$f(c) = \mu e^{\lambda_1 c} + \nu e^{\lambda_2 c} - 2\frac{\lambda\tau^2}{\rho}c - \frac{\pi\rho + 2\lambda\tau^2\eta}{\rho^2} - 2\frac{\lambda\tau^2}{\rho}c\,l$$

with

$$\lambda_i = -\frac{\eta}{\sigma^2} \mp \sqrt{\frac{\eta^2 + 2\sigma^2\rho}{\sigma^4}}$$

(note especially: $\lambda_1 < 0 < \lambda_2$) and where the constants μ, ν depend on l and Δ. For convenience, we introduce the following notations

$$\text{con}_1 := 2\frac{\lambda\tau^2}{\rho}, \quad \text{con}_2 := \frac{\pi\rho + 2\lambda\tau^2\eta}{\rho^2}. \tag{39}$$

Obviously, both these constants are positive. Furthermore, we have

$$v(c) = \mu e^{\lambda_1 (c-\Delta)} + \nu e^{\lambda_2 (c-\Delta)} - \text{con}_1\, c - \text{con}_2 \quad ,$$

and the requirements $v(l) = k$, $v(u) = -k$ lead to the representations

$$\mu = \frac{k(e^{\lambda_2\Delta}+1) - \text{con}_1\left[\Delta + l(1-e^{\lambda_2\Delta})\right] - \text{con}_2\left[1-e^{\lambda_2\Delta}\right]}{e^{\lambda_2\Delta} - e^{\lambda_1\Delta}}, \tag{40a}$$

$$\nu = \frac{-k(e^{\lambda_1\Delta}+1) + \text{con}_1\left[\Delta + l(1-e^{\lambda_1\Delta})\right] + \text{con}_2\left[1-e^{\lambda_1\Delta}\right]}{e^{\lambda_2\Delta} - e^{\lambda_1\Delta}}. \tag{40b}$$

For fixed l, we have the following limiting behaviour of μ, ν as functions in Δ :

$$\mu(\Delta) \xrightarrow{\Delta\to\infty} k + \text{con}_1\, l + \text{con}_2 \qquad \mu(\Delta) \xrightarrow{\Delta\to 0} +\infty \tag{41a}$$

$$\nu(\Delta) \xrightarrow{\Delta\to\infty} 0 \qquad \nu(\Delta) \xrightarrow{\Delta\to 0} -\infty \tag{41b}$$

The existence proof for v(c) is the main part of the proof of the following theorem which states the existence of an optimal qvi-control of the form given in Proposition 39.

Theorem 40 "Existence of an optimal qvi-control"
There exist constants μ, ν, l, u, U with $0 < l < u < U$ such that the function V(c) defined by

$$V(c) = \begin{cases} \frac{\mu}{\lambda_1} e^{\lambda_1 (c-l)} + \frac{\nu}{\lambda_2} e^{\lambda_2 (c-l)} - \frac{\text{con}_1}{2} c^2 - \text{con}_2 c + d, & c \in [0,U] \\ V(U) - k(c-U) & , c > U \end{cases}$$

coincides with the value function $\psi(c)$ of the optimisation problem (33). Here, the constants con_1 and con_2 are defined in (39), the constants μ and ν are defined via relation (40) (with $\Delta := u - l$), and the constant d is chosen such that we have LV = 0 in [0, U], i.e.

$$d = con_2\, l + \tfrac{1}{2}\, con_1 l^2 + [\, 2\rho - \eta(\mu+\nu) - \tfrac{1}{2}\sigma^2[\mu\lambda_1 + \nu\lambda_2]]$$

Proof :
As already said, it is enough to prove the existence of a function v(c) satisfying the relations (37). This will be done in three steps:

Step 1:
Show that for every $l > 0$ there exists a positive number $\Delta = \Delta(l)$ with

$$-K = \phi_1(\, l, \Delta(l)) := \int_u^U (v(x) + k)dx.$$

Step 2:
Show that there exists a pair $(\, l, \Delta(l))$ as in Step 1 that also satisfies

$$K = \phi_2(\, l, \Delta(l)) := \int_0^l (v(x) - k)dx.$$

Step 3:
Show that there exists a triple $(U, l_U, \Delta(l_U))$ with a positive U and $(l_U, \Delta(l_U))$ as in Step 2 satisfying

$$-k = \phi_3(\, U, l_U, \Delta(l_U)) := v(U).$$

We will postpone working out those steps in detail to sub-section v).

□

The main consequence of the theorem is that we have reduced the solution of our impulse control problem to the much simpler task of solving the following system of three non-linear equations for (l, Δ, U) arising from the system (34) :

$$\text{i)}\quad \frac{\mu}{\lambda_1}\left(1 - e^{-\lambda_1 l}\right) + \frac{\nu}{\lambda_2}\left(1 - e^{-\lambda_2 l}\right) - \tfrac{1}{2}\, con_1 \cdot l^2 - (con_2 + k)\, l - K = 0,$$

$$\text{ii)}\quad \mu e^{\lambda_1 (U-l)} + \nu e^{\lambda_2 (U-l)} - con_1\, U - con_2 + k = 0, \qquad (42)$$

$$\text{iii)}\quad \frac{\mu}{\lambda_1}(e^{\lambda_1 (U-l)} - e^{\lambda_1 \Delta}) + \frac{\nu}{\lambda_2}(e^{\lambda_2 (U-l)} - e^{\lambda_2 \Delta})$$
$$- \tfrac{1}{2} con_1\left(U^2 - (l + \Delta)^2\right) - (con_2 - k)(U - l - \Delta) + K = 0$$

where we additionally require

$$0 < l < U, \ 0 < \Delta < U - l$$

and where μ and ν are functions of (Δ, l) given by equations (40). The existence of a unique solution (l, Δ, U) with the required properties is guaranteed by Theorem 40. We will give some numerical examples and a sensitivity analysis of the problem (in dependence of the cost parameters) in the next sub-section.

iv) Some Numerical Examples

As an example, we have solved the system (42) numerically with the following set of parameters:

$$\eta = 0.05, \sigma = 0.2, \rho = 0.06, K = 0.01, k = 0.02, \pi = 0.06, \tau = 0.2, \lambda = 0.2$$

The optimal strategy is given by the upper bound for the cash weight U = 0.527 and the restarting points u = 0.188 (after reaching of U by C_t) and $l = 0.133$ after the total absorption of the cash account by outflows (i.e. reaching of zero by C_t).

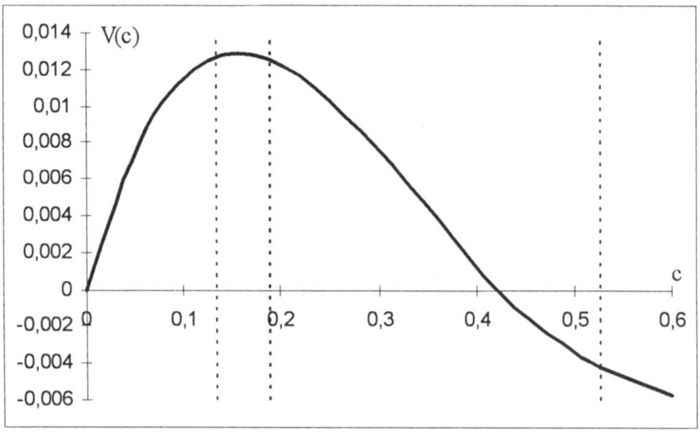

Figure 16: Normalised value function of the cash management problem

Figures 16 and 17 present the value function (normalised to start in zero by subtracting the constant term) and its slope on [0, 0.6], respectively. Recall that the value function is linear on [U, ∞) and that its slope is constant on [U, ∞). The dashed vertical lines represent l, u, U. Note in particular, that l and u have different distances form their corresponding threshold points 0 and U, respectively. This is of course no surprise, because the optimal value for C(t) without transaction costs is negative (note that we assume a positive rate of excess return of the index portfolio compared to the riskless rate). The only reason for shifting the cash weight process away form 0 is to avoid too much transaction costs; the reason for decreasing the cash weight after reaching of U is that it is advantageous to buy additional shares due to the positive excess return Which makes it desirable to take a big step into the direction towards the origin.

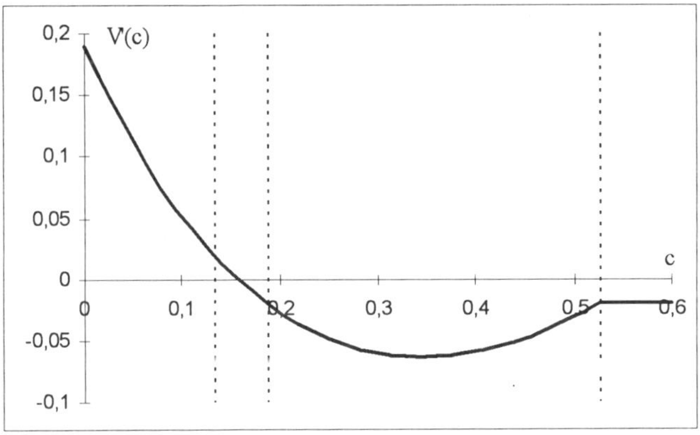

Figure 17: Slope of the value function of the cash management problem

Figures 18 and 19 illustrate the dependence of the optimal strategy given by the triple (l, u, U) plotted as functions on the proportional and fixed cost parameters k and K, respectively (we have used the same data but λ changed to 0.1).

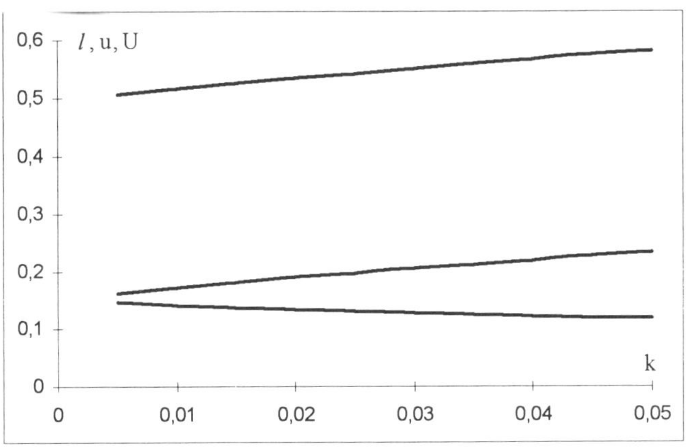

Figure 18: Optimal strategy (l, u, U) as functions of the proportional costs k

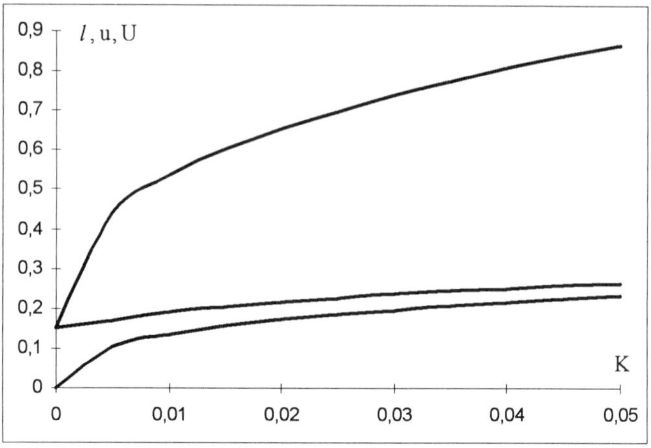

Figure 19: Optimal strategy (*l*, u, U) as functions of the fixed costs K

With increasing proportional cost coefficient k, the lower restarting point *l* will decrease while the upper intervention point U increases. Of course, it is clear that with increasing costs, the no trade interval [0, U] must become larger. Because the lower intervention point of zero is fixed, the upper point U then has to increase. Also there is a tendency that the optimal "actions" *l* and U–u decrease with increasing k which is also plausible due to the rising costs. Figure 19 indicates that if the fixed cost component K tends to zero the optimal strategy will tend to a local time type strategy (i.e. U = u, *l* = 0 and only the minimal actions are taken to prevent C_t from falling below 0 or rising above U). By considering the case of proportional transaction costs, Connor and Leland (1995) indicate by some heuristical arguments that the optimal strategy is of the local time type which supports the above observation. Further, Figure 19 seems to show that with growing K, the optimal restarting points l and u are approaching each other. This can be explained by the vanishing effect of the proportional costs compared to the fixed costs.

v) Proof of the Existence Theorem 40

We will work out each of the steps given in the sketch of the proof of Theorem 40 after another which then completes this proof.

Step 1:

$$\phi_1(l, \Delta(l)) = \mu(\Delta)\,\frac{\left(e^{\lambda_1(U-l)} - e^{\lambda_1\Delta}\right)}{\lambda_1} + \nu(\Delta)\,\frac{\left(e^{\lambda_2(U-l)} - e^{\lambda_2\Delta}\right)}{\lambda_2} - \tfrac{1}{2}\mathrm{con}_1(U^2 - u^2) -$$

$$- (\mathrm{con}_2 - k)(U - u)$$

For (fixed l and) $\Delta \to U - l$, we have $U - u \to 0$ which implies

$$\phi_1(l, (U-l)-) = 0$$

Now consider the case of $\Delta \to 0$. Using the explicit forms of μ, ν, we get

$$\lim_{\Delta\downarrow 0} \frac{\mu(\Delta)\frac{1}{\lambda_1}\left[e^{\lambda_1(U-l)}-e^{\lambda_1\Delta}\right]}{\nu(\Delta)\frac{1}{\lambda_2}\left[e^{\lambda_2(U-l)}-e^{\lambda_2\Delta}\right]} = \frac{2k\frac{e^{\lambda_1(U-l)}-1}{\lambda_1}}{-2k\frac{e^{\lambda_2(U-l)}-1}{\lambda_2}} =: g$$

By noting that $h(x) = (\exp(ax)-1) / a$ is monotonically increasing in a (for positive x) and that we have $\lambda_1 < 0 < \lambda_2$, we can conclude

$$-1 < g < 0\,.$$

Together with the limiting behaviour of $\mu(\Delta)$, $\nu(\Delta)$ for $\Delta \to 0$ (see relation (41)), this results in

$$\phi_1(l, 0+) = -\infty\,.$$

By continuity, we have the existence of a positive number $\Delta(l)$ with $\phi_1(l, \Delta(l)) = -K$.

Step 2:

$$\phi_2(l, \Delta(l)) = \mu(\Delta)\frac{\left(1-e^{-\lambda_1 l}\right)}{\lambda_1} + \nu(\Delta)\frac{\left(1-e^{-\lambda_2 l}\right)}{\lambda_2} - \tfrac{1}{2}\mathrm{con}_1\cdot l^2 - (\mathrm{con}_2 + k)\, l$$

For $l \to 0$, we obtain $\phi_2(0+, \Delta(0+)) = 0$ (Note that (for fixed U) a possible convergence of $\Delta(l)$ against 0 would be a contradiction to the construction of $\Delta(l)$ to satisfy Step 1). For $l \to U$, we have $\Delta(l) \to 0$. If we note

$$\lim_{\Delta\downarrow 0} (\mu(\Delta)/\nu(\Delta)) = -1\,,$$

the limiting behaviour (41) of $\mu(\Delta)$, $\nu(\Delta)$ for $\Delta \to 0$, and that $h(x) = (1-\exp(ax)) / a$ is monotonically decreasing in a for negative x, we obtain $\phi_2(U-, \Delta(U-)) = +\infty$ Thus, by continuity, we have established the existence of a positive l with a corresponding pair $(l, \Delta(l))$, satisfying the requirements of Steps 1 and 2.

Step 3:

$$\phi_3(U, l_U, \Delta(l_U)) = v(U) = \mu e^{\lambda_1(U-l)} + \nu e^{\lambda_2(U-l)} - \mathrm{con}_1 U - \mathrm{con}_2$$

a) $U \to 0$:

For every fixed l, we can choose U sufficiently small such that we will have $\mu > 0$ and $\nu < 0$ (refer to relation (41)). Then, we have

$$\mu e^{\lambda_1(U-l)} + \nu e^{\lambda_2(u-l)} - con_1 U - con_2 < \mu e^{\lambda_1 \Delta} + \nu e^{\lambda_2 \Delta} - con_1 u - con_2 = -k$$

(for the last equality, we have used $v(u) = -k$ which is valid due to the construction of μ, ν). Hence, there exists a positive U with a corresponding triple $(U, l_U, \Delta(l_U))$ such that the pair $(l_U, \Delta(l_U))$ satisfies Steps 1 and 2 and

$$\phi_3(U, l_U, \Delta(l_U)) < -k \tag{43}$$

b) $U \to \infty$:

We have to distinguish between certain subcases to get the existence of a triple $(U, l_U, \Delta(l_U))$ yielding a value of $\phi_3(.)$ which is bigger than $-k$.

Case i) : "$\Delta = \Delta(l_U)$ as chosen in Steps 1 and 2 is unbounded for $U \to \infty$"
Since we are only interested in a value of $\phi_3(.)$ which is bigger than $-k$ (and not in limits of $\phi_3(.)$), we can w.l.o.g. assume that we have

$$\Delta \xrightarrow{U\to\infty} \infty.$$

Using the explicit forms of μ, ν (see the relations (40)), we get the following approximations for $\Delta >> 1$:

$$\nu\, e^{\lambda_2 \Delta} \approx -k + con_1[\Delta + l] + con_2\,,$$

$$\mu e^{\lambda_1 \Delta} \approx con_1 \cdot l \cdot e^{\lambda_1 \Delta}.$$

For $\Delta >> 1$, this will lead to

$$\phi_3(U, l_U, \Delta(l_U)) \approx con_1 \cdot l_U \cdot e^{\lambda_1 \Delta(l_U)} e^{\lambda_1 (U-u)} - con_1 [(U-u) + \Delta(l_U) + l_U] - con_2$$
$$+ [-k + con_1[\Delta(l_U) + l_U] + con_2] e^{\lambda_2 (U-u)}$$
$$= con_1 \cdot l_U [e^{\lambda_2 (U-u)} - 1 + e^{\lambda_1 \Delta(l_U)} e^{\lambda_1 (U-u)}] - k\, e^{\lambda_2 (U-u)}$$
$$+ con_1 \cdot \Delta(l_U) [e^{\lambda_2 (U-u)} - 1] - con_1 (U-u) + con_2 [e^{\lambda_2 (U-u)} - 1]$$

The first term after the equality sign is positive, and thus (for $\Delta >> 1$), we obtain

$$\phi_3(U, l_U, \Delta(l_U)) \geq con_1[\tfrac{1}{2}\Delta(l_U)(e^{\lambda_2 (U-u)} - 1) - (U-u)]$$
$$+ [con_1 \cdot \tfrac{1}{2}\Delta(l_U)(e^{\lambda_2 (U-u)} - 1) - k\, e^{\lambda_2 (U-u)}] + con_2 [e^{\lambda_2 (U-u)} - 1] \geq 0$$

To see the last inequality, note that the first and the last term preceding the inequality sign are obviously positive for large Δ. The second term is surely positive if $U - u$ is bounded. If $U - u$ is unbounded, we can again w.l.o.g. assume that $U - u$ will converge to infinity when Δ converges to infinity. But then, the second term will also be positive for sufficiently large Δ.

<u>Case ii) :</u> "$\Delta = \Delta(l_U) \leq R_1$ for all $U > 0$ (i.e. Δ is bounded by a real constant R_1)"

iia) : "$l \leq R_1$ for all $U > 0$"
Assumptions ii) and iia) imply that u is bounded from above. We have to consider all the different possibilities for the signs of μ and ν in this case.

<u>Assume</u> $\mu > 0$, $\nu < 0$
With the same argument as in Case i), we can now (and in the following subcases) assume that "$\mu > 0$, $\nu < 0$" is valid for all U sufficiently large.

$$\begin{aligned}\phi_3(x, l_U, \Delta(l_U)) &= \mu e^{\lambda_1(x-l_U)} + \nu e^{\lambda_2(x-l_U)} - con_1\, x - con_2 \\ &= \nu\,[e^{\lambda_2(x-l_U)} - e^{\lambda_1(x-l_U)}] + [k + con_2 + con_1\, l_U] e^{\lambda_1(x-l_U)} - con_1 x - con_2 \\ &\leq \varepsilon - con_1 x - con_2 \qquad \text{for } x\ (< U) \text{ sufficiently large)}\end{aligned}$$

where $\varepsilon = \varepsilon(x) > 0$ can be made arbitrarily small by choosing a large value of x (independent of Δ). Note that for the second equality sign, we have used the relation $v(l) = k$ which is satisfied due to the construction of μ, ν. To get the inequality, note that the first term in the line above the inequality sign is negative, and the second one tends to zero for x sufficiently large. For large U, this would be a contradiction to Step 1 (i.e. to $\int_u^U (v(x)+k)dx = -K$).

<u>Assume</u> $\nu > 0$
Note first that $\nu > 0$ implies that we must have $\Delta \geq \varepsilon > 0$ for an appropriate constant ε. But this implies that μ and ν must be bounded. The validity of Step 1 implies the following approximation for large U:

$$\frac{\nu e^{\lambda_2(U-l)}}{\lambda_2} \approx \tfrac{1}{2} con_1 U^2 + (con_2 - k)\, U .$$

This together with the boundedness of μ, ν and the explicit form of $\phi_3(.)$ implies

$$\phi_3(U, l_U, \Delta(l_U)) \xrightarrow{U \to \infty} +\infty.$$

<u>Assume</u> $\mu < 0$, $\nu < 0$
This immediately contradicts

$$\mu + \nu = k + con_1\, l + con_2 > 0$$

which is a consequence of $v(l) = k$.

iib) : "$l \xrightarrow{U\to\infty} \infty$"
This directly implies that u must also tend to infinity with growing U.

<u>**Assume**</u> $\Delta \geq \varepsilon > 0$
This assumption together with the convergence of l towards infinity imply

$$\nu \to \infty, \quad \mu \to \infty .$$

which leads to the following contradiction:

$$K = \phi_2(l, \Delta(l)) = \mu(\Delta)\frac{\left(1-e^{-\lambda_1 l}\right)}{\lambda_1} + \nu(\Delta)\frac{\left(1-e^{-\lambda_2 l}\right)}{\lambda_2} - \tfrac{1}{2}\mathrm{con}_1 \cdot l^2 - (\mathrm{con}_2 + k)\, l$$

$$\geq e^{-\lambda_1 l} - \tfrac{1}{2}\mathrm{con}_1 \cdot l^2 - (\mathrm{con}_2 + k)\, l \xrightarrow{l\to\infty} +\infty .$$

<u>**Assume**</u> $\Delta \xrightarrow{l\to\infty} 0$
1. "$|\nu| \leq S$" (for a real constant S)
From Step 2, we obtain the following approximation for large l:

$$\phi_2(l, \Delta(l)) \approx \mu(\Delta)\frac{\left(1-e^{-\lambda_1 l}\right)}{\lambda_1} - \tfrac{1}{2}\mathrm{con}_1 \cdot l^2 - (\mathrm{con}_2 + k)\, l$$

$$\approx \mathrm{con}_1 \cdot l\,\frac{\left(1-e^{-\lambda_1 l}\right)}{\lambda_1} - \tfrac{1}{2}\mathrm{con}_1 \cdot l^2 \xrightarrow{l\to\infty} +\infty$$

which again is a contradiction to $K = \phi_2(l, \Delta(l))$ for all positive l. For the second approximation, note that from $\nu(l) = k$ and the boundedness of ν, we get

$$\mu \approx \mathrm{con}_1 \cdot l .$$

2. "$\nu \xrightarrow{l\to\infty} -\infty$"
Here, $\nu(l) = k$ leads to the approximation $\mu \approx \mathrm{con}_1 \cdot l - \nu$ for large values of l. Again, using Step 2, we obtain a contradiction :

$$K = \phi_2(l, \Delta(l))$$

$$\approx \mathrm{con}_1 \cdot l\,\frac{\left(1-e^{-\lambda_1 l}\right)}{\lambda_1} - \nu\left[\frac{\left(1-e^{-\lambda_1 l}\right)}{\lambda_1} - \frac{\left(1-e^{-\lambda_2 l}\right)}{\lambda_2}\right] - \tfrac{1}{2}\mathrm{con}_1 \cdot l^2 - (\mathrm{con}_2 + k)\, l \xrightarrow{l\to\infty} +\infty.$$

3. "$\nu \xrightarrow{l\to\infty} +\infty$"
The explicit form of μ shows that for both l and Δ tending to infinity, μ must also tend to (positive) infinity. With the help of Step 2 we obtain a contradiction:

$$K = \phi_2(\,l,\,\Delta(l)) \;= \mu(\Delta)\frac{\left(1-e^{-\lambda_1 l}\right)}{\lambda_1} + \nu(\Delta)\frac{\left(1-e^{-\lambda_2 l}\right)}{\lambda_2} - \tfrac{1}{2}\,\mathrm{con}_1 \cdot l^2 - (\mathrm{con}_2 + k)\,l$$

$$\geq\; e^{-\lambda_1 l} - \tfrac{1}{2}\,\mathrm{con}_1 \cdot l^2 - (\mathrm{con}_2 + k)\,l \xrightarrow{\;l\to\infty\;} +\infty\,.$$

Finally, looking at all the different subcases of case **"b) U → ∞"**, we have the existence of a triple $(U, l_U, \Delta(l_U))$ such that the pair $(l_U, \Delta(l_U))$ satisfies Step 2 and

$$\phi_3(\,U,\,l_U,\,\Delta(l_U)) > -\,k\,. \tag{44}$$

Hence, continuity of ϕ_3, relations (43), and (44) together imply the existence of the required triple $(U, l_U, \Delta(l_U))$ of Step 3.

Putting together all the previous considerations yields the desired existence result for the function v(c) as a solution of the system (37), and we get V(c) by integration. Note in particular that the constant of integration has to be determined such that we have $LV(c) = 0$ in $(0, U)$.

□

6.4 Value Preserving Portfolio Strategies

While from a mathematical point of view the approach of portfolio selection by maximising the expected utility from consumption and terminal wealth is by now well understood, and there are also many successful attempts generalising Merton's problem to make it more realistic (see the foregoing chapters), there are some criticisms arising from a more economical or philosophical viewpoint (see e.g. (Hellwig 1993)).

The most important criticism is that the usual method of discounting future payments to make consumption and terminal wealth time-additive is not acceptable. Discounting makes later time periods less important, and, as a consequence, the "economic ability" of a portfolio decreases with time, because the investor consumes too much in the present which prevents him from preserving the "economic value" of his holdings. An overview for such arguments and a list of corresponding references can be found in (Hellwig 1993).

In this section, we take up an idea of Klaus Hellwig (given in e.g. (Hellwig 1987)) who developed a universally applicable approach to price (economic) resources, the principle of value preservation. The key idea is that their intrinsic value ("value of future returns") should be preserved over time. Therefore, the prices of the relevant goods should be adjusted in such a way that the members of the corresponding market will consume only so much that this intrinsic value will be constant over time ("principle of value preservation"). The main aim of this section is to show how this approach can be applied to portfolio optimisation. While (Hellwig 1987/1993) and (Wiesemann 1995) consider this problem in a discrete-time financial market with a finite state space, we consider continuous-time models with prices given by general diffusion processes (see (Korn 1997a) for a general continuous-time model based on semi-martingales that includes the Black and Scholes model, general diffusion models, jump and jump-diffusion models as special cases).

i) Portfolio Value and Value Preservation

In this sub-section, we will set up the model, give some basic definitions and state some simple consequences of the requirement of value preservation. We look at the standard diffusion type market model consisting of one bond and n stocks with price dynamics given by

$$dP_0(t) = 0, \quad P_0(0) = 1,$$

$$dP_i(t) = P_i(t)\left(b_i(t)dt + \sum_{j=1}^{m} \sigma_{ij}(t)dW_j(t) \right), \quad P_i(0) = p_i, \quad i = 1,\dots, n$$

where b and σ have to satisfy the assumptions of Section 2.1. Especially, $\sigma(t)\sigma(t)'$ is assumed to be uniformly positive definite (as matrix function for $L\otimes P$-almost all $(t, \omega)\in[0, T]\times\Omega$). We allow for $m > n$, i.e. for an incomplete market setting. Recall that a constant bond price means that all stock prices can be regarded as being discounted by the bond price. This feature is notationally very convenient. The following definitions and results can easily be adapted to the case of a non-constant bond price (see the remarks made before Theorem 49).

As we have to modify the definitions of a trading strategy and of a consumption process somewhat for using them in the value preserving approach, we will quickly recall their definitions here.

Definition 41

a) An (n+1)-dimensional, F_t-adapted process $\theta(t) = (\theta_0(t),\dots, \theta_n(t))'$ which is integrable with respect to $P(t) = (P_1(t),\dots, P_n(t))'$ is called a **trading strategy** with wealth process

$$X(t) = \sum_{k=0}^{n} \theta_k(t)P_k(t).$$

b) A (cumulative) **consumption process** C(t) is an Itô-process with C(0) = 0.
c) A pair (θ, C) consisting of a trading strategy and a consumption process is called a **self-financing pair** if the wealth process corresponding to θ satisfies

$$X(t) = x + \int_0^t \theta(s)\,dP(s) - C(t)$$

where in writing $\int \theta dP$, we identify θ with its last n components as we have $dP_0 = 0$.
d) A self-financing pair (θ, C) will be called **admissible** if for every equivalent martingale measure Q (see Section 2.4) for the price process P(t) the "gains process" $\int \theta dP$ is a Q-martingale.

Remark 42

We will shortly comment on the "non-standard" parts of this definition (i.e. parts b) and d)). First, note that a consumption process, as defined above, need not be non-negative. It is thus a mixture of a consumption and an endowment process. By introducing the concept of value preservation below, we will see that a negative consumption rate corresponds to putting additional funds into the portfolio to compensate losses of the whole market. This will be necessary to keep the portfolio value constant in such a situation. Of course, modelling consumption as an Itô-process is not new in the literature. Compare e.g. the CCAPM section in (Back 1991) where consumption is modelled by even more general processes which can attain negative values with positive probability. The requirement that $\int \theta dP$ is a Q-martingale in part d) is slightly stronger than the requirements on a trading strategy in Section 2.2. In a way, as it can be restated as "X(t) + C(t) is a Q-Martingale", it puts a bound on the possible consumption processes C(t). In that sense, this requirement replaces the usual non-negativity constraint on the consumption rate process. Also, it can be interpreted to mean that it should not be possible to make the market advantageous for the investor when it is in equilibrium (represented by the use of an equivalent martingale measure Q) which is a reasonable assumption, because in a market equilibrium a risk-neutral investor is indifferent between investing in different securities. Thus, it is natural to require that he could not turn this equilibrium into a promising market.

The typical definitions used to develop our theory are given in:

Definition 43

Let Q be an equivalent martingale measure for P.
a) The **portfolio value process** (with respect to Q) of an admissible pair (θ, C) is defined as

$$V^Q(t) := E_Q\left(X(T) + \int_t^T dC(s) \middle| F_t \right).$$

b) The **portfolio return process** $R^Q(t)$ (with respect to Q) of an admissible pair (θ, C) with a positive portfolio value process $V^Q(t)$ is defined as

$$dR^Q(t) = \frac{d(C(t) + V^Q(t))}{V^Q(t)}, \quad R^Q(0) = 0$$

c) (θ, C) is called **interest rate oriented** (with respect to Q) if it satisfies

$$dR^Q(t) = H(t) d\left(1/H(t) \right)$$

where H(t) is the Radon-Nikodym-derivative of Q with respect to P.
d) A **value preserving portfolio strategy** is an interest rate oriented pair (θ, C) with a constant portfolio value process with respect to an equivalent martingale measure Q. An equivalent martingale measure Q that allows for a value preserving strategy will be called a **value preserving measure**.

Remark 44

a) The definition of the portfolio value is motivated by option pricing theory. If we interpret the future consumption C(T) – C(t) and the terminal wealth X(T) as payments arising from a contingent claim, the definition of the portfolio value coincides with the usual definition of a possible non-arbitrage price for that contingent claim. Note, that due to our definition of an admissible pair (θ, C), this price is unique and independent of the special form of Q. In particular, (θ, C) represents a replicating strategy for our "contingent claim". As an admissible pair (θ, C) is not uniquely determined by its portfolio value process, simply requiring a constant portfolio value would not be enough to figure out a particular one. Also, why should such a pair be a desirable one ? Therefore, we also specify the form of its return process. In a complete market setting, the requirement of interest rate orientedness means that the return process of (θ, C) should be equal to the wealth process of the growth-optimum portfolio (see Section 2.3, and note that H(t) as defined there coincides with H(t) of Definition 43 because the bond price is assumed to be equal to one in Definition 43). As H(t) can be viewed as a risk-adjusted discount rate (still depending on the chosen equivalent martingale measure !), by requiring a strategy to be interest rate oriented we will ensure that it achieves the market's rate of return (see also Section 2.3 for such an argument). While in the discounted expected utility method the discount factors must be known a priori, we will use a random discount factor by choosing H(t). Such a discount factor takes care of the actual development of the market and is **a priori unknown**. It is given **endogenously** by the market and not exogenously, as is the case of the discounted expected utility method. Of course, the pure bond strategy (which is the natural candidate for a process admitting a constant portfolio value) is not interest rate oriented.

Thus, the choice of the utility functions in the usual expected utility maximisation method is here substituted by the choice of a suitable portfolio return (which models the investor's **intratemporal** preferences) and the requirement of a constant portfolio value process (to model **intertemporal** preferences of the investor). For a more substantial discussion of the economic and decision theoretic aspects aspects of these requirements, we refer to (Hellwig 1993) and (Wiesemann 1995).
b) Note that the problem of existence of a value preserving strategy also includes the problem of existence of a value preserving measure Q.

As a consequence of all our requirements, the consumption process as part of a value preserving strategy is already determined, as will be shown in the next proposition below.

Proposition 45
Let (θ, C) be a value preserving portfolio strategy with a non-zero portfolio value $V^Q(0)$. Then, the consumption process $C(t)$ is given by

$$dC(t) = V^Q(0)\, H(t) d\left(\frac{1}{H(t)} \right)$$

where $H(t)$ is the Radon-Nikodym-derivative of Q with respect to P.

Proof:
Use the two different representations of $dR^Q(t)$ in Definition 43 together with the fact that $V^Q(t)$ is constant over time to obtain the desired representation for $C(t)$.

□

ii) Value Preserving Strategies in General Diffusion Type Markets

In this section, we examine the existence (and uniqueness) of value preserving strategies in some diffusion type markets.

Example 1: "Black and Scholes model"
We start with the classical Black and Scholes case with only a single stock and one bond. As usual, we assume constant market coefficients b and σ. As we look at discounted prices, we assume a constant bond price and a stock price which follows a

geometric Brownian motion. The unique equivalent martingale measure Q is given by its Radon-Nikodym derivative H(t) with respect to P,

$$H(t) = e^{-\frac{1}{2}\left(\frac{b}{\sigma}\right)^2 t - \frac{b}{\sigma} W(t)}.$$

Consider an investor with an initial capital of x. We have the following theorem:

Theorem 46
The unique value preserving portfolio strategy (θ, C) with initial portfolio value $V^Q(0) = x$ is given by

$$\theta_0(t) = x\left(1 - \frac{b}{\sigma^2}\right), \quad \theta_1(t) = x\,\frac{b}{\sigma^2}\frac{1}{P_1(t)},$$

$$dC(t) = x\left(\left(\frac{b}{\sigma}\right)^2 dt + \frac{b}{\sigma} dW(t)\right) =: x\,\frac{b}{\sigma}\, dW^Q(t).$$

By means of the Girsanov theorem, $W^Q(t)$, constructed as above, is a Brownian motion with respect to Q, the unique equivalent martingale measure for P(t).

Proof :
The wealth process X(t) corresponding to (θ, C), as given above, satisfies

$$X(t) = \theta_0(t)P_0(t) + \theta_1(t)P_1(t) \equiv x.$$

from which it can directly be verified that, (θ, C) is self-financing. Using this and the explicit form of C(t) yields that (θ, C) is value preserving with portfolio value x. Moreover, by Definition 41, the portfolio value of an admissible pair cannot exceed x. Next, consider any other value preserving strategy (θ, C) with a constant portfolio value of x. By Proposition 45, this strategy has to satisfy

$$\begin{aligned} x = V^Q(0) = X(T) &= x + \int_0^T \theta_1(t)dP_1(t) - C(T) \\ &= x + \int_0^T \theta_1(t)P_1(t)\sigma dW^Q(t) - \int_0^T x\frac{b}{\sigma} dW^Q(t) \\ &= x + \int_0^T \left(\theta_1(t)P_1(t)\sigma - x\frac{b}{\sigma}\right) dW^Q(t). \end{aligned} \tag{45}$$

Because the left side of this equation is constant, so is the right one. Thus, the quadratic variation of the stochastic integral must be zero which implies

$$\theta_1(t)P_1(t)\sigma = x\frac{b}{\sigma^2} \quad \forall\, t\in[0, T] \text{ a.s..}$$

Hence, $\theta_1(t)$ must have the desired form. As we must further have

$$\theta_0(t)P_0(t) = X(t) - \theta_1(t)P_1(t) = x - \theta_1(t)P_1(t),$$

uniqueness of the value preserving strategy follows.

□

Remark 47

The characteristic features of the value preserving portfolio approach can be seen by comparing its resulting strategy to the results of the portfolio problem

$$\max_{(\pi,c)\in A'(x)} E\left(\int_0^T \ln(c(t))dt + \ln(X(T))\right).$$

From Example 3.19, we know that in this problem, the optimal portfolio and consumption rate processes are given by

$$\pi(t) = \frac{b}{\sigma^2}, \quad c(t) = \frac{x}{T+1}\, e^{\frac{1}{2}\left(\frac{b}{\sigma}\right)^2 t + \frac{b}{\sigma}W(t)}$$

which lead to a strictly increasing, non-negative consumption process of

$$C(t) = \frac{x}{T+1}\int_0^T \frac{1}{H(t)}dt = \frac{x}{T+1}\int_0^T e^{\frac{1}{2}\left(\frac{b}{\sigma}\right)^2 t + \frac{b}{\sigma}W(t)}dt$$

and a non-constant, non-negative final wealth of

$$X(T) = \frac{x}{T+1}\,\frac{1}{H(T)} = \frac{x}{T+1}e^{\frac{1}{2}\left(\frac{b}{\sigma}\right)^2 T + \frac{b}{\sigma}W(T)}.$$

Remarkably, the optimal portfolio process in the log-utility example and that of the value preserving portfolio strategy, $(\theta_1(t)P_1(t))/X(t)$, coincide (note that in the value preserving case, the wealth process is constant with $X(t) \equiv x$). But the "cost" of the non-negativity of the consumption process in this case is a non-constant portfolio value process

$$V^Q(t) = \frac{x}{H(t)}\left(1 - \frac{t}{T+1}\right)$$

with a strictly decreasing (risk-neutral) mean of

$$E_Q\left(V^Q(t)\right) = x\left(1-\frac{t}{T+1}\right).$$

Thus, in the risk-neutral sense, the investor is (in the mean) continuously absorbing parts of his future economic power by consuming too much (This is not necessarily correct with respect to the subjective measure P). Moreover, comparison of the value preserving results and that of the logarithmic case gives us a new interpretation of value preservation. The identity of the portfolio processes shows that value preservation can be interpreted as maximising the growth rate of the wealth at every time instant (which is the interpretation of the logarithmic case). In this sense, it is a myopic strategy (compare also (Wiesemann 1995) for the discrete time case). But the different consumption strategy in the value preserving case yields an intertemporal rebalancing between different time instants. One is not allowed to consume more than that part of the instantaneous return which is not needed for value preservation. On the negative side, one has to place additional funds into the portfolio if the "whole market is going down" (which is the interpretation of a negative consumption rate dC(t) in the value preserving approach). Of course, having both value preservation **and** a non-negative non-zero consumption process would be an arbitrage opportunity. Therefore, we could not require both these features at the same time. In sub-section iii), we will give some numerical examples that will further enlighten the relationship between the value preserving and the logarithmic case.

Example 2 : "An incomplete Black and Scholes model (and general diffusion type models)"

We still consider a market with only one bond (with constant price) and one stock with constant market coefficients. But here we will model more sources of uncertainty by replacing the one-dimensional Brownian motion W(t) by an m dimensional one with $m > 1$. Therefore, we have the following form of the stock price

$$dP_1(t) = P_1(t)\left(bdt + \sum_{i=1}^{m}\sigma_i dW_i(t)\right)$$

where $W(t) = (W_1(t),\ldots, W_m(t))'$ is an m-dimensional Brownian motion, and the (m-dimensional) row vector σ satisfies

$$\sigma\sigma' > 0.$$

In this case, we have an infinite number of equivalent martingale measures Q^Y with a general form given by

$$Q^Y(A) = E\left(1_A Z_T(Y)\right) \quad \forall\ A \in F_T,$$

$$Z_t(Y) = \exp\left(-\sum_{i=1}^{m} Y_i W_i(t) - \tfrac{1}{2}\|Y\|^2 t\right) =: H^Y(t)$$

for every $Y \in \mathbf{R}^m$ with

$$\sigma Y = b. \tag{46}$$

This can directly be verified with the help of the Girsanov Theorem B23. For every such Y, the corresponding "market rate of return" $d(1/H^Y(t))$ (which in particular determines the consumption process of a value preserving strategy) must have the form

$$H^Y(t)\, d\left(\frac{1}{H^Y(t)}\right) = \|Y\|^2 dt + Y' dW(t) =: Y' dW^Y(t)$$

where $W^Y(t)$ is an m-dimensional Brownian with respect to Q^Y. So, it is not at all clear if there exists a unique (or maybe many) value preserving measure(s).

Theorem 48
There is a unique value preserving measure Q^Y which is determined by

$$Y = \frac{b}{\sigma\sigma'}\sigma' .$$

The corresponding value preserving strategy with a portfolio value of x is given as

$$\theta_0(t) = x\left(1 - \frac{b}{\sigma\sigma'}\right), \quad \theta_1(t) = x\frac{b}{\sigma\sigma'}\frac{1}{P_1(t)},$$

$$dC(t) = x\, Y' dW^Y(t) = x\, \|Y\|^2 dt + xY' dW(t) .$$

Proof:
The characteristics of (θ, C) as a value preserving strategy can be shown as in the proof of Theorem 46. As in this proof, we can deduce

$$\theta_1(t) P_1(t) \sigma = x\, Y' \quad \forall\, t \in [0, T] \text{ a.s.}$$

with the help of an analogue of equation (45). Hence, we must have

$$Y' = a\sigma$$

for some $a \in \mathbf{R}$. Then, equation (46) implies

$$a = \frac{b}{\sigma\sigma'} .$$

Thus, Y has the desired form, and we obtain uniqueness of (θ, C) as in the proof of Theorem 46.

□

The foregoing results in the complete and in the incomplete case generalise directly to a general diffusion type model as given in Chapter 2. The market coefficients $r(t)$, $b(t)$, $\sigma(t)$ should all be uniformly bounded ($L\otimes P$-a.s.), F_t-adapted processes of appropriate dimensions. The sources of uncertainty are modelled as an m-dimensional Brownian motion, and these sources should not be degenerate in the sense that we require $\sigma(t)\sigma(t)'$ to be uniformly positive definite $L\otimes P$-a.s.. By taking into account that now the bond price is non-constant, we modify the definition of the portfolio value process to have the following form:

$$V^Q(t) := E_Q\left(\frac{P_0(t)}{P_0(T)} X(T) + \int_t^T \frac{P_0(t)}{P_0(s)} dC(s) \middle| F_t \right).$$

For the same reason, $H^Y(t)$ is now given as the product of the inverse of the bond price and the Radon-Nikodym-derivative of an equivalent martingale measure Q with respect to P, i.e.

$$H^Y(t) = \exp\left(-\int_0^t \left(r(s) + \tfrac{1}{2}\|Y(s)\|^2 \right) ds - \sum_{i=1}^m \int_0^t Y_i(s) dW_i(s) \right)$$

for some F_t-adapted, $\mathbf{R}^m$-valued process $Y(t)$ with

$$E\left(\tfrac{1}{2} \int_0^T \|Y(s)\|^2 ds \right) < \infty .$$

Then, by analogous considerations as in the proofs of Theorem 46 and 48, we get the following general form of the existence theorem:

Theorem 49
There is a unique value preserving measure Q^Y determined by

$$Y(t) = \sigma(t)' (\sigma(t)\sigma(t)')^{-1}(b(t) - r(t) \underline{1}) .$$

The corresponding unique value preserving strategy is given as

$$dC(t) = x\,((r(t) + \|Y(t)\|^2)\, dt + Y(t)' dW(t))$$

$$\theta_0(t) = \left(1 - \underline{1}'(\sigma(t)\sigma(t)')^{-1}(b(t) - r(t)\underline{1})\right)\frac{x}{P_0(t)},$$

$$\theta_i(t) = \left((\sigma(t)\sigma(t)')^{-1}(b(t) - r(t)\underline{1})\right)_i \frac{x}{P_i(t)}, \quad i = 1,\ldots, n.$$

Remark 50

a) The unique value preserving measure of Theorems 48 and 49 is well known from the theory of option hedging in incomplete markets. It is the so-called minimal martingale measure of (Föllmer and Schweizer 1990). The relationship between value preserving measures and minimal martingale measures for general continuous price processes $P_i(t)$ (which are not necessarily Itô-processes) is studied in (Korn 1996).

b) The effect of additional constraints on the portfolio strategies is examined in (Wiesemann 1995) and (Korn 1997a).

iii) Some Numerical Examples

In this part, we present some simulation results to illustrate the performance of the value preserving strategy and to compare it to the log-optimal strategy as given in the situation of Example 1. For the case of T = 1, we simulate a geometric Brownian motion for the price process $P_1(t)$ for various values of b/σ in a realistic range (i.e. $b/\sigma = 0.1, 0.33, 0.5, 0.8, 1$). Because the whole situation is symmetric in b/σ, we do not consider negative values. Recall that the performance of the value preserving and the log optimal strategies depend only on this ratio. For every case presented below, we have performed 500 simulations. In Table 8 we give the mean, minimum, maximum and expected value of the total consumption C(T) of the value preserving strategy and of the total consumption $C_0(T)$ and terminal wealth $X_0(T)$ of the log optimal strategy, respectively. Comparing the sum of consumption and terminal wealth between the two types of strategies, we see that the value preserving one performs better for values of b/σ up to 0.5. The log-optimal one is superior for higher values of b/σ. Thinking of the value preserving approach as one that is developed for long time investment the above behaviour is reasonable, because in the long run, high values of b/σ are not realistic (remember that we have assumed r= 0). On the other hand, the range of the total consumption of the value preserving strategy exceeds that of the log-optimal one. Of course, this fact must be expected, because in the logarithmic case the market uncertainty influences both the consumption and the terminal wealth while in the value preserving approach the consumption process has to carry the whole uncertainty alone.

	b/σ	0.1	0.33	0.5	0.8	1
Value preserving strategy	$C(T)_{mean}$	1.26	10.84	26.37	61.00	93.79
	$C(T)_{min}$	–34.82	–115.97	–142.92	–184.38	–256.08
	$C(T)_{max}$	31.15	119.39	181.64	298.13	390.08
	$E(C(T))$	1.00	11.11	25.00	64.00	100.00
	$X(T)$	100.00	100.00	100.00	100.00	100.00
Log-optimal strategy	$C_0(T)_{mean}$	50.35	53.18	57.02	69.91	81.78
	$C_0(T)_{min}$	41.81	31.09	22.04	19.54	19.42
	$C_0(T)_{max}$	61.18	98.73	127.83	346.23	425.51
	$E(C_0(T))$	50.25	52.88	56.81	70.04	85.91
	$X_0(T)_{mean}$	50.61	55.79	65.50	91.29	127.55
	$X_0(T)_{min}$	35.12	14.83	10.57	5.74	2.34
	$X(T)_{max}$	67.93	156.09	271.35	715.71	1499.43
	$E(X_0(T))$	50.50	55.88	64.20	94.83	135.91

Table 8 : Performance of value preserving and log-optimal strategy

To compare the typical behaviour of both strategies, we compare them pathwise by looking at two simulated paths. In both examples we have chosen $b = 0.1245$ and $\sigma = 0.2195$. Further, we normalised the stock price to start at 1. For visual purposes, we always give pictures comparing the "market return" $1/H(t)$, the actual and the expected stock price (in Figures 20 and 23), the value preserving and the log-optimal total consumption, $C(t)$ and $C_0(t)$, (in Figures 21 and 24), and the value preserving and the log-optimal wealth process, $X(t) \equiv x$ and $X_0(t)$ (in Figures 22 and 25) .

In the first example (compare figures 20–22), the market performs very well. The stock price and the market return $1/H(t)$ are above the expected value of the stock price after $t \approx 0.3$, and from this time on, the market return is above the stock price, too. Also, from this time on, the value preserving total consumption stays above the log-optimal total consumption until $t \approx 0.7$, and they are approximately equal from 0.7 to the terminal time. In contrast to this, the log-optimal wealth lies below the constant wealth of the value preserving approach up to $t \approx 0.3$. Up to $t \approx 0.7$, it is above x, and then goes down to a lower level. From the value preserving point of view, one can interpret this as that the log-optimal agent consumes too much in the first third of the time period. Then, until 0.7, he saves more money than necessary which results in the higher total consumption of the value preserving one.

In the last part of the period, the log-optimal agent once again "lives too well". This means, he consumes more than his economic situation would allow him.

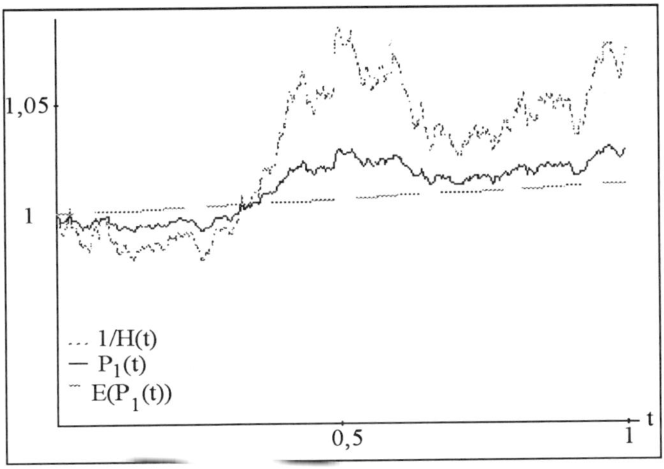

Figure 20 : Stock price, market return, expected stock price

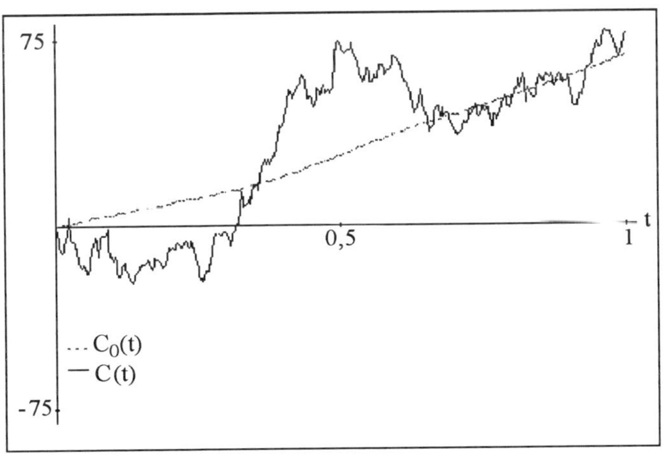

Figure 21 : Total consumption

In the second example (see Figures 23 – 25), the market performance is very poor. The stock price and the market return have decreasing tendency (more or less) throughout the whole period. In particular, the market return is below the stock price.

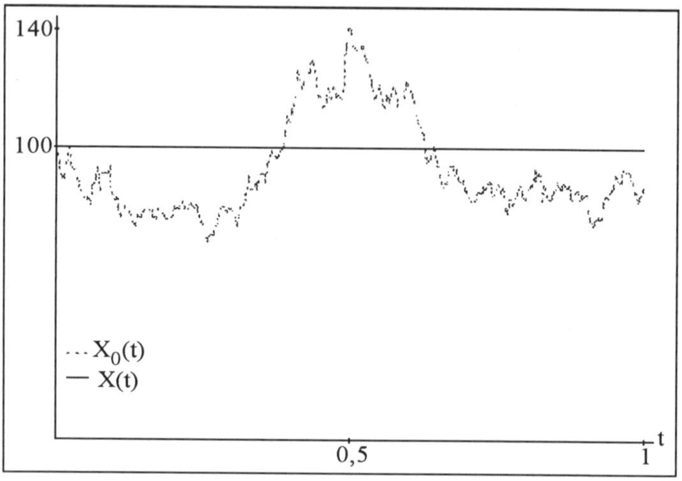

Figure 22: Wealth process

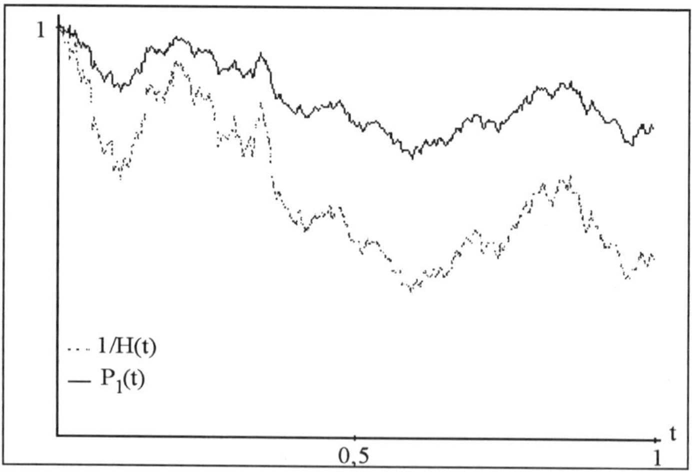

Figure 23 : Stock price, market return

However, the total consumption corresponding to the log-optimal strategy is strictly increasing, but only slowly. On the other hand, the total consumption in the value preserving case is always negative (with a minimum of around –95). Thus, staying at the same economic level in such a bad situation requires additional funds which is expressed in negative consumption rates in our framework. The value preserving agent has to "work" for holding his portfolio value constant.

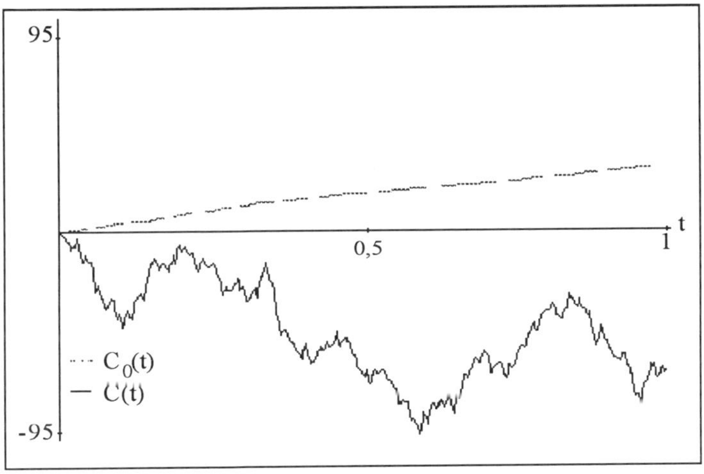

Figure 24 : Total consumption

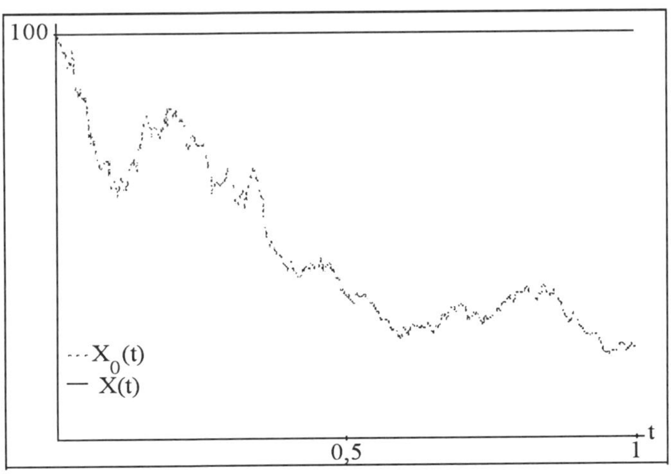

Figure 25 : Wealth process

In contrast to that, the log-optimal agent consumes parts of his future economic power by following a strictly increasing consumption process which is also reflected by the decreasing tendency of his wealth process. Thus, we see in both examples that there is a certain symmetry between the two strategies. On one hand, we have the strictly increasing consumption of the log-optimal strategy (that behaves relatively smooth) and the high variety in the value preserving case. On the other hand we can compare the high variety in the log-optimal wealth process against the constant wealth of the value preserving agent. Also, we see that the value preserving strategy delivers a positive total consumption when the market performs well. If the market performance is very poor, one cannot expect a positive consumption if the goal is value preservation. In such a case, the value preserving investor is required to place additional funds into the portfolio to hold his position constant in terms of the portfolio value.

Appendix A
Normal Distribution, Conditional Expectation, Stochastic Processes

A.1 Normally Distributed Random Variables

As the normal distribution (also called Gaussian distribution) will play a very prominent role throughout the book, it is appropriate to recall its form and some basic characteristics of normal (or Gaussian) random variables.

Definition A1

a) A real-valued random variable X is said to be **normally distributed** with mean μ and variance $\sigma^2 > 0$ (we will write $X \sim N(\mu, \sigma^2)$ for brevity) if it posseses a probability density function $\varphi_{\mu,\sigma^2}(x)$ of the form

$$\varphi_{\mu,\sigma^2}(x) = \frac{1}{\sqrt{2\pi\sigma^2}} e^{-\frac{(x-\mu)^2}{2\sigma^2}}, \quad x \in \mathbf{R}.$$

Such an X a **normal random variable**. It is called standard normally distributed if we have $(\mu, \sigma^2) = (0, 1)$. In this case, we will denote its density by $\varphi(x)$ and its distribution function by $\Phi(x)$.

b) An $\mathbf{R}^n$-valued random variable $X = (X_1,\ldots, X_n)'$ is said to have a **multivariate normal distribution** if for every vector $a \in \mathbf{R}^n \setminus \{0\}$ the real-valued random variable $Y = a'X$ is normally distributed. We will write $X \sim N(\mu, \Sigma)$ if $\mu \in \mathbf{R}^n$ denotes the vector of (component wise) expectations and $\Sigma \in \mathbf{R}^{n,n}$ the covariance matrix of X, i.e.

$$\mu_i = E(X_i), \; \sigma_{ij} = Cov(X_i, X_j), \; i, j = 1,\ldots, n$$

where we have set $\Sigma = (\sigma_{ij})_{1 \le i, j \le n}$.

Remark A2 "Some properties of normal random variables"

a) If X and Y are independent random variables with $X \sim N(\mu_1, \sigma_1^2)$, $Y \sim N(\mu_2, \sigma_2^2)$ then we have $X+Y \sim N(\mu_1+\mu_2, \sigma_1^2+\sigma_2^2)$, i.e. the sum of two independent normal random variables is again a normal random variable.

b) Let X be a multivariate normal, $\mathbf{R}^n$-valued random variable. Then we have:

i) All components X_i are normally distributed, more precisely, $X_i \sim N(\mu_i, \sigma_{ii})$.

ii) $X_1,\ldots, X_n$ are independent. $\Leftrightarrow$ The covariance matrix of X has diagonal form.

(i.e. for multivariate normal variables, independence of the components is equivalent to the components being uncorrelated).

c) A very useful inequality relating the probability density and the distribution function of the standard normal is the following (for a proof see p. 105 of (Gänssler and Stute 1977)):

$$(x^{-1} - x^{-3})\varphi(x) < \Phi(-x) < x^{-1}\varphi(x) \quad \forall\, x > 0.$$

Definition A3

A positive random variable Z is said to have a **lognormal distribution** with parameters (μ, σ^2) if its logarithm is normally distributed, more precisely if we have

$$\ln(Z) \sim \mathcal{N}(\mu + \tfrac{1}{2}\sigma^2, \sigma^2)\,.$$

A.2 Conditional Expectation

The probabilistic tool for modelling the expected future price of a security, given some prior information on that security (e.g. past prices), is the conditional expectation. To define it, let (Ω, F, P) be a complete probability space, X a real-valued random variable defined on Ω.

Definition A4

Let $E(|X|) < \infty$ and H be a sub-σ-algebra of F. Then the **conditional expectation of X given *H*** is the (a.s. unique) real-valued random variable $E(X \mid H)$ on Ω satisfying

i) $E(X \mid H)$ is H-measurable, ii) $\int_A E(X \mid H)dP = \int_A XdP \quad \forall\, A \in H$.

Remark A5

a) As already indicated in the definition, a random variable, satisfying the two conditions above, is only almost surely unique. Thus, we should better talk of the conditional expectation as the class of almost surely equal random variables satisfying these conditions. However, as is usually done, we will identify all the members of this class and talk of "the conditional expectation".

b) If X is an $\mathbf{R}^n$-valued random variable then the definition of the conditional expectation can easily be extended to that case if the integrals in part ii) have the appropriate multidimensional meanings.

d) If Y is another real-valued random variable then we will write $E(X \mid Y)$ and call this the conditional expectation of X **given Y** which is defined as

$$E(X \mid Y) = E(X \mid \sigma(Y))$$

where $\sigma(Y)$ is the σ-algebra generated by Y, i.e. $\sigma(Y) = \sigma(Y^{-1}(A) \mid A \in B)$ where B is the Borel-σ-algebra on $\mathbf{R}$.

We will give a simple example where we can explicitly compute the conditional expectation $E(X \mid H)$. Although this method of computation has no meaning in the general case, it is instructive to keep this example in mind.

Example A6
If the σ-algebra H in Definition A4 is generated by a finite partition $H_1, \ldots, H_k$ of Ω with $P(H_i) > 0$ for all $i = 1, \ldots, k$ then we have

$$E(X \mid H) = \sum_{i=1}^{k} \frac{1}{P(H_i)} E(X\, 1_{H_i}) 1_{H_i} \quad \text{a.s.,} \tag{1}$$

(where 1_A denotes the indicator of the set A) i.e. on the sets H_i the conditional expectation $E(X \mid H)$ is (a.s.) equal to the mean of X over H_i.

We will not go deeper into the theory of conditional expectation. For our purposes, it is enough to collect some simple properties and rules to compute conditional expectations. Such a collection is given in the following proposition.

Proposition A7
Let X and H be as in Definition A4. Then, we have:
a) If X is H-measurable then we have $E(X \mid H) = X$ a.s..
b) For the choice of $H = \{\emptyset, \Omega\}$, we have $E(X \mid H) = E(X)$ a.s..
c) The conditional expectation is linear, i.e. for constants a, b and a random variable Y on Ω with $E(|Y|) < \infty$ we have $E(aX+bY \mid H) = a\,E(X \mid H) + b\,E(Y \mid H)$.
d) $E(E(X \mid H)) = E(X)$ a.s., i.e. the expectation of the conditional expectation is the unconditional one.
e) If X is independent of H then we have $E(X \mid H) = E(X)$ a.s..
f) For $X \geq Y$ a.s., we have $E(X \mid H) \geq E(Y \mid H)Y$ a.s..
g) If Z is bounded and H-measurable then we have $E(ZX \mid H) = Z\,E(X \mid H)$ a.s..
h) $E(E(X \mid H) \mid G) = E(E(X \mid G) \mid H) = E(X \mid G)$ a.s. for all sub-σ-algebras G of H.

A.3 Stochastic Processes, Local Martingales and Martingales

We will first recall the notion of a stochastic process (most of the following material can be found in (Karatzas and Shreve 1988) or in (Lamberton and Lapeyre

1996). As a global assumption valid for the whole book, we will always assume that all random variables are defined on a complete probability space (Ω, F, P).

Definition A8
A (d-dimensional) **stochastic process** X is a collection of ($\mathbf{R}^d$-valued) random variables $\{X_t, t\in I\}$ where I is an interval contained in the non-negative real numbers. For fixed $\omega\in\Omega$, the set $\{X_t(\omega), t\in I\}$ is called a **path** (or a **realisation**) of X.

We will sometimes write X(t) instead of X_t when it is convenient and mostly suppress the dependence on ω of the random variables. We could of course give a more general definition of a stochastic process than the one above, but this one contains all relevant cases for our applications. The usual interpretation of a stochastic process is that it describes the evolution of random phenomena over time. Thus, the index set I is a synonym for time. In the following, we will for notational convenience often state definitions and results only for the case of $I = [0, \infty)$. However, it should always be clear how to modify them for an interval $I = [a, b]$ with non-negative real numbers a, b.

To describe the flow of information over time that is given by observing stochastic processes or other random events, we introduce the notion of a filtration.

Definition A9
a) A **filtration** $\{F_t\}_{t\in[0,\infty)}$ is a non-decreasing sequence of sub-σ-fields of F. The quadruple $(\Omega, F, P, \{F_t\}_{t\in[0,\infty)})$ will be called a **filtered probability space**.
b) A filtration $\{F_t\}_{t\in[0,\infty)}$ is called **right-continuous** if it satisfies

$$F_t = F_{t+} := \bigcap_{\varepsilon>0} F_{t+\varepsilon} \quad \forall\, t \geq 0 .$$

It is called **left-continuous** if it satisfies

$$F_t = F_{t-} := \sigma\{\bigcup_{s<t} F_s\} \quad \forall\, t > 0 .$$

It is called **continuous** if it is both left- and right-continuous.
c) We say that a filtration $\{F_t\}_{t\in[0,\infty)}$ satisfies the **usual conditions** if it is right-continuous and if F_0 contains all P-zero sets.
d) A stochastic process $\{X_t\}_{t\in[0,\infty)}$ is called **measurable** if for all $A\in B(\mathbf{R}^d)$ (where $B(\mathbf{R}^d)$ denotes the Borel-σ-algebra on $\mathbf{R}^d$) we have

$$\{(t, \omega) \mid X_t(\omega)\in A\} \in B([0, \infty))\otimes F.$$

e) A measurable stochastic process $\{X_t\}_{t\in[0,\infty)}$ is called **adapted** to the filtration $\{F_t\}_{t\in[0,\infty)}$ (for short: F_t-adapted) if X_t is F_t-measurable for all $t\in[0,\infty)$.

f) Two stochastic processes X and Y are called **indistinguishable** if almost all their paths coincide, i.e. if we have

$$P(X_t = Y_t \ \forall\ t\in[0, \infty)) = 1.$$

Remark A10 (Important !)
a) By using the notation $\{(X_t, F_t);\ t\in[0, \infty)\}$ for a stochastic process, we always mean that it is F_t-adapted.
b) The good reason for assuming that F_0 contains all P-zero sets (as in part c) of Definition A9) is that if we have Z = V a.s. and Z is F_t-measurable for some $t\in[0,\infty)$ then V is also F_t-measurable. This fact is sometimes used implicitly in the book.
c) If a stochastic process is measurable then its paths are measurable functions in t. It is usually not necessary to require that an F_t-adapted should also be measurable as we have done above. This means that as a **general assumption** throughout this book, we assume our processes to be measurable (without explicitly requiring it). Examples of measurable processes are adapted processes having solely left-continuous paths (or only right-continuous paths).
d) If we talk of a "unique" stochastic process in this book then we usually mean uniqueness up to indistinguishability.

The obvious filtration to think of in connection with the stochastic process X is the filtration $\{F_t^X\}_{t\in[0,\infty)}$ that is generated by the process itself, i.e.

$$F_t^X := \sigma(X_s\ ;\ s \in[0, t)).$$

However, this filtration can be defective in the following sense: Even if a process X is right-continuous (i.e. the paths $X_.(\omega)$ as a function of time t for fixed $\omega\in\Omega$ are right-continuous), the corresponding filtration $\{F_t^X\}_{t\in[0,\infty)}$ can fail to be so (see (Karatzas and Shreve 1988), page 89). In contrast, if a process X is left-continuous then so is $\{F_t^X\}_{t\in[0,\infty)}$. Right-continuity of the underlying filtration is a fundamental ingredient in the theory of stochastic integrals and stochastic differential equations. As is shown in (Karatzas and Shreve 1988), there is a possibility of obtaining a right-continuous filtration (for a certain class of stochastic processes) by augmenting $\{F_t^X\}_{t\in[0,\infty)}$, i.e. by considering its P-augmentation

$$F_t := F_t^P := \sigma(\{F_t^X \cup \mathcal{N}\})$$

where $\mathcal{N}$ are the P-zero sets.

The most important example of a stochastic process in our context of continuous-time finance is, of course, the Brownian motion (or Wiener process) which is the subject of the following example.

Example A11 "Brownian motion"
a) The continuous, real-valued process $\{W_t\}_{t\in[0,\infty)}$ with $W_0 = 0$ a.s. and increments $W_t - W_s$ which are independent of F_s^W and satisfy $W_t - W_s \sim \mathcal{N}(0, t-s)$ for $0 \le s < t$ is called a (one-dimensional, standard) **Brownian motion**.
It is obvious from this definition that $\{W_t\}_{t\in[0,\infty)}$ is also a Brownian motion with respect to the P-augmentation $\{F_t\}_{t\in[0,\infty)}$ of $\{F_t^W\}_{t\in[0,\infty)}$. Since this filtration satisfies the usual conditions (see (Karatzas and Shreve 1988), page 91), we will call it the "natural filtration" of Brownian motion. If not otherwise stated, we will always use this filtration in connection with Brownian motion and call it the **Brownian filtration**.
b) The n-dimensional process $W(t) = (W_1(t),\ldots, W_n(t))'$ where the n components $W_i(t)$ are independent one-dimensional, standard Brownian motions is called a (standard) **n-dimensional Brownian motion**.
c) Among the most remarkable properties of the paths of Brownian motion are the nowhere differentiability of almost all paths and the fact that almost all paths have infinite variation on every finite time interval.

The following class of processes is the fundamental building block of the theory of stochastic calculus and its applications to finance.

Definition A12
A real-valued stochastic process $\{(X_t, F_t);\ t\in[0,\infty)\}$ with $E(|X_t|) < \infty$ for all non-negative t is called a **martingale** if it satisfies

$$E\,(X_t \mid F_s) = X_s \quad \text{a.s.} \quad \forall\ 0 \le s < t\,. \tag{2}$$

It is called a **submartingale** if we have "$\geq$" and a **supermartingale** if we have "$\leq$" in the characterising relation (2).

Remark A13
A simple consequence of relation (2) is that we have

$$E(X_t) \overset{\geq}{\underset{\leq}{=}} E(X_s) \ \ \forall\, 0 \le s < t \ \text{ if X is a } \begin{cases} \text{submartingale} \\ \text{martingale} \\ \text{supermartingale} \end{cases}. \tag{3}$$

In discrete time (i.e. when the time index set is $\mathbf{N}_0$ rather than $[0, \infty)$), martingales are often used to model fair games. If one thinks of X_n as the sequence of the amount of money a gambler owns after taking part in n rounds of a game, his expected wealth after playing round n+1 (given all the results of previous games) should equal X_n if the game is fair, i.e. we should have

$$E(X_{n+1} \mid F_n^X) = X_n \text{ a.s.}$$

which means that X_n should be a martingale. The interpretation of a martingale as a model for (the outcomes of) a fair game is often very useful in interpreting the meaning of security prices or wealth processes that are martingales. We will also give some continuous-time examples for (sub-,super-) martingales.

Example A14

a) A one-dimensional Brownian motion $\{(W_t, F_t); t\in[0,\infty)\}$is a martingale. To see this, note

$$\begin{aligned} E(W_t \mid F_s) &= E(W_s + W_t - W_s \mid F_s) = W_s + E(W_t - W_s \mid F_s) \\ &= W_s + E(W_t - W_s) = W_s \quad \text{a.s. } \forall\ 0 \le s < t\,. \end{aligned}$$

For the second but last equation we have used the independence of the increments of F_s. As this and the fact that the increments $W_t - W_s$ have zero mean are the only special properties of Brownian motion that were actually used to derive the martingale property, we can conclude that all processes X with independent, zero-mean increments are martingales.

b) The process $\{(X_t, F_t); t\in[0,\infty)\}$ with $X_t = W_t + \mu t$ where W is a one-dimensional Brownian motion is called a **Brownian motion with drift** μ. By using part a), it is obvious that X is a submartingale if μ is positive and a supermartingale if μ is negative.

A direct consequence of Jensen's inequality is the following useful proposition.

Proposition A15

Let $\{(X_t, F_t); t\in[0,\infty)\}$ be a martingale (a submartingale) and $g: \mathbf{R}\to\mathbf{R}$ a convex (a convex and non-decreasing) function with $E(|g(X_t)|) < \infty$ for all $t\in[0,\infty)$.
Then $\{(g(X_t), F_t); t\in[0,\infty)\}$ is a submartingale.

Important special cases of this proposition are given by the choices $g(x) = x^2$ or $g(x) = |x|$. The proposition can also be generalised to the case of a d-dimensional vector of martingales and convex functions g: $\mathbf{R}^d \to \mathbf{R}^d$.

In some applications in finance, we need a somewhat weaker, localised version of the term martingale, so-called local martingales which will be defined below.

Definition A16

a) A non-negative random variable τ is called a **stopping time** with respect to a filtration $\{F_t\}_{t\in[0,\infty)}$ if for every $t \geq 0$ we have $\{\tau \leq t\} \in F_t$.

b) A stochastic process $\{(X_t, F_t);\ t\in[0,\infty)\}$ with $X_0 = 0$ a.s. and $\{F_t\}_{t\in[0,\infty)}$ satisfying the usual conditions is called a **local martingale** if there exists a non-decreasing sequence $\{T_n\}_{n\in\mathbf{N}}$ of stopping times with $T_n \to \infty$ a.s. as $n \to \infty$ such that the stopped processes $\{(X_t^{(n)}, F_t);\ t \in [0,\infty)\}$ given by

$$X_t^{(n)} := X_{t\wedge T_n}$$

(where $t\wedge T_n := \min\{t, T_n\}$) are martingales for every $n\in\mathbf{N}$. $\mathcal{M}^{loc}$ (resp. $\mathcal{M}^{c,loc}$) will denote the sets of local martingales (resp. continuous local martingales).

Hence, there are always embedded martingales in a local martingale. This fact will prove to be very useful in dealing with stochastic integrals. For example, by choosing $T_n = \inf\{t\geq 0 \mid X_t \geq n\}$ for a continuous process X(t), one can force the existence of moments of every order for the embedded martingales. This localisation argument will be used in a number of proofs throughout the book (as long as X_t is a.s. finite for every t, we have $T_n \to \infty$ a.s. as $n \to \infty$). Of course, every martingale (starting in 0) is a local martingale. On the other hand, a local martingale need not be a martingale (see Exercises 3.36, 3.37 in (Karatzas and Shreve 1988) for some counterexamples). The following proposition gives two simple but often useful cases where a local martingale is also a supermartingale respectively a martingale. It can easily be proved with the help of Fatou's lemma (part a)) and the bounded convergence theorem (part b)).

Proposition A17

a) Every non-negative local martingale is a supermartingale.

b) A bounded local martingale is a martingale.

We will close this section by formulating the optional sampling theorem which in particular states that the martingale property of a sufficiently regular martingale is preserved if we randomly stop the process.

Theorem A 18 "Optional sampling theorem"
Let $\{(X_t, F_t); t\in[0,\infty]\}$ be a right-continuous submartingale with a last element X_∞. Let S, T be two stopping times with $S \leq T$. Then we have

$$E(X_T \mid F_S) \geq X_S \text{ a.s.,}$$

and we have equality in this relation if moreover X is a martingale with a last element X_∞.

Appendix B
Introduction to Stochastic Integrals and the Itô-Calculus

B.1 Construction of Itô-Integrals

As stochastic calculus is the mathematical basis underlying the whole theory of continuous-time finance, we will give a short outline of the construction of stochastic integrals and the corresponding calculus. We will not give any proof but instead refer to standard textbooks on stochastic calculus such as (Karatzas and Shreve 1988), (Oksendal 1992) or (Rogers and Williams 1987) for a more detailed and rigorous introduction to the field. Further, we will only look at stochastic integrals with respect to Brownian motion (so called **Itô-integrals**) which saves us some technicalities.

Let us consider a complete probability space (Ω, F, P) on which there is defined a one-dimensional standard Brownian motion W(t). As always, let $\{F_t\}_{t\in[0,\infty)}$ be the Brownian filtration. Our goal will be to define the **Itô-integral**

$$\int X\,dW .$$

The first problem in doing this is that two naive approaches do not work. As the path of W(t) are nowhere differentiable (see Remark A11 c)), it is not possible to define this integral pathwise as

$$\int_0^T X(t,\omega)\frac{dW(t,\omega)}{dt}\,dt .$$

Moreover, we cannot even define it pathwise (for almost every $\omega\in\Omega$) as a Stieltjes integral (see (Protter 1992)). The main reason for that is that the Brownian paths are simply "too long". More precisely, for $\omega\in\Omega$ as a function of time t, the path $W(t,\omega)$ has infinite variation on every compact interval [t, s] with $t < s$ (see Remark A11 c)). But finite variation of the integrator is a necessary condition to define a

Stieltjes integral. It is therefore necessary to think of another approach to introduce the stochastic integral with respect to Brownian motion. We will follow the classical L^2-approach of Itô as given in the above mentioned textbooks. This approach is similar to the standard way of introducing the Lebesgue integral in analysis:

> " Start by defining $\int XdW$ for simple integrands. Then extend the definition to more general ones."

We will do this in several steps.

Step 1: "Itô-integral for simple processes"
If $X(t)$ is a stochastic process that is constant on finite intervals $]t_{i-1}, t_i]$ with $t_0 = 0$, $t_{i-1} < t_i$ and equal to ϕ_i on $]t_{i-1}, t_i]$ then it is natural to define $\int XdW$ by

$$\int_0^t X(s)dW(s) := \sum_{i=1}^{k} \phi_i (W(t_i) - W(t_{i-1})) + \phi_{k+1}(W(t)-W(t_k)) \qquad (1)$$

for $t\in]t_k, t_{k+1}]$. As we want $\int XdW$ to be F_t-adapted, we have to require ϕ_i to be $F_{t_{i-1}}$-measurable. As we further want to ensure the existence of (at least) the expectation of $\int XdW$, we also require ϕ_i to be bounded. More formally :

Definition B1
a) A stochastic process $\{X(t); t\in[0, T]\}$ such that there exist real numbers $0 = t_0 < t_1 < \ldots < t_p = T$ and $F_{t_{i-1}}$-measurable, bounded random variables ϕ_i with

$$X(t,\omega) = \sum_{i=1}^{p} \phi_i(\omega) 1_{]t_{i-1}, t_i]}(t), \quad t\in[0, T]$$

is called a **simple process**.
b) The **Itô-integral** for a simple process $\{X(t); t\in[0, T]\}$ is defined by representation (1) for $t\in[0, T]$.

The following property is a simple consequence of the above definition, but plays the crucial role in the extension of the Itô-integral to more general integrands.

Proposition B2
Let $\{X(t); t\in[0, T]\}$be a simple process.

a) $\{I_t(X), F_t\}_{t\in[0,T]} := \left\{ \int_0^t X(s)dW(s), F_t \right\}_{t\in[0,T]}$ is a continuous martingale.

b) $$E\left(\int_0^t X(s)dW(s)\right)^2 = E\left(\int_0^t X^2(s)dW(s)\right) \tag{2}$$

Remark B3

a) By defining

$$\int_t^T X(s)dW(s) := \int_0^T X(s)dW(s) - \int_0^t X(s)dW(s),$$

we obtain

$$\int_0^T 1_A X(s) 1_{\{t<s\}} dW(s) = 1_A \int_t^T X(s)dW(s) \quad \forall\, A \in F_t\,. \tag{3}$$

b) As it is also easy to see that the stochastic integral for simple processes is linear, part b) of Proposition B2 states that with respect to the norms

$$\|I(X)\|_{L[0,T]} := \sqrt{E\left(\int_0^T X(s)dW(s)\right)^2} = \sqrt{E\left(\int_0^T X^2(s)ds\right)} =: \|X\|_T, \tag{4}$$

the correspondence $X \to I(X)$ is an isometry (i.e. a linear mapping that preserves distances) between the space of simple processes and that of the corresponding Itô-integrals. It is therefore referred to as the **Itô-isometry**. We will use this isometry in the next step to extend the Itô-integral.

Step 2: "Itô-integral for processes in $L^2[0, T]$ "

The idea behind the extension of the Itô-integral for simple processes is to use the Itô-isometry (4) as the starting point. Indeed, we consider exactly these of processes X which are F_t-adapted and have a finite norm $\|X\|_T$. This set of processes will be called $L^2[0, T]$. The construction of $\int XdW$ for $X \in L^2[0, T]$ relies on the following facts:

- $X \in L^2[0, T]$ can be approximated (in a suitable way) by a sequence $\{X_n\}_{n \subset \mathbf{N}}$ of simple processes with $X_n \in L^2[0, T]$.
- For an approximating sequence $\{X_n\}_{n \in \mathbf{N}}$ of X, as above, we can define the Itô- integrals $I(X_n)$.
- The sequence $\{I(X_n)\}_{n \in \mathbf{N}}$ of Itô-integrals converges (in a suitable sense) towards a process I(X) (independent of the approximating sequence for X). We finally define the Itô-integral for X to be equal to this process : $\int XdW := I(X)$.

We also extend the $L[0, T]$-norm to the space M of all continuous martingales $\{M(t), F_t;\, t\in[0, T]\}$ that satisfy $M(0) = 0$ a.s. and

$$\|M\|_{L[0,T]} := \sqrt{E(M(T))^2} < \infty.$$

With this extension, we are ready to state the necessary facts for the construction of the Itô-integral for processes $X\in L^2[0, T]$ in a rigorous way:

Proposition B4

a) For every process $X\in L^2[0, T]$ there exists a sequence $\{X_n\}_{n\in\mathbf{N}}$ of simple processes with $X_n\in L^2[0, T]$ satisfying

$$\lim_{n\to\infty} \|X - X_n\|_T = 0 .$$

b) For any sequence $\{X_n\}_{n\in\mathbf{N}}$ of simple processes as in part a), we have

$$\|I(X_m) - I(X_n)\|_{L[0,T]} < \varepsilon \quad \forall n, m > N(\varepsilon)$$

for all $\varepsilon > 0$ (where $N(\varepsilon)$ is a suitable positive integer), i.e. the corresponding sequence of Itô-integrals is a Cauchy-sequence in the space M.

c) Independent of the approximating sequence $\{X_n\}_{n\in\mathbf{N}}$ for X the corresponding sequence $\{I(X_n)\}_{n\in\mathbf{N}}$ of Itô-integrals converges towards a process $I(X)\in M$ (with respect to the $L[0, T]$-norm), i.e. we have

$$\lim_{n\to\infty} \|I(X) - I(X_n)\|_{L[0,T]} = 0 ,$$

and we define the Itô-integral for $X\in L^2[0, T]$ to be equal to this process, i.e.

$$\int X dW := I(X).$$

It is important to know that this so defined stochastic integral has the same properties as the stochastic integral for simple processes:

Proposition B5 "Properties of the Itô-integral for processes of $L^2[0, T]$ "

For every $X\in L^2[0, T]$ the Itô-integral $I(X)$ is a continuous square integrable martingale with

$$E(I_t(X))^2 = E\left(\int_0^t X(s)^2\, ds\right), \quad t\in[0, T],$$

$$E((I_t(X)-I_s(X))^2 | F_s) = E\left(\int_s^t X(u)^2 \, du \middle| F_s \right) \text{ a.s., } 0 \le s \le t \le T,$$

$$I(\alpha X + \beta Y) = \alpha I(X) + \beta I(Y)$$

for $Y \in L^2[0, T]$, α, $\beta \in \mathbf{R}$. Further, for any two stopping times $\upsilon \le \tau$ a.s. and any number $t \in [0, T]$, we have

$$E((I_{t\wedge\tau}(X) | F_\upsilon) = I_{t\wedge\upsilon}(X) \text{ a.s.,}$$

$$I_{t\wedge\tau}(X) = I_t(\tilde{X})$$

where $\tilde{X}(t) := X(t)1_{\{t \le \tau\}}$.

Step 3: "Itô-integral for processes in $H^2[0, T]$ "
For modelling purposes in finance, it is necessary to extend the stochastic integral to an even wider class of integrands which we will call $H^2[0, T]$. It is given by

$$H^2[0, T] := \{ X \mid X(t) \; F_t\text{-adapted, } \int_0^T X(s)^2 \, ds < \infty \text{ a.s. } \}.$$

As processes in $H^2[0, T]$ do not necessarily have a finite $\| \, . \, \|_T$-norm, we cannot approximate them by simple process in the same way as we did in Step 2 for processes in $L^2[0, T]$. But we can localise processes in $H^2[0, T]$ by using suitable stopping times. More precisely, define the sequence T_n of stopping times by

$$T_n(\omega) := T \wedge \inf\left\{ 0 \le t \le T \middle| \int_0^t X(s,\omega)^2 \, ds \ge n \right\}$$

and define the stopped process $X^{(n)}$ by

$$X^{(n)}(t,\omega) := X(t,\omega)1_{\{T_n(\omega) \ge t\}}.$$

Thus, by construction, these stopped processes $X^{(n)}$ are members of $L^2[0, T]$. Hence, we can define the stochastic integral $I(X^{(n)})$. Note further that due to the finiteness of $\int X^2 ds$ the stopping times T_n as defined above converge to infinity almost surely as n goes to infinity. We define the Itô-integral I(X) of X as

$$I_t(X) := I_t(X^{(n)}) \quad \text{on } \{0 \le t \le T_n\}.$$

Due to the stopping time property of Proposition B5, this is a consistent definition as we have

$$I_t(X^{(n)}) = I_t(X^{(m)}) \quad \text{on } \{0 \le t \le T_n\}.$$

By this way of defining I(X), we directly have that the Itô-integral for $X \in H^2[0, T]$ is a **local** martingale (the corresponding stopping times T_n in the definition of a local martingale are obviously given by the stopping times T_n appearing in the construction of $X^{(n)}$). It is very important to keep in mind that I(X) is in general no martingale. Thus all the properties of the Itô-integral given in Proposition B5 which involve expectations are in general no longer valid. However, linearity of the integral, continuity of its path (with respect to t) and the stopping property remain valid.

Having made the one-dimensional definition, we can now also define the **multi-dimensional Itô-integral** in the following sense:
Let $\{(W(t), F_t);\ t \in [0,\infty)\}$ be a d-dimensional Brownian motion and $X(t) = (X_{ij}(t))$ be an (n, d)-matrix of stochastic processes $X_{ij}(t) \in H^2[0, T]$. We define the d-dimensional Itô-integral of X as the following vector of (sums) of one-dimensional Itô-integrals:

$$\int_0^t X(s)dW(s) := \begin{pmatrix} \sum_{j=1}^{d} \int_0^t X_{1j}(s)dW_j(s) \\ \vdots \\ \sum_{j=1}^{d} \int_0^t X_{nj}(s)dW_j(s) \end{pmatrix}.$$

Note that for this definition to work, it is necessary that the definition of the one-dimensional Itô-integral can be extended to the case where the underlying filtration $\{F_t\}_{t \in [0,\infty)}$ is not necessary the natural Brownian filtration. This is possible as long as all the Brownian motions involved are still Brownian motions (of suitable dimensions) with respect to $\{F_t\}_{t \in [0,\infty)}$ (see (Oksendal 1992) for a short discussion of this problem). Otherwise, an expression such as

$$\int_0^t W_2(s)dW_1(s)$$

which clearly is a special case of the above defined multi-dimensional Itô-integral would not be defined.

At the end of this section, we state a simple but useful result on the distribution of an Itô-integral for deterministic functions.

Proposition B6
Let $X \in H^2[0, T]$ be a deterministic real-valued function. Then the Itô-integral $I_t(X)$ is $\mathcal{N}(0, \int_0^t X(s)^2\, ds)$-distributed.

B.2 The Itô-Formula

The fundamental tool of Itô-calculus is Itô's formula (more precisely its various variants). In this section, we will give a version of it for the special class of Itô-processes which is sufficiently general to cover most of the applications in this book.

Definition B7
Let $\{(W(t), F_t);\ t \in [0,\infty)\}$ be a d-dimensional Brownian motion.
a) $\{(X(t), F_t);\ t \in [0,\infty)\}$ is called a real-valued **Itô-process** if it admits the (unique) representation

$$X(t) = X(0) + \int_0^t K(s)ds + \int_0^t H(s)dW(s) \tag{5}$$

where $X(0)$ is a.s. finite and F_0-measurable. $K(t)$ and $H(t)$ are F_t-adapted real-valued respectively $\mathbf{R}^d$-valued processes with

$$\int_0^t |K(s)|ds < \infty \text{ a.s.,} \quad \int_0^t (H_i(s))^2 ds < \infty \text{ a.s.} \quad \forall\, t \in [0, \infty),\ i = 1,\ldots, d.$$

b) An **n-dimensional Itô-process** $X = (X_1,\ldots, X_n)$ is a vector of real-valued Itô-processes.
c) If X and Y are Itô-processes (with respect to the same filtration) where X has the representation (5) and Y is given by

$$Y(t) = Y(0) + \int_0^t L(s)ds + \int_0^t M(s)dW(s)$$

(with the processes $L(t)$, $M(t)$ satisfying the appropriate conditions of part a)) then the **quadratic covariation process** $\langle X, Y \rangle$ of X and Y is defined as

$$\langle X, Y \rangle(t) := \sum_{i=1}^{d} \int_0^t H_i(s) M_i(s) ds\ .$$

Especially, $\langle X \rangle := \langle X, X \rangle$ is called the **quadratic variation** of X.

Remark B8

a) It can be shown that the representation (5) of an Itô-process is unique. By this we mean that if X(t) would have a second representation of the form (5),

$$X(t) = Y(0) + \int_0^t L(s)ds + \int_0^t M(s)dW(s),$$

then Y(0) would be a.s. equal to X(0) and K and L respectively H and M would be indistinguishable. Therefore, the quadratic variation and covariation as in the preceding definition are well-defined. In particular, if X would be the "zero process", i.e. $X(t) = 0 \quad \forall\ t \in [0, \infty)$ a.s. then X(0), K and H would be indistinguishable from zero (in the appropriate meanings !).

b) A motivation for the name quadratic variation and its relation to the quadratic variation of ordinary calculus can be found in Section 1.5 of (Karatzas and Shreve 1988). By construction, the quadratic covariation is bilinear and symmetric in X, Y.

b) The representation (5) of the Itô-process X will often be stated in differential notation:

$$dX(t) = K(t)dt + H(t)dW(t).$$

c) Now, we can also define the stochastic integral of Y with respect to a real-valued Itô-process X admitting representation (5) as

$$\int_0^t Y(s)dX(s) := \int_0^t Y(s)K(s)ds + \int_0^t Y(s)H(s)dW(s)$$

for all Y with $Y \cdot H \in H^2$. The extension to multi-dimensional Itô-processes is straight forward but requires more complicated notation.

Theorem B9 "Itô's formula in one dimension"

Let X be an Itô-process with representation (5), let f: $\mathbf{R} \to \mathbf{R}$ be a twice continuously differentiable function. Then $f(X_t)$ is again an Itô-process with representation

$$f(X_t) = f(X_0) + \int_0^t f'(X(s))dX(s) + \tfrac{1}{2}\int_0^t f''(X(s))d\langle X\rangle(s)$$

$$= f(X_0) + \int_0^t \left(f'(X(s))K(s) + \tfrac{1}{2} f''(X(s))\|H(s)\|^2\right)ds + \tfrac{1}{2}\int_0^t f'(X(s))H(s)dW(s).$$

Examples B10

a) For X(t) = t, we recover the following well-known formula as a special case of Itô's formula

$$f(t) = f(0) + \int_0^t f'(s)ds,$$

and more general for X(t) = h(t), where h is a C^1-function, we see that the chain rule

$$(f\circ h)(t) = (f\circ h)(0) + \int_0^t f'(h(s))h'(s)ds$$

is also a special case of in Itô's formula.

b) For X(t) = W(t) (where W(t) is a one-dimensional standard Brownian motion) and $f(x) = x^2$, we obtain

$$W(t)^2 = 2\int_0^t W(s)dW(s) + t$$

which demonstrates that the rules of Itô-calculus differ from that of usual calculus. The additional term "t" in this expression (compared to usual calculus) has its origin in the non-vanishing quadratic variation of the integrator W(t).

c) By choosing $X(t) = (b-\frac{1}{2}\sigma^2)t + \sigma W(t)$ and f(x) = exp(x) it can be directly verified using Theorem B11 that Z(t) = exp(X(t)) satisfies the following **stochastic differential equation**

$$dZ(t) = Z(t)\,[\,bdt + \sigma dW(t)],\quad Z(0) = 1\ .$$

We will also state a multi-dimensional version of Itô's formula.

Theorem B11 "Itô's formula in multi dimensions"

Let $X = (X_1,\ldots, X_n)$ be an n-dimensional Itô-process with representations

$$X_i(t) = X_i(0) + \int_0^t K_i(s)ds + \sum_{j=1}^{d}\int_0^t H_{ij}(s)dW_j(s)$$

where $W = (W_1,\ldots, W_d)$ is a d-dimensional Brownian motion. Let f: $[0,\infty)\times\mathbf{R}^n \to \mathbf{R}$ be a $C^{1,2}$- function. Then $f(t, X_t)$ is again an Itô-process with representation

$$\begin{aligned} f(t, X_1(t),\ldots, X_n(t)) = f(0, X_1(0),\ldots, X_n(0)) &+ \int_0^t \frac{\partial f}{\partial t}(s, X_1(s),\ldots,X_n(s))ds \\ &+ \sum_{i=1}^{n}\int_0^t \frac{\partial f}{\partial x_i}(s, X_1(s),\ldots,X_n(s))dX_i(s) \\ &+ \frac{1}{2}\sum_{i,j=1}^{n}\int_0^t \frac{\partial^2 f}{\partial x_i \partial x_j}(s, X_1(s),\ldots,X_n(s))d\langle X_i, X_j\rangle(s)\ . \end{aligned}$$

By choosing f(x, y) = xy in Theorem B11 we get the useful product rule for Itô-integrals:

Corollary B12 "Product rule"
Let X, Y be two Itô-processes (as in part c) of Definition B7). We then have :

$$X(t)Y(t) = X(0)Y(0) + \int_0^t X(s)dY(s) + \int_0^t Y(s)dX(s) + \langle X, Y\rangle(t).$$

Some further generalisations of Itô's formula will be given in section B.7.

B.3 Some Basics on Stochastic Differential Equations

In Example B10 c), we saw a first example of a **stochastic differential equation** of the form

$$dX(t) = b(t, X(t))\, dt + \sigma(t, X(t))\, dW(t)\,. \tag{6}$$

This form is in deed only a symbolic representation of the corresponding relation between stochastic integrals

$$X(t) = X(0) + \int_0^t b(s, X(s))ds + \int_0^t \sigma(s, X(s))dW(s) \tag{7}$$

where $\{W(t), F_t\}_{t\geq 0}$ is an m-dimensional Brownian motion and b: $[0,\infty)\times\mathbf{R}^n \to \mathbf{R}^n$, σ: $[0,\infty)\times\mathbf{R}^n \to \mathbf{R}^{n,m}$ are deterministic functions. For convenience, we always state a stochastic differential equation (for brevity: sde) in the notation (6), but keep its real meaning (7) in mind.

We will give conditions for the existence of an F_t-adapted process X(t) satisfying this sde (see e.g. (Karatzas and Shreve 1988)). Such a solution is called a **strong solution**. As for our purposes it is enough to consider strong solutions, we will not introduce the concept of a **weak solution** of an sde (excellent references for the treatment of this topic can be found in e.g. (Karatzas and Shreve 1988), (Revuz and Yor 1991) or (Rogers and Williams 1987)).

Theorem B13
Let T > 0 and b: $[0, T]\times\mathbf{R}^n \to \mathbf{R}^n$, σ: $[0, T]\times\mathbf{R}^n \to \mathbf{R}^{n,m}$ be measurable functions satisfying

$$\|b(t,x)\| + \|\sigma(t,x)\| \leq C(1 + \|x\|) \tag{8}$$

$$\|b(t,x)-b(t,y)\| + \|\sigma(t,x)-\sigma(t,y)\| \le D\|x-y\| \qquad (9)$$

for x, $y \in \mathbf{R}^n$ and some constants C, D (where $\|.\|$ are the appropriate Euclidean norms, i.e.

$$\|x\|^2 = \sum_{i=1}^{n} x_i^2, \quad \|\sigma\|^2 = \sum_{i,j=1}^{n} \sigma_{ij}^2 \text{).}$$

Then, for every $x \in \mathbf{R}^n$ the stochastic differential equation

$$dX(t) = b(t, X(t))\, dt + \sigma(t, X(t))\, dW(t), \quad X(0) = x$$

has a unique solution X(t) which is continuous in t and satisfies the relation

$$E\left(\sup_{0 \le s \le T} \|X_s\|^2\right) < \infty .$$

Remark B14

a) It is also possible to generalise the foregoing theorem by having a random initial condition X(0) = Z. All the claims of the theorem remain valid if we assume that Z is independent of the Brownian filtration and satisfies $E(\|Z\|^2) < \infty$.

b) In the autonomous case (i.e. if the coefficient functions b, σ only depend on X(t) but not on time), condition (8) can be dropped in the theorem above.

The following theorem resembles the well-known variations of constants formula for linear ordinary differential equations.

Theorem B15 "Variation of constants":

Let $\{W(t), F_t\}_{t \ge 0}$ be an m-dimensional Brownian. Let A, a, S_j, σ_j, j = 1,..., m, be F_t-adapted, real valued processes satisfying

$$\int_0^t |A(s)| ds < \infty, \quad \int_0^t |a(s)| ds < \infty, \text{ a.s.}$$

$$\int_0^t S_j(s)^2\, ds < \infty, \quad \int_0^t \sigma_j(s)^2\, ds < \infty, \text{ a.s.}, \quad j = 1,\dots, m .$$

Then there exists a unique solution to the (general) linear sde

$$dX(t) = [A(t)X(t) + a(t)]\, dt + \sum_{j=1}^{m} \left[S_j(t)X(t) + \sigma_j(t)\right] dW_j(t), \quad X(0) = x$$

which is given by

$$X(t) = Z(t)\,[\,x + \int_0^t \frac{1}{Z(u)}(a(u) - \sum_{j=1}^{m} S_j(u)\sigma_j(u))du + \sum_{j=1}^{m}\int_0^t \frac{\sigma_j(u)}{Z(u)} dW_j(u) \quad (10)$$

with Z(t) defined as

$$Z(t) = \exp(\int_0^t (A(u) - \tfrac{1}{2}\sum_{i=1}^{m} S_j(u)^2)du + \sum_{j=1}^{m} S_j(u)dW_j(u)). \qquad (11)$$

Remark B16

a) Setting m =1, $a(t) = \sigma_j(t) = 0$ and choosing A(t), $S_j(t)$ equal to constants, we rediscover Example B10 c) as special case of this theorem. Remark B14 a) on random initial conditions stays valid for Theorem B15, too. It is important to note that Theorem B15 is not a special case of Theorem B13, because the coefficients A, a, S, σ in the linear sde are not necessarily bounded, random processes.

b) Note that Z(t) is the unique solution to the homogenous equation

$$dZ(t) = Z(t)\,[A(t)dt + \sum_{j=1}^{m} S_j(t)dW_j(t)],\ Z(0) = 1.$$

Therefore, representation (10) of X(t) is in total analogy with the deterministic variation of constants formula.

c) It is possible to consider a more general form of an sde than the one in equation (6). One could replace the coefficient functions b(t, X(t)), σ(t, X(t)) by functionals depending depending on the values of the path of X(s), $s \leq t$. As we will not need such a generalisation, we refer to (Revuz and Yor 1991) or (Rogers and Williams 1987).

An important feature of a (strong) solution X(t) to the sde (6) is that it has the Markov property, i.e. for any bounded Borel function f and $t \leq s$, we have

$$E(f(X(s)) \mid F_t) = E(f(X(s)) \mid X(t)).$$

This relation has the interpretation that the future development of the X(s) process only depends on its current level X(t) and not on the whole past information, given by F_t. X(s) will be called a **Markov process**. For simplicity, we will formulate the corresponding theorem only in dimension one.

Theorem B17 "Markov property of solutions of an sde"

Let n = 1. A solution X(s) to the sde (9) is a Markov process. Further, for any bounded Borel function f, we have

$$E(f(X(s)) \mid F_t) = g(X(t))$$

with $g(x) = E(f(X_s^{t,x}))$ where by $X_s^{t,x}$ we denote the solution of the sde (9) starting at time s with a value of x, i.e. it satisfies

$$X_s^{t,x} = x + \int_t^s b(u, X_u^{t,x})du + \int_t^s \sigma(u, X_u^{t,x})dW(u).$$

Remark B18
In the spirit of this theorem, we will often use the notation $E^{t,x}(f(X(s))$ which denotes the expectation of the process $X_s^{t,x}$. The Markov property (and this notation) plays an essential role in the next section and in Part C, the part on controlled stochastic differential equations.

B.4 Partial Differential Equations and Stochastic Differential Equations — The Feynman–Kac Representation

It is definitely beyond the scope of this book to give an introduction to pde theory. We will only state two results relating solutions of an sde with that of a pde. To state these results, assume that the continuous functions f: $\mathbf{R}^n \to \mathbf{R}$, g: $[0, T]\times\mathbf{R}^n \to \mathbf{R}$ and k: $[0, T]\times\mathbf{R}^n \to [0, \infty)$ satisfy the conditions

$$f(x) \geq 0 \;\forall\; x\in\mathbf{R}^n \;\;\textbf{or}\;\; |f(x)| \leq L\left(1+\|x\|^{2\lambda}\right) \;\forall\; x\in\mathbf{R}^n, \tag{12}$$

$$g(t, x) \geq 0 \;\forall\; 0 \leq t \leq T, x\in\mathbf{R}^n \;\;\textbf{or}\;\; |g(t, x)| \leq L\left(1+\|x\|^{2\lambda}\right) \;\forall\; 0 \leq t \leq T, x\in\mathbf{R}^n \tag{13}$$

for suitable constants $L > 0$, $\lambda \geq 1$. The problem to find a function v, belonging to class $C^{1,2}$ on $[0, T)\times\mathbf{R}^n$, continuous on $[0, T]\times\mathbf{R}^n$, satisfying

$$-v_t + kv = g + \tfrac{1}{2}\sum_{i=1}^{n}\sum_{j=1}^{n} a_{ik}(t,x)v_{ik} + \sum_{i=1}^{n} b_i(t,x)v_i, \;\; (t, x)\in[0, T)\times\mathbf{R}^n$$

$$v(T, x) = f(x), \;\; x\in\mathbf{R}^n,$$

where the matrix a(t, x) is given by $a(t, x) = \sigma(t, x)\sigma(t, x)'$, is called the **Cauchy problem** for v. The relation between it and the sde (9) is the subject of the well-known Feynman–Kac representation Theorem.

Theorem B19 "Feynman–Kac Representation "
Let the functions b, σ satisfy the assumptions of Theorem B13 and assume further that the matrix a(t, x) is positive definite uniformly in (t, x). If there exists a solution v to the Cauchy problem (with the above smoothness properties) that satisfies the growth condition

$$\max_{0\le t\le T} | v(t, x) | \le M\left(1+\|x\|^{2\mu}\right) \quad \forall\, x\in\mathbf{R}^n \tag{14}$$

for some $M > 0$, $\mu \ge 1$. Then v(t, x) is given by the stochastic representation

$$v(t, x) = E^{t,x}\left(f(X(T))\exp\left(-\int_t^T k(u, X(u))du\right) + \int_t^T g(s, X(s))\exp\left(-\int_t^s k(u, X(u))du\right)ds\right).$$

where X(s) is a solution to the sde (9). In particular implies that such a solution is unique.

Hence, if we know (by whatever results !) that there exists a solution to the Cauchy problem, satisfying condition (14) then we can obtain it by calculating the expected value of an appropriate functional of the unique solution of sde (9). The following theorem (which is given as Lemma 7.1 in (Karatzas, Lehoczky and Shreve 1987)), although somewhat specialised, includes an existence part for the solution of the Cauchy problem. For its formulation, consider let X(t) be the unique solution of the sde

$$dX(t) = -X(t)\,[\,(\mu - r)dt - \theta\,'dW(t)] \tag{15}$$

with μ, $r > 0$, $\theta\in\mathbf{R}^n$.

Theorem B20
Let the real-valued function q(t, y) be continuous on $[0, T]\times(0, \infty)$, as well as Hölder continuous in y uniformly with respect to (t, y) on compact subsets of its domain. Further, let f: $(0, \infty)\to\mathbf{R}$ be continuous. Both q and f are assumed to satisfy the growth condition

$$\max_{0\le t\le T} | u(t, y) | \le K\left(1+y^{\alpha}+y^{-\alpha}\right), \quad 0<y<\infty \tag{16}$$

for some $K > 0$, $\alpha > 0$. Then the function

$$v(t, y) = E^{t,y}\left(e^{-\mu(T-t)} f(X(T)) + \int_t^T e^{-\mu(s-t)} q(s, X(s))ds \right)$$

(where X(s) satisfies the sde (15)) is the unique solution of the Cauchy problem

$$v_t + \tfrac{1}{2}\|\theta\|^2 y^2 v_{yy} + (\mu - r)y v_y - \mu v + q(t, y) = 0\,, \quad 0 \le t < T, 0 < y < \infty,$$
$$v(T, y) = f(y)\,,\ 0 < y < \infty$$

with the required smoothness properties which also satisfies a growth condition of the type of condition (16).

B.5 Itô's Martingale Representation Theorem

The theoretical result that makes the martingale method of portfolio optimisation (see Section 3.4) work is the martingale representation theorem for Brownian martingales. It states that every martingale with respect to a Brownian filtration is in fact simply an Itô-integral (with respect to the corresponding Brownian motion) of a suitable integrand.

Definition B21
Let $\{W(t), F_t\}_{t\ge 0}$ be an m-dimensional Brownian where $\{F_t\}_{t\ge 0}$ is the natural filtration. Then a (local) martingale $\{M(t), F_t\}_{t\ge 0}$ is called a (local) **Brownian martingale**.

Theorem B22 "Martingale representation theorem"
Let $\{M(t), F_t\}_{t\in[0,T]}$ be a square-integrable Brownian martingale (i.e. it satisfies $E(M(t)^2) < \infty\ \forall\ t\in[0, T]$). Then there exists a unique F_t-adapted, $\mathbf{R}^m$-valued process $\psi(t), t\in[0, T]$, with

$$E\left(\int_0^T \|\psi(t)\|^2 dt \right) < \infty,$$

$$M(t) = M(0) + \int_0^t \psi(s)' dW(s)\ .$$

Especially, M(t) is a continuous martingale. If M(t) is only a local Brownian martingale then the above assertion still holds true, but $\psi(t)$ is only a member of $H^2[0, T]$.

B.6 Change of Measure and The Girsanov Theorem

A fundamental result for option pricing (and also for the Sections 2.4, 3.5, 6.4) is Girsanov's theorem. It tells us how to construct one Brownian motion out of another via a change of probability measure. Therefore, let W(t) be an n-dimensional Brownian motion on a complete probability space (Ω, F, P) equipped with a filtration $\{F_t\}_{t\in[0,\infty)}$ satisfying the usual conditions. Let further X(t) be an n-dimensional process with

$$P\left(\int_0^T (X_i(t))^2\,dt < \infty\right) = 1, \quad 1 \le i \le n, \; 0 \le T < \infty .$$

Then, we set

$$Z_t(X) = \exp\left(\sum_{i=1}^n \int_0^t X_i(s)\,dW_i(s) - \tfrac{1}{2}\int_0^t \|X(s)\|^2\,ds\right).$$

If $Z_t(X)$ is a martingale (it is at least a supermartingale under our assumptions)then we can define a probability measure Q_T on F_T for each $0 \le T < \infty$ by

$$Q_T(A) = E(1_A Z_T(X)) \quad \forall\, A \in F_T.$$

By the martingale property of $Z_t(X)$, we also have the consistency condition

$$Q_T(A) = Q_t(A) \quad \forall\, A \in F_t.$$

The positivity of $Z_T(X)$ yields that P and Q_T are equivalent (as measures) on F_T.

Theorem B23 "Girsanov's Theorem"
If the process Z(X), as given above, is a martingale then the process $\tilde{W} = \{\tilde{W}(t), F_t;\ 0 \le t \le T\}$ defined componentwise as

$$\tilde{W}_i(t) = W_i(t) - \int_0^t X_i(s)\,ds, \quad i = 1,\dots, n,$$

is an n-dimensional Brownian motion on (Ω, F_T, Q_T).

A condition, ensuring that Z(X) is a martingale, is the following one which is known as the **Novikov condition**:

$$E\left(\exp\left(\tfrac{1}{2}\int_0^T \|X(s)\|^2\,ds\right)\right) < \infty.$$

B.7 Two Generalisations of Itô's Formula

This section contains two generalisations of Itô's formula which are very useful for applications in stochastic control (in particular in Sections 5.1 and 6.3). The first one is from Chapter IV of (Rogers and Williams 1987).

Theorem B24

Let X be a one-dimensional Itô-process, and let f be a C^1-function on **R** such that there exists a measurable function h, integrable on each interval [−a, a] with

$$f'(y) - f'(x) = \int_x^y h(z)dz$$

(i.e. we have f'' = h in the distributional sense). Then, we have

$$f(X(t)) = f(X(0)) + \int_0^t f'(X(s))dX(s) + \tfrac{1}{2}\int_0^t h(X(s))d\langle X\rangle(s) .$$

This theorem is extremely useful as it also covers the case where f''(x) exists on $\mathbf{R} \setminus \{a_1, \ldots, a_k\}$ and is continuous there. Another type of generalisation of Itô's formula will be needed in Section 5.1 where the process X not only fails to be an Itô-process, but also could have jumps. More precisely, let X(t) have the form

$$X(t) = V(t) + \int_0^t K(s)dW(s) \tag{17}$$

with $K(s) \in H^2[0, T]$ for all T > 0 and V(t) an F_t-adapted, right-continuous process with paths of finite variation. Further, the left-limits V(t−) of V should exist for all t > 0 almost surely, and V should have at most countably many jumps (almost surely). Denote by A the continuous process

$$A(t) = V(t) - \sum_{0<s\leq t} \Delta V(s)$$

with $\Delta V(s) = V(s) - V(s-)$. Due to the assumption that V has sample paths of finite variation, we know

$$\sum_{0<s\leq t} |\Delta V(s)| < \infty .$$

Hence, A(t) is well-defined. We have now set the scene for the next generalisation of Itô's formula which can be found in (Harrison 1985), p. 71.

Theorem B25

Let f: $\mathbf{R}\to\mathbf{R}$ be twice continuously differentiable. With the above notation (in particular with a process X(t) of the form (17)), we have

$$f(X(t)) = f(X(0)) + \int_0^t f'(X(s))K(s)dW(s) + \int_0^t f'(X(s))dA(s)$$
$$+ \tfrac{1}{2}\int_0^t f''(X(s))K(s)^2\,ds + \sum_{0<s\le t}\Delta f(X(s)).$$

Remark B26

Further generalisations of Itô's formula include versions for convex functions (see e.g. (Karatzas and Shreve 1988)), versions with generalised derivatives (see (Krylov 1980) or (Haussmann 1994)) or versions based on generalisations of the quadratic variation process (see (Föllmer, Protter and Shiryaev 1995)).

Appendix C
Controlled Stochastic Differential Equations

A stochastic differential equation of the form

$$dX(t) = \mu(t, X(t), u(t))dt + \sigma(t, X(t), u(t))dW(t) \tag{1}$$

where u(t) is a stochastic process that can be "chosen by a controller" is called a **controlled stochastic differential equation**. However, to give equation (1) a rigorous meaning, various points have to be made more precise, and some assumptions have to be put on the coefficient functions $\mu(t, x, u)$, $\sigma(t, x, u)$ and the "control process" u(t), $t \ge 0$. The first problem that arises is, of course, the existence and uniqueness of a solution X(t) of the stochastic differential equation (1) in dependence of the control process u. To emphasise this dependence, we will write $X^u(t)$ for such a solution corresponding to the control u. As existence and uniqueness for the general form (1) of such a controlled stochastic differential equation is a non-trivial problem (and requires additional mathematical tools that we will not introduce here), we will specialise to control equations of the following affine linear form

$$dX^u(t) = \left([\mu_1'u_1(t) + \mu_2]\, X^u(t) + \nu_1'u_2(t) + \nu_2\right)dt$$
$$+ \left([u_1(t)'\sigma_1 + \sigma_2]\, X^u(t) + u_2(t)'\eta_1 + \eta_2\right)dW(t) \tag{2}$$

where $\mu_1 \in \mathbf{R}^n$, $\nu_1 \in \mathbf{R}^k$, μ_2, $\nu_2 \in \mathbf{R}$, $\sigma_1 \in \mathbf{R}^{n,m}$, $\eta_1 \in \mathbf{R}^{k,m}$, σ_2, $\eta_2 \in \mathbf{R}^m$ are constants of appropriate dimensions, W(t) is a k-dimensional Brownian motion (together with its natural filtration $\{F_t\}_{t \geq 0}$), and $u_1(t)$, $u_2(t)$ are F_t-adapted processes of suitable dimensions (for a treatment of controlled stochastic equations of the general form (1), the monographs by (Fleming and Rishel 1975), (Fleming and Soner 1993) or (Gihman and Skorohod 1979) are recommended). The requirement that the controls have to be F_t-adapted, has the interpretation that the controller should base his decision on past and present values of W(t) and thus also on past and present values of $X^u(t)$. He is not allowed to look into the future.

From the variation of constants formula B15, we directly obtain that equation (2) has a unique strong solution $X^u(t)$ if the components of the control processes u(t) = $(u_1(t), u_2(t))$ satisfy

$$\int_0^T u_{1i}(t)^2 \, dt < \infty, \quad \int_0^T u_{2i}(t)^2 \, dt < \infty \text{ a.s.} \tag{3}$$

for i = 1,..., n, j = 1,..., k. Therefore, the set of adapted stochastic processes satisfying condition (3) would be a natural candidate for the set of admissible controls. However, to apply some standard results from stochastic control theory (and to maximise expected utility later on), we have to restrict this class a bit further.

Definition C1

Let $U \subseteq \mathbf{R}^{n+k}$ be a closed set (not necessarily bounded!). An F_t-adapted process u(t) = $(u_1(t), u_2(t))$ with $u(t) \in U$ for all $t \in [0, T]$ a.s. is called an **admissible control** (for equation (2)) if we have

$$\|u_1(t)\|^2 < C \quad \forall\, t \in [0, T],$$

$$E\left(\int_0^T \|u(s)\|^m \, ds \right) < \infty \quad \forall\, m \in \mathbf{N}$$

(where $u_1(t)$ denote the first n components of u(t), $\|.\|$ denotes Euclidean norm of appropriate dimension, and C is a given constant).

As already mentioned, this condition is stronger than condition (3). In particular, it implies that $E(|X^u(t)|^m)$ is finite for all m (see (Fleming and Soner 1993), Appendix D).

The goal of the controller will be to maximise his utility of a "good position" of $X^u(t)$ on [0, T) and at the final time T while not incurring excessive control costs. This task is expressed by the optimisation problem

$$\max_{u \in \mathcal{U}(x)} E\left(\int_0^T L(s, X^u(s), u(s))ds + \psi(X^u(T)) \right) =: \max_{u \in \mathcal{U}(x)} J(0,x,u) \qquad (4)$$

where $\mathcal{U}(x)$ denotes the set of admissible controls on [0, T] when the process $X^u(t)$ starts in $X^u(0) = x$. L(s, x, u) and $\psi(x)$ are continuous, real valued functions that satisfy the polynomial growth conditions

$$| L(t, x, u) | \leq C (1 + |x|^p + \|u\|^p),$$

$$| \psi(x) | \leq C (1 + |x|^p)$$

for all $(t, x, u) \in [0, T] \times \mathbf{R} \times U$ and suitable constants $C > 0$ and $p \in \mathbf{N}$. We will later comment on the case where these two functions are only defined for values of x in a restricted region (such as $[0, \infty)$).

To solve the above problem, we introduce the corresponding **value function**

$$v(t, x) := \sup_{u \in \mathcal{U}(t,x)} E^{t,x}\left(\int_t^T L(s, X^u(s), u(s))ds + \psi(X^u(T)) \right) = \sup_{u \in \mathcal{U}(t,x)} J(t,x,u)$$

which describes the behaviour of the maximum utility as a function in its initial parameters (t, x), i.e. v(t, x) is the maximum possible utility if the maximisation problem starts at time t with X(t) = x. $\mathcal{U}(t, x)$ denotes the set of admissible controls corresponding to these initial conditions. If we are able to determine v(t, x) then we obtain the maximum possible utility of the original maximisation problem as v(0,x). Even more, the method, we will develop below, also yields an optimal control process (if the method works at all !). First, we will derive it formally via some heuristic computations (and justify this derivation afterwards by citing some suitable results of stochastic control theory). To do so, in the following, we assume to have all required smoothness properties, permission to interchange limits, and the validity of the so-called Bellman principle. The Bellman principle consists of the following representation of the value function

$$v(t, x) = \sup_{u \in \mathcal{U}([t,s],x)} E^{t,x}\left(\int_t^s L(r, X^u(r), u(r)dr + v(s, X^u(s)) \right) \qquad (5)$$

where $\mathcal{U}([t, s], x)$ denotes the subset of strategies in $\mathcal{U}(t, x)$ restricted to the time interval [t, s]. The expression in brackets on the right hand side of equation (5) can be interpreted as the utility gained from following the strategy u on [t, s] (which results in the integral and in the value $X^u(s)$ at time s) and behaving optimally on the remaining time interval [s, T] (which results in v(s, $X^u(s)$)). The Bellman principle says that taking the supremum over the expected value of these expressions yields the value function. If we apply Itô's formula to v(s, $X^u(s)$) in equation (5), we obtain

$$v(t,x) = v(t,x) + \sup_{u \in \mathcal{U}([t,s],x)} E^{t,x}\Bigg(\int_t^s \Big(L(r, X^u(r), u(r)) + v_t(r, X^u(r)) +$$

$$+ \big[(\mu_1' u_1(r) + \mu_2) X^u(r) + \nu_1' u_2(r) + \nu_2 \big] v_x(r, X^u(r)) +$$

$$+ \tfrac{1}{2} \big\| (u_1(r)'\sigma_1 + \sigma_2) X^u(r) + u_2(r)'\eta_1 + \eta_2 \big\|^2 v_{xx}(r, X^u(r)) \Big) dr +$$

$$+ \int_t^s \big((u_1(r)'\sigma_1 + \sigma_2) X^u(r) + u_2(r)'\eta_1 + \eta_2 \big) v_x(r, X^u(r)) dW(r) \Bigg)$$

By assuming that the stochastic integral in this equation is a martingale and by subtracting v(t, x) on both sides, we obtain

$$0 = \sup_{u \in \mathcal{U}([t,s],x)} E^{t,x}\Bigg(\int_t^s \Big(L(r, X^u(r), u(r)) + v_t(r, X^u(r)) +$$

$$+ \big[(\mu_1' u_1(r) + \mu_2) X^u(r) + \nu_1' u_2(r) + \nu_2 \big] v_x(r, X^u(r)) +$$

$$+ \tfrac{1}{2} \big\| (u_1(r)'\sigma_1 + \sigma_2) X^u(r) + u_2(r)'\eta_1 + \eta_2 \big\|^2 v_{xx}(r, X^u(r)) \Big) dr \Bigg).$$

Dividing both sides by s–t and taking the limit $s \downarrow t$, we arrive at:

$$0 = \sup_{u \in \mathcal{U}} \Big(L(t,x,u) + v_t(t,x) + \big[(\mu_1' u_1) + \mu_2) x + \nu_1' u_2 + \nu_2 \big] v_x(t,x) +$$

$$+ \tfrac{1}{2} \big\| (u_1'\sigma_1 + \sigma_2) x + u_2'\eta_1 + \eta_2 \big\|^2 v_{xx}(t,x) \Big). \tag{6}$$

Note that the expectation operator has vanished, because at time t the values of $X^u(t)$ and of u(t) are known. Equation (6) together with the obvious boundary condition

$$v(T, x) = \psi(x) \tag{7}$$

constitute the Hamilton-Jacobi-Bellman-Equation (for short: HJB-Equation) of the optimisation problem (4). One of the main results of stochastic control theory consists of so called verification theorems, i.e. theorems stating that a sufficiently regular solution of the HJB-Equation coincides with the value function **and** that an optimal control u*(t) can be constructed by looking at the values that yield the supremum in equation (6). More precisely :

Theorem C2 "Verification theorem"
Let $\tilde{v}(t, x)$ be a polynomially bounded solution to the HJB-Equation (C6/7) satisfying $\tilde{v} \in C^{1,2}([0, T)\times\mathbf{R})$ and $\tilde{v}$ continuous on $[0, T]\times\mathbf{R}$. Then we have:

a) $$\tilde{v}(t, x) \geq v(t, x) \ \ \forall\ (t, x)\in[0, T)\times\mathbf{R}\,.$$

b) If there exists an admissible control u*(t) satisfying

$$\begin{aligned} u^*(t) \in \underset{\pi \in U}{\arg\max}\Big(& L(t, X^*(t), \pi) + v_t(t, X^*(t)) + \\ & + \big[(\mu_1'\pi_1) + \mu_2)X^*(t) + \nu_1'\pi_2 + \nu_2\big] v_x(t, X^*(t)) + \\ & + \tfrac{1}{2}\left\|(\pi_1'\sigma_1 + \sigma_2)X^*(t) + \pi_2'\eta_1 + \eta_2\right\|^2 v_{xx}(t, X^*(t))\Big) \end{aligned}$$

(a.s. with respect to $L\otimes P$) where X*(t) denotes the corresponding solution of the stochastic differential equation (2) then we have

$$\tilde{v}(t, x) = v(t, x) = J(t, x, u^*) \ \ \forall\ (t, x)\in[0, T)\times\mathbf{R}\,.$$

In particular, u*(t) is an optimal control.

This theorem (which is a special case of Theorem IV.3.1 in (Fleming and Soner 1993)) suggests the following algorithm to solve our optimisation problem:

Step 1:
Solve the optimisation problem inside the HJB-Equation to obtain $u^* = (u^*_1, u^*_2)$ (depending on the **yet unknown** value function $\tilde{v}$ and its relevant partial derivatives).

Step 2:
Insert u* into the HJB-Equation, drop the supremum operator, and solve the resulting partial differential equation.

However, as the above verification theorem only presents a sufficient condition for optimality, we cannot expect this method to work in the general case. Moreover, as the partial differential equation, occurring in the second step of the above algorithm, is usually highly non-linear, it is a rare occasion that we can solve it explicitly. We will present some examples for this case in Section 3.3.

Before closing this section, we will state some further useful variants of our optimisation problem and the corresponding HJB-Equations. First, we look at the problem given by its value function of the form

$$v(t,x) = \sup_{u \in U(t,x)} E^{t,x}\left(\int_t^T e^{-\rho(s-t)} L(X^u(s), u(s))ds + e^{-\rho(T-t)} \psi(X^u(T)) \right).$$

Then, the corresponding variant of Theorem C2 stays correct if in the HJB-Equation we replace

$$L(t, x, u) \quad \text{by} \quad L(x, u) - \rho\, v(t, x)$$

and if we make the corresponding change in the maximisation problem of part b) of Theorem C2. Further, we will look at the infinite time horizon case given by the value function

$$v(x) = \sup_{u \in \mathcal{U}(x)} E^x\left(\int_0^\infty e^{-\rho s} L(X^u(s), u(s))ds \right).$$

The set $\mathcal{U}(x)$ then consists of all controls that satisfy the requirements of Definition C1 (uniformly) for all finite T. Further, the members of $\mathcal{U}(x)$ are assumed to satisfy

$$E^x\left(\int_0^\infty e^{-\rho s} L(X^u(s), u(s))ds \right) < \infty .$$

The HJB-Equation related to this problem has the form

$$0 = \sup_{u \in \mathcal{U}} \Big(L(x,u) - \rho v(x) + \left[(\mu_1' u_1) + \mu_2 \right) x + \nu_1' u_2 + \nu_2 \big] v_x(x) + \\ + \tfrac{1}{2} \left\| (u_1' \sigma_1 + \sigma_2) x + u_2' \eta_1 + \eta_2 \right\|^2 v_{xx}(x) \Big), \qquad (8)$$

and the corresponding verification theorem (which is a special case of Theorem IV.5.1 in (Fleming and Soner 1993)) reads:

Theorem C3

Let $\tilde{v}(x)$ be a polynomially bounded solution to the HJB-Equation (C8) satisfying $\tilde{v} \in C^2([0,\infty)\times\mathbf{R})$. Then we have:

a)
$$\tilde{v}(x) \geq J(x, u)$$

for all $x \in \mathbf{R}$ and all admissible controls u that satisfy

$$\liminf_{t\to\infty} e^{-\rho t} E^x\left(\tilde{v}(X^u(t))\right) \geq 0. \tag{9}$$

b) If there exists an admissible control u*(t) satisfying

$$u^*(t) \in \arg\max_{\pi \in U}\Big(L(X^*(t),\pi)) + \left[(\mu_1'\pi_1) + \mu_2)X^*(t + \nu_1'\pi_2 + \nu_2\right]v_x(X^*(t)) +$$
$$-\rho v(X^*(t)) + \tfrac{1}{2}\left\|(\pi_1'\sigma_1 + \sigma_2)X^*(t) + \pi_2'\eta_1 + \eta_2\right\|^2 v_{xx}(X^*(t))\Big)$$

(a.s. with respect to $\lambda\times P$) and

$$\liminf_{t\to\infty} e^{-\rho t} E^x\left(\tilde{v}(X^*(t))\right) \geq 0. \tag{9*}$$

where X*(t) denotes the corresponding solution of the stochastic differential equation (2) then we have

$$\tilde{v}(x) = v(x) = J(x, u^*) \;\; \forall\, (t, x) \in [0, T)\times\mathbf{R}\,.$$

Especially, u*(t) is an optimal control.

Remark C4

The additional condition (9) in Theorem C3 appears for technical reasons. If $\tilde{v}(x)$ is always positive (which must be the case if L(x, u) is) then this condition is trivially satisfied.

Important Remark: "Controlled stochastic differential equations with a restricted state space"

If the utility functions L, ψ are only defined for $x \in \overline{O}$ where O is an open set strictly contained in the real line then we will stop the process $X^u(t)$ at the first time it reaches the boundary of O. In the finite horizon problems we will modify the formulation of the problems to obtain the value functions

$$v(t, x) := \sup_{u \in \mathcal{U}(t,x)} E^{t,x}\left(\int_t^\tau L(s, X^u(s), u(s))ds + \psi(\tau, X^u(\tau))\right),$$

$$v(t, x) = \sup_{u \in U(t,x)} E^{t,x}\left(\int_t^\tau e^{-\rho(s-t)} L(X^u(s), u(s))ds + e^{-\rho(\tau-t)}\psi(\tau, X^u(\tau))\right),$$

respectively. Here, τ is the minimum of T and the first exit time of $X^u(t)$ from O. As this exit time could be strictly smaller than T, it is reasonable to introduce an explicit time dependence into the function ψ. The verification theorem C2 stays correct if we replace T by τ in their statements. Then, the HJB-Equations (6) and (8) are only required to hold on $[0, \tau)\times O$ with boundary data

$$v(t, x) = \psi(t, x) \;\; \forall\, (t, x) \in \{[0, T)\times\partial O\} \cup \{\{T\}\times\overline{O}\}$$

In the infinite horizon problem, we have to introduce a further utility function g(.) which is continuous on **R** and represents the utility obtained from reaching the boundary of O. If O is an interval of the form (a, b) with $a \in \mathbf{R}\cup\{-\infty\}$, $b \in \mathbf{R}\cup\{+\infty\}$ then, with τ denoting the first exit time of $X^u(t)$ from O, the infinite horizon problem is given by its value function

$$v(x) = \sup_{u \in U(x)} E^x\left(\int_0^\tau e^{-\rho s} L(X^u(s), u(s))ds + 1_{\{\tau<\infty\}}\, g(X^u(\tau))e^{-\rho\tau}\right),$$

and we have to add the boundary condition

$$v(x) = g(x)\,,\; x \in \partial O$$

to the HJB-Equation. Of course, in Theorem C3, we have to replace ∞ by τ in part b), and the technical condition (9) must be changed to

$$\liminf_{t\to\infty} e^{-\rho t} E^x\left(1_{\{\tau\geq t\}}\, \tilde{v}(X^u(t))\right) \geq 0$$

(clearly, condition (9*) has to be changed in the obvious way, too).

In recent years, the stochastic control method has been greatly improved by the introduction of so-called viscosity solutions to the HJB-Equation. The definition of such a solution requires substantially weaker conditions on the value function (especially no smoothness requirements) as in our verification theorems. However, the introduction of viscosity solutions is clearly beyond the scope of this book, and we refer the interested reader to the monograph (Fleming and Soner 1993) and the sources cited therein.

Appendix D
Laplace's Method of Integration

To prove some asymptotic results in Chapter 6, we needed the following result on the asymptotic behaviour of an integral. For a proof, see (Müller 1995). A similar theorem is given on p. 277 of (Widder 1946).

Theorem D1 "Laplace's method of integration"
Let g: $[a, b] \to \mathbf{R}$, $a, b \in \mathbf{R}$, be a twice continuously differentiable function with a maximum point $u^* \in (a, b)$. Assume further that we have $g''(x) \leq -K < 0$ for $x \in (a, b)$. Then we have

$$\int_a^b e^{ng(u)} du \sim e^{ng(u^*)} \sqrt{\frac{2\pi}{n|g''(u^*)|}}$$

(where $x_n \sim y_n$ means that we have $(x_n / y_n) \to 1$ for $n \to \infty$).

We also cite a multidimensional version of this theorem. Its proof can again be found in (Müller 1995).

Theorem D2 "Laplace's method of integration"
Let W: $C \to \mathbf{R}$ be a twice continuously differentiable function with a uniformly negative definite Hessian J. Assume further that the minimum of W over C will be attained in $c^* \in \text{int}(C)$ (where C is the (m−1)-dimensional simplex). Then we have

$$\int_C e^{nW(c)} dc \sim e^{nW(c^*)} \left(\frac{2\pi}{n}\right)^{\frac{m-1}{2}} \frac{1}{\sqrt{\det(-J^*)}}$$

where J^* is the Hessian in c^*.

Appendix E
Optimisation of Convex Functionals

In this section, we will give a short survey on optimisation of convex functionals over a convex set. Our presentation is based on Chapter 7 of (Luenberger 1969). However, we adapt it to our setting of maximising concave functionals.

Definition E1
a) Let f be a convex functional defined on a convex subset C of a normed space X. The set

$$C^* = \left\{ x^* \in X^* \,\middle|\, \sup_{x \in C}\left[\langle x, x^* \rangle - f(x)\right] < \infty \right\}$$

(where X* is the dual space to X) is called the **conjugate set** (to (f, C)), and the functional

$$f^*(x^*) = \sup_{x \in C}\left[\langle x, x^* \rangle - f(x)\right], \quad x^* \in C^*,$$

is called the **convex conjugate functional** to f on C*.

b) Let g be a concave functional defined on a convex subset D of a normed space X. The set

$$D^* = \left\{ x^* \in X^* \,\middle|\, \inf_{x \in D}\left[\langle x, x^* \rangle - g(x)\right] > -\infty \right\}$$

is called the **conjugate set** (to (g, D)), and the functional

$$g^*(x^*) = \inf_{x \in D}\left[\langle x, x^* \rangle - g(x)\right], \quad x^* \in D^*,$$

is called the **concave conjugate functional** to g on D*.

Remark E2

a) For f = 0, we obtain

$$f^*(x) = \sup_{x \in C}\left[\langle x, x^* \rangle\right],$$

i.e. f*(x) then coincides with the support function of C (see also Definition 4.10).

b) The choices of $X = C = \mathbf{R}^n$ and

$$f(x) = \frac{1}{p}\sum_{i=1}^{n}|x_i|^p$$

with p > 1 fixed leads to $X^* = C^* = \mathbf{R}^n$ and

$$f^*(x^*) = \frac{1}{q}\sum_{i=1}^{n}|x_i^*|^q$$

with $p^{-1} + q^{-1} = 1$.

The main application of the theory of conjugate sets and functionals is the role they play in duality relations for optimisation problems. Therefore, consider two convex subsets C and D of X, a functional f which is convex over C and a functional g which is concave over D. We seek to solve the maximisation problem

$$\max_{C \cap D} [g(x) - f(x)] . \tag{1}$$

The fundamental result relating this problem to its dual (that still has to be defined !) is the Fenchel Duality Theorem.

Theorem E3 "Fenchel Duality Theorem"
Let f, g, C, D be as above. Assume that $C \cap D$ contains points in both the relative interior of C and D. Further, let there exist a pair $(r, x) \in \mathbf{R} \times (C \cap D)$ with either $f(x) < r$ or $g(x) > r$. Finally, assume that

$$\mu = \sup_{C \cap D} [g(x) - f(x)]$$

is finite. Then, we have

$$\mu = \sup_{C \cap D} [g(x) - f(x)] = \inf_{C^* \cap D^*} [f^*(x) - g^*(x)] . \tag{2}$$

Remark E4
The minimisation problem given by the right hand side of equation (2) is called the **dual problem** to the maximisation problem (1). In some cases, the dual problem is much easier to solve than the primal one. In particular, it is sometimes possible that the dual to a constrained problem is an unconstrained one. We also refer to the dual optimisation problems in Sections 3.5, 4.2, and 4.4.

REFERENCES

AKIAN M., SÉQUIER P., SULEM A. (1995) *A finite horizon multidimensional portfolio selection problem with singular transaction costs*, Proceedings of the 34th Conference on Decision and Control, 2193–2198.

ATKINSON C., PLISKA S.R., WILMOTT P. (1995) *Portfolio mangament with transaction costs*, preprint.

ATKINSON C., WILMOTT P. (1995) *Portfolio mangament with transaction costs: An asymptotic analysis of the Morton and Pliska model*, Mathematical Finance **5** (4), 357–367.

BACHELIER L. (1900) *Théorie de la speculation*, Annales Scientifique de l' École Normal Superieure **17,** 21–86.

BACK, K. (1991) *Asset pricing for general processes*, Journal of Mathematical Economics **20** (4), 371–395.

BAXTER M., RENNIE A. (1996) *Financial Calculus*, Cambridge University Press.

BENSOUSSAN A., LIONS J.L. (1984) *Impulse Control and Quasi-Variational Inequalities*, Gauthier-Villars, Paris.

BLACK F., SCHOLES M. (1973) *The pricing of options and corporate liabilities*, Journal of Political Economics **81**, 637–659.

BREIMAN L. (1961) *Optimal gambling systems for favourable games*, in *Fourth Berkeley Symposium on Mathematical Statistics and Probability*, I, 65–78.

BUCKLEY I.R.C., KORN R. (1997) *Optimal cash management for equity index tracking in the presence of fixed and proportional transaction costs*, Berichte zur Stochastik und verwandten Gebieten 97-3, Johannes Gutenberg-Universität Mainz.

CONNOR G., LELAND H. (1995) *Cash management for index tracking*, Financial Analysts Journal **51**, 75–80.

COVER T. (1991) *Universal portfolios*, Mathematical Finance **1** (1), 1–29.

COX J., HUANG C.F. (1989) *Optimal consumption and portfolio policies when asset prices follow a diffusion process* , Journal of Economic Theory **49**, 33–83.

COX J., HUANG C.F. (1991) *A variational problem arising in financial economics*, Journal of Mathematical Economics **20**, 465–487.

CVITANIC J., KARATZAS I. (1992) *Convex duality in constrained portfolio optimization*, Annals of Applied Probabiltiy **2**, 767–818.

DAVIS M.H.A. (1993) *Markov Models and Optimization*, Chapman and Hall, London.

DAVIS M.H.A., NORMAN, A.R. (1990) *Portfolio selection with transaction costs*, Mathematics of Operations Research **15** (4), 676–713.

DAVIS M.H.A., PANAS V.G., ZARIPHOPOLOU T. (1993) *European option pricing with transaction costs*, SIAM Journal on Control and Optimization **31**, 470–493.

DOLÉANS-DADE C. (1976) *On the existence and unicity of solutions of stochastic integral equations*, Zeitschrift für Wahrscheinlichkeitstheorie und verwandte Gebiete **36**, 93–101.

DUFFIE D. (1992) *Dynamic Asset Pricing Theory*, Princeton.

DUFFIE D., RICHARDSON H.R. (1991) *Mean variance hedging in continuous time*, Annals of Applied Probability **1**, 1–15.

EASTHAM J.E., HASTINGS K.J. (1988) *Optimal impulse control of portfolios*, Mathematics of Operations Research **13** (4), 588–605.

FLEMING W.H., RISHEL R.W. (1975) *Deterministic and stochastic optimal control*, Springer , Berlin.

FLEMING W.H., SONER M.H. (1993) *Controlled Markov Processes and Viscosity Solutions* , Springer , Berlin.

FLEMING W.H., ZARIPHOPOULOU T. (1991) *An optimal investment/consumption problem model with borrowing*, Mathematics of Operations Research **16**, 802–822.

FLETCHER R. (1981) *Practical Methods of Optimization, Volume II: Constrained Optimization*, Wiley, New York.

FÖLLMER H., PROTTER P., SHIRYAEV A. (1995) *Quadratic covariation and an extension of Itô's formula*, Bernoulli **1**, 149–170.

FÖLLMER H., SCHWEIZER M. (1990) *Hedging of contingent claims under incomplete information*, in *Applied stochastic analysis* eds. M.H.A. Davis and R.J. Elliot, Gordon and Breach, London.

FÖLLMER H., SCHWEIZER M. (1993) *A microeconomic approach to diffusion models for stock prices*, Mathematical Finance **3** (1), 1-23.

GÄNSSLER P., STUTE W. (1977) *Wahrscheinlichkeitstheorie*, Springer, Berlin.

GIHMAN I.I., SKOROHOD A.V. (1979) *Controlled Stochastic Processes*, Springer, Berlin.

GILL P.E., MURRAY W. (1978) *Numerically stable methods for quadratic programming*, Mathematical Programming **14**, 349–372.

GOLDFARB D., IDNANI A. (1983) *A numerically stable dual method for solving strictly convex quadratic programs*, Mathematical Programming **27**, 1–33.

HARRISON J.M.(1985) *Brownian Motion and Stochastic Flow Systems*, Wiley, New York.

HARRISON J.M., KREPS D. (1979) *Martingales and arbitrage in multiperiod securities markets*, Journal of Economic Theory **20**, 381–408.

HARRISON J.M., PLISKA S.R. (1981) *Martingales and stochastic integrals in the theory of continuous trading*, Stochastic Processes and Applications **11**, 215–260.

HARRISON J.M., PLISKA S.R. (1983) *A stochastic calculus model of continuous trading: Complete markets*, Stochastic Processes and Their Applications **15**, 313–316.

HARRISON J.M., T.SELKE, A.TAYLOR (1983) *Impulse control of a Brownian motion,* Mathematics of Operations Research **8**, 454–466.

HAUGAN R. (1993) *The New Finance*, Prentice Hall, Englewood Cliffs.

HAUSSMANN, U.G. (1994) *Generalized solutions of the Hamilton-Jacobi-Bellman equation*, SIAM Journal on Control and Optimization **32** (3), 728–743.

HE H., PEARSON N. (1991) *Consumption and portfolio policies with incomplete markets and short-sale constraints: The infinite dimensional case*, Journal of Economic Theory **54**, 259–304.

HELLWIG K. (1987) *Bewertung von Ressourcen* , Physica Verlag, Heidelberg.

HELLWIG K. (1993) *Intertemporal choice reexamined*, Discussion Papers in Economics, University of Ulm.

HULL J. (1993) *Introduction to Futures and Options Markets*, Prentice Hall, Englewood Cliffs.

IOFFE A.D., TICHOMIROFF V.M. (1979) *Theorie der Extremalwertaufgaben*, VEB Deutscher Verlag der Wissenschaft, Berlin.

JAMSHIDIAN F. (1992) *Asymptotically optimal portfolios*, Mathematical Finance **2** (2), 131–150.

JEANBLANC-PIQUÉ M. (1993) *Impulse control method and exchange rate*, Mathematical Finance **3**, 161–177.

KALLIANPUR G., STRIEBEL C. (1969) *Stochastic differential equations occurring in the estimation of continuous parameter stochastic processes*, Theory of Probability and Its Applications **14** (4), 567–594.

KARATZAS I. (1989) *Optimization problems in continuous trading*, SIAM Journal on Control and Optimization **27,** 1221–1259.

KARATZAS I., LEHOCZKY J.P., SETHI S., SHREVE S.E. (1986) *Explicit solutions of a general consumption/investment problem*, Mathematics of Operations Research **11**, 261–294.

KARATZAS I., LEHOCZKY J.P., SHREVE S.E. (1987) *Optimal portfolio and consumption decisions for a "small investor" on a finite horizon*, SIAM Journal on Control and Optimization **27,** 1157 – 1186.

KARATZAS I., LEHOCZKY J.P., SHREVE S.E., XU G.L. (1991) *Martingale and duality methods for utility maximisation in an incomplete market*, SIAM Journal on Control and Optimization **29,** 702–730.

KARATZAS I., SHREVE S.E. (1988) *Brownian Motion and Stochastic Calculus*, Springer, Berlin.

KIESEL R. (1996) *Pricing in incomplete markets: A quadratic utility approach*, preprint, Birkbeck College, London.

KORN R. (1992) *Optimierungsprobleme bei Wertpapierhandel in stetiger Zeit*, doctoral dissertation, Johannes Gutenberg-Universität Mainz.

KORN R. (1994) *Portfolio optimization with strictly positive transaction costs,* Berichte zur Stochastik 94-1, Johannes Gutenberg-Universität Mainz, to appear in: Finance and Stochastics.

KORN R. (1996) *Value preserving portfolio strategies and the minimal martingale measure*, Berichte zur Stochastik 96-6, to appear in: Mathematical Methods of Operations Research **47** (2).

KORN R. (1997a) *Value preserving portfolio strategies in continuous-time models*, Mathematical Methods of Operations Research **45**(1), 1–43.

KORN R. (1997b) *Some applications of L^2-hedging with a non-negative wealth process*, Applied Mathematical Finance **4** (1), 1-15.

KORN R. (1997c) *Generalised Impulse Control and Value Preserving Control of Continuous-Time Stochastic Processes with Applications to Finance*, Habilitationsschrift, Johannes Gutenberg-Universität Mainz.

KORN R. (1997d) *Optimal impulse control when control consequences are random*, to appear in : Mathematics of Operations Research.

KORN R., TRAUTMANN S. (1995) *Continuous-time portfolio optimization under terminal wealth constraints* , ZOR **42** (1), 69–93.

KRYLOV N. (1980) *Controlled Diffusion Processes*, Springer, Berlin.

LAMBERTON D., LAPEYRE B. (1996) *Introduction to Stochastic Calculus Applied to Finance*, Chapman and Hall, London.

LAUE S. (1998) *Approximation unerreichbarer Ziele in der Portfolio-Optimierung*, Diploma thesis, Johannes Gutenberg-Universität Mainz.

LIONS P.L., SZNITMAN A.S. (1984) *Stochastic differential equations with reflecting boundary conditions*, Communications on Pure and Applied Mathematics **37**, 511–537.

LIOUI A. (1997) *On mean variance optimization in continuous-time*, preprint.

LUENBERGER D. (1969) *Optimization by Vector Space Methods*, Wiley, New York.

MAGILL M.J.P., CONSTANTINIDES G.M. (1976) *Portfolio selection with transaction costs*, Journal of Economic Theory **13**, 245–263.

MARKOWITZ H. (1952) *Portfolio selection*, Journal of Finance **7**, 77–91.

MARKOWITZ H. (1959) *Portfolio Selection: Efficient Diversification of Investment*, Wiley, New York.

MERTON R.C. (1969) *Lifetime portfolio selection under uncertainty: the continuous case*, Reviews of Economical Statistics **51**, 247–257.

MERTON R.C. (1971) *Optimum consumption and portfolio rules in a continuous time model*, Journal of Economic Theory **3**, 373–413.

MERTON R.C. (1990) *Continuous-Time Finance*, Blackwell, Cambridge MA.

MORTON A.J., PLISKA S.R. (1995) *Optimal portfolio management with fixed transaction costs*, Mathematical Finance **5** (4), 337–356.

MÜLLER P. (1995) *Universelle Portfolios*, Diploma thesis, Johannes Gutenberg-Universität Mainz.

NAYFEH A.H. (1981) *Introduction to Perturbation Techniques*, Wiley, New York.

OCCONE D., KARATZAS I. (1991) *A generalized Clark representation formula, with application to optimal portfolios*, Stochastics and Stochastics Reports **34,** 187–228.

OKSENDAL B. (1992) *Stochastic Differential Equations*, 3rd edition, Springer, Berlin.

PLISKA S.R. (1986) *A stochastic calculus model of continuous trading: Optimal portfolios*, Mathematics of Operations Research **11**, 371–382.

PRESMAN E., SEHTI S. (1991) *Risk-aversion behaviour in consumtion/investment problems*, Mathematical Finance **1**, 100–124.

PROTTER P. (1992) *Stochastic Integrals and Differential Equations*, Springer, Berlin.

REVUZ D., YOR M. (1991) *Continuous Martingales and Brownian Motion*, Springer, Berlin.

RICHARDSON H.R. (1989), *A minimum variance result in continuous trading portfolio optimization*, Management Science **35**, 1045–1055.

ROCKAFELLAR R.T. (1970) *Convex Analysis*, Princeton University Press.

ROCKAFELLAR R.T. (1976) *Integral functionals, normal intregrands and measurable selection*, eds. J.P. Gossez e.a., Springer, Berlin.

ROGERS L.C.G., WILLIAMS D.W. (1987) *Diffusions, Markov Processes and Martingales , Volume II,* Wiley, Chichester.

SAMUELSON P.A. (1973) *Mathematics of speculative price*, SIAM Reviews **15,** 1–42.

SCHÄL M. (1974) *A selection theorem for optimization problems*, Archive der Mathematik **25**, 219–224.

SCHWEIZER M. (1992) *Mean-variance hedging for general claims*, Annals of Applied Probability **2**, 171–179.

SCHWEIZER M. (1995) *On the minimal martingale measure and the Föllmer-Schweizer decomposition*, Stoch. Anal. Appl. 13, 573–599

SEHTI S., TAKSAR M. (1988) *A note on Merton's 'Optimum consumption and portfolio rules in a continuous-time model'*, Journal of Economic Theory **46**, 395–401.

SHIRYAEV A. (1978) *Optimal Stopping Rules*, Springer, Berlin.

SHREVE S.E., SONER H.M. (1994) *Optimal investment and consumption with transaction costs*, Annals of Applied Probability **4** (3), 609–692.

SULEM A. (1986) *A solvable one-dimensional model of a diffusion inventory system*, Mathematics of Operations Research **11**, 125–133.

TAKSAR M., KLASS M.J., ASSAF, D. (1988) *A diffusion model for optimal portfolio selection in the presence of brokerage fees* , Mathematics of Operations Research **13** (2), 277–294.

TOBIN J. (1958) *Liquidity preference as behaviour towards risk*, Review of Economic Studies **25**, 68–65.

TOBIN J. (1965) *The theory of portfolio selection*, in *The Theory of Interest Rates*, eds. F.B.R. Brechling and F.H. Hahn, Macmillan, London.

VARADHAN S.R.S., WILLIAMS R.J. (1985) *Brownian motion in a wedge with oblique reflection*, Communications on Pure and Applied Mathematics **38**, 405–433.

WHALLEY A.E., WILMOTT P. (1994) *Optimal hedging of options with small but arbitrary transaction cost structure*, preprint, OCIAM Working paper, Mathematical Institute Oxford.

WIDDER D.V. (1946) *The Laplace Transform*, Princeton University Press, Princeton.

WIESEMANN T. (1995) *Wertorientiertes Portfoliomanagement : Ein Modell zur intertemporalen Portfoliowerterhaltung* , Doctoral Dissertation, University of Ulm.

WILLINGER W., TAQQU M.S. (1991) *Towards a convergence theory for continuous stochastic securities market models*, Mathematical Finance **1**, 55–100.

WILMOTT P., DEWYNNE J.N., HOWISON S.D. (1993) *Option Pricing : Mathematical Models and Computation* , Oxford Financial Press.

XU G.L. (1990) *A Duality Method for Optimal Consumption and Investment under Short-Selling Prohibition*, Ph.D. dissertaion, Carnegie Mellon University.

XU G.L., SHREVE S.E. (1992a) *A duality method for optimal consumption and investment under short-selling prohibitions I : General market coefficients*, Annals of Applied Probability **2**, 87–112.

XU G.L., SHREVE S.E. (1992b) *A duality method for optimal consumption and investment under short-selling prohibitions I : Constant coefficients*, Annals of Applied Probability **2**, 314–328.

INDEX